Hydatid Disease

Sami Akbulut
Editor

Hydatid Disease

Diagnosis, Treatment and Follow Up Strategies

Springer

Editor
Sami Akbulut
Department of Surgery and Liver Transplantation Institute
Inonu University Faculty of Medicine
Malatya, Türkiye

Department Biostatistics and Medical Informatics
Inonu University Faculty of Medicine
Malatya, Türkiye

ISBN 978-3-031-97276-8 ISBN 978-3-031-97277-5 (eBook)
https://doi.org/10.1007/978-3-031-97277-5

This Springer imprint is published by the registered company Springer Nature Switzerland AG
The registered company address is: Gewerbestrasse 11, 6330 Cham, Switzerland

Preface

Echinococcosis, also known as hydatidosis or hydatid disease, is caused by the larval stages of cestodes (tapeworms) from the genus *Echinococcus*. *Echinococcus granulosus*, which is the most commonly encountered form, causes cystic echinococcosis. Another species, E. *multilocularis*, causes alveolar echinococcosis, and is becoming increasingly more common. Two exclusively New World species, *E. vogeli* and *E. oligarthrus*, are associated with "Neotropical echinococcosis"; *E. vogeli* causes a polycystic form whereas *E. oligarthrus* causes the extremely rare unicystic form.

Hydatid disease continues to pose a significant global health challenge, particularly in regions where the disease remains endemic. Despite notable advancements in medical science, the diagnosis, treatment, and management of hydatid disease are still fraught with difficulties. These challenges stem from the parasite's complex life cycle, its wide spectrum of clinical presentations, and the potential for severe complications. Addressing hydatid disease requires a multidisciplinary approach, integrating clinical expertise with public health initiatives to mitigate its impact.

This book is a comprehensive compilation of the most current knowledge on hydatid disease, covering its epidemiology, pathophysiology, clinical manifestations, diagnostic methods, therapeutic strategies, and emerging challenges. It brings together the expertise of leading researchers and clinicians from various disciplines, aiming to bridge the gap between basic science and clinical practice.

The primary objective of this book is to serve as a detailed reference for healthcare professionals, researchers, and students who encounter hydatid disease in their clinical practice or academic pursuits. By addressing the disease from a multidisciplinary perspective, we hope to enhance understanding and facilitate more effective diagnosis and management strategies.

The journey of preparing this book was marked by profound challenges. In the midst of writing and compiling, two devastating earthquakes struck Turkey, claiming tens of thousands of lives and leaving an indelible mark on our community. These tragic events not only shattered lives but also disrupted the progress of this work in unimaginable ways. Additionally, the ongoing conflict in Gaza posed further difficulties, as some of our contributors were directly affected and unable to complete their chapters. Yet, through unwavering dedication, collective perseverance, and a shared commitment to advancing medical knowledge, we succeeded in completing this book despite the unprecedented hardships.

This book brings together the perspectives and expertise of authors from diverse backgrounds and countries. As a result, variations in terminology, data presentation, and clinical approaches may be observed throughout the chapters. These differences are a natural outcome of the global and multidisciplinary collaboration, reflecting the unique insights of each contributor. We believe that acknowledging these differences enriches the content, offering a more comprehensive and multifaceted view of hydatid disease.

A special emphasis has been placed on novel diagnostic approaches, including the use of artificial intelligence, as well as the latest surgical techniques and postoperative follow-up protocols. Furthermore, the book explores the public health implications of hydatid disease, with a focus on prevention and control strategies in endemic areas.

I would like to express my deepest gratitude to all the contributors whose dedication and expertise have made this book possible. I am also thankful to Springer Nature for providing the platform to share this knowledge with the global medical community.

I dedicate this book to the memory of my beloved parents, the late Zurab Akbulut and the late Bahri Akbulut, who made every sacrifice to support my education. My thoughts and prayers are always with them. Their unwavering love and support have been the foundation of my academic journey. Additionally, I would like to extend my heartfelt gratitude to my wife and children, Bahri Yusuf Akbulut and Ali Çınar Akbulut, whose unwavering support and encouragement have been a constant source of strength, especially during the challenging times following the earthquake. Even in the most challenging moments, their unwavering support and love have remained my guiding light, providing the strength to overcome adversity.

I sincerely hope that this book will serve as a valuable resource for medical professionals and researchers, fostering a deeper understanding of hydatid disease and ultimately contributing to better patient outcomes. By highlighting the most recent advances and addressing the multifaceted challenges of hydatid disease, this book aspires to shape future research and enhance clinical practices worldwide. We truly believe that the insights presented in this book will ignite further research, spark innovative solutions, and ultimately contribute to more effective strategies in the global fight against hydatid disease.

Department of Surgery and Liver Transplantation Institute
Inonu University Faculty of Medicine
Malatya, Türkiye

Department Biostatistics and Medical Informatics
Inonu University Faculty of Medicine
Malatya, Türkiye

Sami Akbulut

Introduction: Understanding Hydatid Disease

Introduction

Echinococcosis, commonly referred to as hydatidosis or hydatid disease, is an infectious condition caused by the larval stages of tapeworms belonging to the genus *Echinococcus* [1]. These parasites belong to the class Cestoda, order Cyclophyllidea, and family Taeniidae [2]. Among the species responsible for the disease, *Echinococcus granulosus* is the most prevalent, causing cystic echinococcosis, characterized by the formation of one or more hydatid cysts. These cysts most commonly occur in the liver (52–77%) and lungs (5–47%), but they may also be found in other organs such as bones (0.5–2.5%), kidneys (1–4%), spleen (0.5–11.3%), soft tissue and muscles (1–4%), thyroid gland (0–3.4%), breast (0.27%), adrenal gland (<1%), and the central nervous system (1–3%) [3–9]. The disease often remains asymptomatic for many years, as hydatid cysts typically grow slowly, averaging approximately 1–5 cm per year [7, 10, 11]. Clinical symptoms generally develop only when cysts attain substantial size. Approximately half of patients requiring medical intervention present within a few years after initial infection, reflecting variability in growth rates influenced by parasite species, host immunity, and cyst localization [10]. Alveolar echinococcosis, caused by *Echinococcus multilocularis*, has an asymptomatic long incubation period, presenting as a primary tumor-like lesion, typically in the liver. Its increasing frequency underscores the importance of early diagnosis and therapeutic interventions [3, 12]. Additionally, two lesser-known New World species, *Echinococcus vogeli* and *Echinococcus oligarthrus*, cause Neotropical echinococcosis, manifesting as polycystic and unicystic forms, respectively [13].

Life Cycle

The life cycle of *Echinococcus* species varies between different species but generally involves definitive hosts (typically carnivores like dogs and foxes) and intermediate hosts (herbivores such as sheep, cattle, goats, and humans). Humans act as accidental intermediate hosts. The adult *Echinococcus* tapeworms reside in the intestines of definitive hosts, where they produce eggs that are excreted into the environment through the host's feces. These eggs contaminate soil, water, and vegetation, posing a significant risk of transmission. The adult worms of *E. granulosus* live in the small intestine of canids. Eggs are passed with feces and can survive in the environment. Herbivores and humans ingest the eggs, which hatch in the small intestine, releasing oncospheres. These oncospheres penetrate the intestinal wall and migrate to the liver, lungs, or other organs where they develop into hydatid cysts. The cycle completes when canids consume infected viscera containing viable cysts. Definitive hosts of *E. multilocularis* are wild canids (like foxes) and domestic dogs. Intermediate hosts include small mammals (like rodents) and humans. Eggs released in feces are ingested by intermediate hosts, where they hatch, penetrate the intestinal wall, and develop primarily in the liver. The cycle completes when the definitive host consumes infected intermediate hosts. *E. vogeli* and *E. oligarthrus* are primarily found in Central and South America. *E. vogeli* is associated with wild carnivores and rodents, while *E. oligarthrus* primarily infects felids. The life cycle is similar, with eggs excreted by definitive hosts and ingested by intermediate hosts. Cysts develop in the liver or lungs, and the cycle completes when definitive hosts consume infected tissues [14–21].

Role of Humans in the Life Cycle

Humans, as accidental intermediate hosts, become infected by ingesting eggs from contaminated food, water, or soil, or through direct contact with infected animals. In humans, the cysts typically develop in the liver and lungs, but they can also form in other organs, including the spleen, kidneys, heart, bones, and central nervous system. If cysts rupture, they may cause severe allergic reactions or anaphylactic shock. Understanding the distinct life cycles of each *Echinococcus* species is crucial for developing targeted public health measures to control transmission and reduce the risk of human infection. The transmission cycle of *E. granulosus* is closely associated with sheep farming and dog husbandry, particularly in rural and pastoral regions. The presence of stray dogs significantly increases the risk of transmission. In contrast, *E. multilocularis* has a more complex cycle, involving wildlife reservoirs such as foxes and rodents, leading to human exposure through contaminated environments. The lifecycle and transmission dynamics vary significantly among regions, depending on factors such as climate, animal husbandry practices, and public awareness. Implementing targeted interventions, such as dog deworming programs and public health education, can significantly reduce the prevalence of hydatid disease [16, 17, 19, 22, 23].

Geographical Distribution

Hydatid disease exhibits a global distribution, predominantly in regions where livestock farming and close human-animal interactions occur frequently. The distribution of hydatid disease is directly related to the species of *Echinococcus* involved and ecological interactions between definitive and intermediate hosts. *E. granulosus* is primarily found in pastoral regions, including Eurasia, Africa, South America, and the Middle East [21, 24]. According to recent studies, the Eastern Mediterranean region, including countries such as Iran, Iraq, and Tunisia, represents one of the major global hotspots for cyctic echinococcosis, with approximately 688 million people at risk [14]. Notably, regions within the Mediterranean, Central Asia, and parts of China have significant endemicity [25]. South America, notably Argentina, Peru, and Chile, represents critical endemic zones due to pastoral practices and high exposure to domestic dogs [4, 25, 26].

E. multilocularis primarily occurs in the northern hemisphere. Endemic regions include Central and Northern Europe (particularly Germany, France, and Switzerland), parts of Asia (notably western China, Mongolia, and Russia), and regions of North America (Alaska and northern Canada) [27–29]. Wild carnivores, especially foxes, play crucial roles in maintaining transmission cycles, with rodents serving as intermediate hosts. Human infection is relatively less common than cystic echinococcosis but presents significant health concerns due to severe outcomes.

Less common species such as *E. vogeli* and *E. oligarthrus* are predominantly found in Central and South America [13]. *E. vogeli*, associated with wild canids and rodents, has been identified primarily in forested regions of Brazil, Colombia, and Venezuela. Human infections, although rare, can result in polycystic echinococcosis, posing serious health risks. *E. oligarthrus* primarily affects felids such as jaguars and pumas, with human cases being rare and associated mainly with wildlife interactions in tropical forests. The geographical variability in echinococcosis prevalence underscores the necessity for targeted surveillance, localized public health interventions, and control measures adapted to specific ecological and socio-economic conditions. Understanding these geographical and ecological dynamics is crucial for designing effective prevention and control programs to mitigate the public health impacts of echinococcosis [16, 17].

Global, Regional, and National Burden

Echinococcosis, including both cystic and alveolar forms, continues to represent a substantial global health burden. According to the Global Burden of Disease Study 2021, echinococcosis caused approximately 148,521 new cases, 633,404 prevalent cases, 1364 deaths, and 105,072 disability-adjusted life years (DALYs) worldwide [18]. The age-standardized incidence rate was 1.82 per 100,000 population, and the age-standardized prevalence rate was 7.69 per 100,000 population [18]. Despite advancements in diagnostic and therapeutic methods, mortality rates remain significant, particularly in regions with low socio-demographic indexes where healthcare infrastructure is inadequate.

The economic burden of hydatid disease, particularly cystic echinococcosis, is considerable. The global annual costs exceed 3 billion USD, driven by healthcare expenses, loss of productivity, and financial impacts on livestock farming [14, 30]. In hyperendemic areas of South America, the prevalence among slaughtered livestock can range from 20% to 95%, significantly affecting the meat and dairy industries due to reduced carcass weight, decreased milk production, and fertility loss [30]. Similarly, in the Eastern Mediterranean region, the economic impact remains high due to the persistence of the disease and insufficient healthcare resources. Implementing comprehensive surveillance, public awareness, and One Health approaches is essential to mitigate both health and economic burdens, especially in countries with low socio-demographic indexes [31, 32].

Clinical Presentations

Hydatid disease presents with a wide spectrum of clinical manifestations, primarily influenced by the size, number, and anatomical location of the cysts [3, 33]. Cystic echinococcosis often remain asymptomatic for years before the cysts grow large enough to cause symptoms in the affected organs. Hepatic involvement often results in right upper quadrant, jaundice, nausea, vomiting, and abnormal liver function tests, while pulmonary cysts may present with chest pain, shortness of breath, cough, and hemoptysis [34, 35]. Rupture of the cysts can trigger a host immune response, presenting with symptoms such as fever, urticaria, eosinophilia, and in severe cases, anaphylactic shock [36, 37]. Additionally, rupture may result in cyst dissemination, leading to secondary infection or new cyst formation in distant organs. Other nonspecific signs include anorexia, weight loss, and general weakness. Alveolar echinococcosis primarily affects the liver as a slow-growing, destructive tumor. The incubation period can last 5–15 years before clinical manifestations arise [1, 3, 19]. The disease typically presents with abdominal pain, biliary obstruction, weight loss, and signs of hepatic failure. In addition to the liver and lungs, hydatid cysts can rarely localize in other organs, leading to organ-specific symptoms. For instance, pancreatic cysts may present with pancreatitis, while renal cysts can cause hematuria or hydatiduria [38, 39]. *E. vogeli* primarily involves the liver, typically presenting as a slow-growing, tumor-like mass, with frequent secondary cystic development [40]. In contrast, *E. oligarthrus* remains poorly characterized due to the limited number of reported cases.

Diagnostic Modalities

The diagnosis of hydatid disease relies on a comprehensive approach that includes clinical evaluation, imaging studies, and serological tests [41]. Accurate classification of cysts is crucial for determining the most suitable treatment strategy. The World Health Organization Informal Working Group on Echinococcosis (WHO-IWGE) classification categorizes hydatid cysts into active (CE1, CE2), transitional (CE3a, CE3b), and inactive (CE4, CE5) stages, which directly influence management decisions [20, 42, 43]. Gharbi's classification, based on ultrasonographic (US) findings, categorizes hydatid cysts into five types, ranging from purely cystic (Type I) to completely calcified (Type V) [44].

The WHO-IWGE PNM classification system was developed specifically for alveolar echinococcosis to provide a standardized approach for evaluating diagnostic and therapeutic strategies. Based on imaging findings, this system categorizes the disease by assessing three key factors: the extent of the parasitic mass within the liver (P), the involvement of adjacent organs (N), and the presence of metastases (M). Once the P, N, and M components are determined, they are combined to assign an overall stage ranging from I to IV. To further specify disease severity, Stage III is divided into subcategories IIIA and IIIB. This structured approach ensures consistency in disease assessment and helps guide clinical management [3, 45].

US is the primary imaging modality for diagnosing cystic echinococcosis due to its availability, cost-effectiveness, and ability to characterize cyst morphology and detect complications like rupture or superinfection [43]. US is particularly valuable in distinguishing active cysts (CE1, CE2) from inactive or calcified cysts (CE4, CE5). However, factors like obesity, intestinal gas, or previous surgery can limit its effectiveness [46]. Computed tomography (CT) is employed when US findings are inconclusive or when detailed anatomical mapping is necessary [46]. CT is particularly effective for detecting calcified cyst walls (CE5) and evaluating complex anatomical relationships [47]. It is useful in preoperative planning and assessing the extent of organ involvement. Magnetic resonance imaging (MRI) offers superior soft tissue contrast and is especially useful for characterizing cyst contents. T2-weighted sequences enhance the visualization of liquid components, making MRI valuable for differentiating cyst types and detecting internal septations or daughter cysts [47, 48]. Magnetic resonance cholangiopancreatography (MRCP) is indicated when biliary involvement is suspected. It noninvasively visualizes the biliary tree, helping to identify biliary fistulas and potential cyst-biliary communications [42, 49, 50]. Endoscopic retrograde cholangiopancreatography (ERCP) is recommended when there is strong suspicion of biliary obstruction or communication, often indicated by MRCP. Besides diagnostic capabilities, ERCP also allows therapeutic interventions, such as biliary drainage or stent placement when necessary [51, 52].

The use of PET/CT in the diagnosis of alveolar echinococcosis holds significant importance in the literature, particularly for evaluating disease activity and monitoring treatment progress. PET/CT serves as an imaging modality to assess the metabolic activity of alveolar echinococcosis lesions. This technique is especially valuable in inoperable patients, as it aids in monitoring the effectiveness of benzimidazole therapy and determining the duration of treatment. When combined with serological tests, metabolic activity measurements obtained through PET/CT facilitate a more comprehensive understanding of disease progression. Clinical applications and findings support the utility of PET/CT in alveolar echinococcosis diagnosis and management. Firstly, there is a notable correlation between metabolic activity measured by PET/CT and EgHF antibody levels, indicating that this relationship can be utilized to monitor disease activity throughout treatment. Secondly, the morphological classification of hepatic alveolar echinococcosis lesions (EMUC-CT) is associated with FDG uptake, where type IV lesions, in particular, demonstrate significantly lower FDG accumulation compared to other types. Lastly, PET/CT findings are instrumental in guiding the duration of benzimidazole therapy. Negative PET/CT results, coupled with low serological activity, may indicate the potential cessation of treatment [53–55].

Serological tests complement imaging by confirming the diagnosis, especially when radiological findings are ambiguous. Commonly used tests include: enzyme-linked immunosorbent assay (ELISA), indirect hemagglutination (IHA), latex agglutination (LA), immunofluorescence assay (IFA), and immunoelectrophoresis (IEP) [42]. ELISA and IHA are the most widely used due to their high sensitivity for liver cysts (85–98%) [42]. However, false-negative results may occur in calcified cysts or in patients previously treated. To improve accuracy, combining at least two serological methods is recommended. Western Blot (WB) serves as a confirmatory test with specificity approaching 100% [42, 56]. Serology plays a crucial role both in the diagnosis and in the post-treatment follow-up of hydatid disease. While it supports initial diagnosis, its primary value lies in monitoring recurrence or residual disease after therapeutic interventions.

Immunological markers play a crucial role in differentiating active from inactive cysts, especially in alveolar echinococcosis. EgHF-ELISA, Em2-ELISA, recEm18-ELISA, and Em-immunoblotting are essential serological tests in the diagnosis and follow-up of alveolar echinococcosis [3, 11, 42, 57–59]. EgHF-ELISA, utilizing *E. granulosus* hydatid fluid, shows high diagnostic sensitivity (95% for alveolar echinococcosis and 91% for cystic echinococcosis), making it suitable for initial screening. Em2-ELISA, based on the specific Em2 antigen from *E. multilocularis*, provides enhanced specificity for alveolar echinococcosis, with a sensitivity of 86%, although this decreases to 64% after radical surgery [59]. RecEm18-ELISA, which utilizes the recombinant Em18 antigen, is valuable in post-treatment monitoring, showing an 80% sensitivity, but may be less effective after curative surgery (73% sensitivity). Em-immunoblotting, leveraging *E. multilocularis* vesicular fluid antigens, exhibits the highest sensitivity (98%), even in patients who have undergone radical surgery, but its application is limited by cost and technical complexity [59]. The combination of these tests improves diagnostic accuracy and monitoring in clinical practice [3, 11, 42]. Diagnostic aspiration and biopsy may be considered for differential diagnosis, especially when the lesion's nature is uncertain or malignancy is suspected [42]. However, due to the risk of cyst rupture and secondary dissemination, it is generally avoided when noninvasive methods can provide sufficient diagnostic information [42]. A multimodal approach that integrates imaging, serological, and molecular diagnostics offers the most reliable method for identifying and characterizing hydatid cysts. This comprehensive strategy ensures accurate diagnosis, optimal treatment planning, and appropriate follow-up.

Therapeutic Approaches

Therapeutic approaches to cystic echinococcosis vary depending on cyst characteristics, location, and the patient's clinical condition. The treatment strategy is determined based on the WHO-IWGE classification, which categorizes cysts according to their stage and complexity. The main therapeutic modalities include medical treatment, percutaneous interventions, surgical approaches (conservative or radical), and a "watch and wait" strategy for asymptomatic or inactive cysts [11, 42, 60].

(i) Medical Treatment: Medical therapy is mainly indicated for uncomplicated cysts (CE1 and CE3a), small cysts (<5 cm), inoperable cases, and as an adjunct to surgery or percutaneous interventions [3, 42, 61]. The primary medical treatment for hydatid disease is the use of benzimidazole derivatives, particularly albendazole and mebendazole. Albendazole is currently the drug of choice to treat hydatid disease, either alone or together with interventional treatments. It is given orally at a dosage of 10–15 mg/kg/day, divided into two doses, with a fat-rich meal to increase its bioavailability [15, 42, 62]. As an alternative to albendazole, mebendazole, the first benzimidazole successfully tested against hydatid disease, can be used when albendazole is unavailable or not tolerated. The recommended dosage is 40–50 mg/kg/day, divided into three doses. There is no clear consensus on the duration of benzimidazole treatment for hydatid disease. However, it is generally recommended to administer therapy cyclically or continuously for 1–6 months [15, 63, 64]. During treatment, regular monitoring of liver function tests is essential to detect potential hepatotoxicity. Praziquantel at a dose of 40 mg/kg once a week in combination with albendazole seems more effective in killing protoscoleces compared to albendazole alone [3]. Although promising, the effectiveness of praziquantel in preventing secondary echinococcosis still requires further research. Additionally, combining albendazole (400 mg twice daily) with praziquantel (25–40 mg/kg/day) may enhance therapeutic efficacy, but the routine use of this combination remains debated [3, 11, 15, 63, 64]. Long-term therapy is essential when surgery is not feasible or in cases of recurrent disease [42].

(ii) Percutaneous Treatments: The PAIR (puncture, aspiration, injection, reaspiration) technique is the preferred method for CE1 and CE3a cysts due to its high efficacy and minimal invasiveness. This technique is most effective for cysts with liquid content and is contraindicated in calcified or thick-walled cysts (CE4 and CE5). Pre-treatment with albendazole for 1–2 weeks prior to PAIR decreases infection risks and cyst viability. For cysts where PAIR is not feasible, modified catheterization technique (MoCaT) is an alternative, especially for CE2 and CE3b cysts, though recurrence rates are a concern [3, 11, 15, 42, 65, 66].

(iii) Surgical Treatment: Surgery is indicated for large cysts (>10 cm), CE2-CE3b cysts with multiple daughter vesicles, complicated cysts (e.g., infected or ruptured), and those at high risk of rupture [7, 67]. Radical surgery, including pericystectomy and hepatic resection, is preferred when complete excision is feasible to minimize bile leakage risks. Conservative surgery, including cystotomy and partial cystectomy, is used when radical procedures are impractical. The use of scolicidal agents to sterilize the cyst cavity followed by cavity management techniques such as unroofing, omentoplasty, or capitonnage is essential [3, 11, 42, 67, 68]. Laparoscopic approaches are increasingly preferred for their reduced morbidity and quicker recovery, particularly for anterior liver segment cysts [3, 11, 42, 67, 68].

(iv) Watch and Wait Strategy: For inactive or calcified cysts (CE4 and CE5), a conservative approach with regular monitoring is suitable. These cysts have a low risk of complications, and the "watch and wait" strategy is appropriate when imaging and serological tests confirm inactivity [3, 11, 15, 22, 69].

(v) Alveolar echinococcosis primarily affects the liver and typically presents as slow-growing, tumor-like lesions. The preferred treatment approach is radical surgical resection, when feasible, as it offers the best chance for disease control. However, due to the infiltrative and invasive nature of alveolar echinococcosis, many patients present at advanced stages where complete resection is not possible. In such cases, palliative surgery or long-term medical therapy becomes necessary to manage symptoms and control disease progression [11]. Long-term benzimidazole treatment for several years is mandatory in all inoperable alveolar echinococcosis patients and following surgical resection of the parasite lesions. Since residual parasite tissue may remain undetected even after radical surgery, including liver transplantation, benzimidazole should be administered for at least 2 years, with a follow-up period of at least 10 years to monitor for possible recurrence [11]. Preoperative benzimidazole administration is generally not recommended except in the case of liver transplantation. Liver transplantation has been performed worldwide in more than 100 reported cases, primarily for patients with inoperable lesions and/or chronic liver failure due to alveolar echinococcosis (rarely E. granulosus infection). Liver transplantation is particularly indicated when there is severe liver disease (e.g., secondary biliary cirrhosis or Budd-Chiari syndrome), recurrent life-threatening cholangitis, or when radical liver resection is not feasible [70–72]. Importantly, patients should not have extra-hepatic alveolar echinococcosis locations, as residual alveolar echinococcosis in the lung or abdominal cavity significantly complicates outcomes. Immunosuppression following liver transplantation may favor the re-growth of larval remnants or the formation/increase in size of metastases [71]. Given the complexity of these cases, multidisciplinary care and ongoing monitoring are crucial. Limited literature suggest that despite the continuous use of immunosuppressive drugs after liver transplantation, benzimidazole can effectively inhibit the growth of alveolar echinococcosis metastases. Therefore, the combination of liver transplantation and sustained benzimidazole therapy can prolong patient survival, both in the presence and absence of metastases in other organs [3, 73]. However, there is still no consensus in the literature regarding the optimal use, dosage, and duration of benzimidazole therapy post-transplant [71].

Recent studies are exploring the use of anti-angiogenic drugs, antitumor agents, antibiotics, immunosuppressants, anti-infective agents, antimalarial drugs and insecticides, synthetic compounds, and drugs carriers to manage alveolar echinococcosis [74]. Moreover, innovative therapies involving natural compounds from traditional Chinese and Tibetan medicine are being evaluated for their potential to interfere with parasitic metabolic pathways [74]. These novel approaches aim to increase therapeutic efficacy while minimizing the adverse effects commonly associated with long-term albendazole use. In conclusion, the choice of treatment for echinococcosis should be individualized based on cyst type, patient condition, and available medical resources. Ongoing follow-up with imaging and serological tests is crucial to detect recurrence and manage complications.

Follow-Up

Follow-up after the treatment of cystic and alveolar echinococcosis is essential to detect potential recurrences or residual disease [42, 75]. The primary goal of follow-up is to identify residual cysts or early recurrence, which may still occur despite successful treatment interventions [75]. The follow-up strategy should be individualized based on the type of intervention performed, the characteristics of the cyst, and the patient's clinical condition.

(i) Postoperative Imaging: The recommended imaging follow-up starts with the first assessment at 1 month post-treatment using US. Subsequent checks should be performed at the third, sixth, and ninth months within the first year. After the first year, follow-up should continue every 6 months for at least 3 to 5 years, especially in endemic areas. For high-risk patients or those with complicated cysts, additional imaging with CT is advisable to confirm uncertain US findings [3, 42, 76].

(ii) Serological Monitoring: Serological tests, such as ELISA and IHA, play a critical role in follow-up. It is important to note that serological markers may take years to become negative, and any re-positivity after a period of negativity should be carefully evaluated as it might indicate recurrence. Therefore, periodic serological follow-up is recommended, particularly in high-risk or endemic areas [3, 42].

(iii) Recurrent Cases and Treatment: In cases of recurrence after percutaneous treatment, the primary approach should be re-attempting percutaneous therapy. If percutaneous methods fail, conservative surgery should be considered, and if the patient is eligible, radical surgery should be performed.

(iv) Multidisciplinary Approach: A comprehensive follow-up strategy should involve a multidisciplinary team, including hepatobiliary surgeons, infectious disease specialists, and radiologists. Patient education regarding recurrence risks and the importance of consistent follow-up is crucial. Additionally, implementing public health measures, such as regular deworming of livestock and dogs, is vital to minimize transmission.

The Role of Artificial Intelligence in Hydatid Disease Management

In recent years, the integration of artificial intelligence into the management of hydatid disease has shown significant potential. AI algorithms, particularly those based on machine learning and deep learning, have demonstrated remarkable capabilities in automating cyst detection, predicting complications, and assisting in surgical planning. These technologies leverage large datasets from imaging studies, including US, CT, and MRI, to enhance diagnostic accuracy and reduce the likelihood of human error in interpreting complex cases. One of the primary applications of artificial intelligence in hydatid disease management is automated cyst detection. By training convolutional neural networks (CNNs) on vast imaging datasets, artificial intelligence systems can identify cyst characteristics with high sensitivity and specificity. These systems assist radiologists in differentiating between active, transitional, and inactive cysts, thereby facilitating early diagnosis and treatment planning. Recent studies have shown that deep learning algorithms, such as U-Net and CNN, have achieved high accuracy in staging liver hydatid cysts, particularly in differentiating between CE1 and CE2 stages.

Artificial intelligence models have also been developed to predict potential complications associated with cyst rupture, superinfection, or biliary communication. By analyzing patient data, cyst morphology, and radiological features, these models provide risk assessments that allow clinicians to take proactive measures. For instance, artificial intelligence-driven predictive analytics can estimate the likelihood of cyst rupture based on size, location, and internal structure. Models integrating clinical and serological data have shown promising results in predicting postoperative complications and recurrence risks. Integrating artificial intelligence into surgical planning involves using machine learning algorithms to model different surgical scenarios and predict outcomes. Artificial intelligence-based decision support systems recommend surgical techniques based on cyst type, size, anatomical location, and potential intraoperative challenges. Moreover, artificial intelligence algorithms trained on historical surgical data can predict postoperative recovery patterns and complication rates, guiding surgical decision-making. By analyzing patient-specific factors and cyst characteristics, artificial intelligence can tailor treatment plans, optimizing the choice between percutaneous techniques, surgical excision, or conservative management. Decision support algorithms that incorporate real-time data enable clinicians to choose the most suitable intervention while minimizing complications. Hybrid AI models combining imaging data with serological markers have shown potential in refining treatment strategies. Ongoing research focuses on improving the interpretability and robustness of artificial intelligence models. Challenges include the need for large, high-quality datasets and the implementation of data fusion techniques to effectively integrate multimodal data (clinical, imaging, serological) to enhance predictive accuracy. Data fusion methodologies, such as feature-level fusion and decision-level fusion, can synergistically combine heterogeneous data sources, improving the robustness and reliability of artificial intelligence-driven diagnostic and predictive models. Future artificial intelligence systems aim to offer real-time guidance during surgery, postoperative monitoring, and long-term follow-up. Multicenter validation studies are essential to standardize artificial intelligence applications, ensuring their safe and effective integration into clinical practice [77–93].

The Hydatid Disease Book: From Fundamentals to Future Innovations

Overall, this book aims to provide a comprehensive resource that covers all facets of hydatid disease, from basic pathophysiology and epidemiology to advanced therapeutic approaches and innovative technologies. By synthesizing the latest evidence and clinical expertise, the book strives to support healthcare professionals in making informed decisions and advancing patient care in regions burdened by this challenging parasitic infection. Additionally, it emphasizes the importance of using accurate and consistent terminology when discussing hydatid disease, as precise language is crucial for effective communication among healthcare professionals and for the standardization of diagnosis and treatment approaches.

Editor Note The ongoing conflict in Gaza has created a profound public health crisis, with devastating consequences for vulnerable populations. Disruption of healthcare infrastructure, shortage of clean water, and limited access to food and medicines have resulted in a surge of preventable child deaths and a dramatic rise in morbidity among older adults. Overcrowded shelters and destruction of sanitation systems have facilitated the rapid spread of infectious diseases, including diarrheal outbreaks and respiratory infections, further overwhelming the already fragile healthcare system. These conditions have not only increased acute mortality but have also amplified the long-term burden of malnutrition, chronic disease complications, and psychosocial trauma within the affected population.

From a public health and epidemiological perspective, the rise in famine-related deaths over the past few months—largely due to the obstruction of humanitarian aid—highlights the urgent need for coordinated international action. Global institutions, including WHO and UNICEF, bear significant responsibility to respond decisively rather than remain passive observers in the face of such a crisis. There is no moral or rational justification for the ongoing suffering of civilians anywhere in the world, and the immediate establishment of secure humanitarian corridors must be considered an ethical imperative.

Sami Akbulut

Department of Surgery
Inonu University Faculty of Medicine
Malatya, Turkey

Department of Public Health
Inonu University Faculty of Medicine
Malatya, Turkey

Department of Biostatistics and Medical Informatics
Inonu University Faculty of Medicine
Malatya, Turkey

References

1. Kakamad FH, Anwar KA, Ahmed HK, Habibullah IJ, Kaka Ali HH, Nasralla HA, Abdullah HO, Tahir SH, Kareem HO, Hasan AH, Gharib DT, Asaad HR, Mohammed AA, Abdalla BA, Esmaeil DA, Rashid RJ, Hamahussein KF. Risk factors associated with human echinococcosis: a systematic review and meta-analysis. Front Vet Sci. 2024;11:1480579. https://doi.org/10.3389/fvets.2024.1480579. PMID: 39654835 PMCID: PMC11625768.
2. Crotti S, Brustenga L, Cruciani D, Bonelli P, D'Avino N, Felici A, Morandi B, Sebastiani C, Spina S, Gobbi M. Molecular screening of Echinococcus spp. and other cestodes in wild carnivores from Central Italy. Vet Sci. 2023;10(5):318. https://doi.org/10.3390/vetsci10050318.
3. Brunetti E, Kern P, Vuitton DA. Expert consensus for the diagnosis and treatment of cystic and alveolar echinococcosis in humans. Acta Trop. 2010;114(1):1–16. https://doi.org/10.1016/j.actatropica.2009.11.001. PMID: 19931502.
4. Tian T, Miao L, Wang W, Zhou X. Global, regional and national burden of human cystic echinococcosis from 1990 to 2019: A systematic analysis for the global burden of disease study 2019. Trop Med Infect Dis. 2024;9(4) https://doi.org/10.3390/tropicalmed9040087. PMID: 38668548 PMCID: PMC11054543.
5. Gun E, Etit D, Buyuktalanci DO, Cakalagaoglu F. Unusual locations of hydatid disease: a 10-year experience from a tertiary reference center in Western Turkey. Ann Diagn Pathol. 2017;29:37–40. https://doi.org/10.1016/j.anndiagpath.2017.01.011. PMID: 28807340.
6. Kumar S, Gautam S, Prakash R, Kumar S, Srivastava VK. Simultaneous primary hydatid cysts of liver and spleen with spontaneous intraperitoneal rupture of liver cyst. J Clin Diagn Res. 2014;8(11):Nj01. https://doi.org/10.7860/jcdr/2014/9097.5105. PMID: 25584268 PMCID: PMC4290289.
7. Sozuer E, Akyuz M, Akbulut S. Open surgery for hepatic hydatid disease. Int Surg. 2014;99(6):764–9. https://doi.org/10.9738/intsurg-d-14-00069.1. PMID: 25437585 PMCID: PMC4254238.
8. Akbulut S, Demircan F, Sogutcu N. Hydatid cyst disease of the thyroid gland: report of two cases. Int Surg. 2015;100(4):643–7. https://doi.org/10.9738/intsurg-d-14-00124.1. PMID: 25598424 PMCID: PMC4400932.
9. Bannour B, Romdhani M, Chiba D, Bannour I, Abdelkader AB, Mokni M, Boughizane S. Isolated hydatid cyst of the breast: A rare pseudotumor of the breast. Eur J Breast Health. 2025;21(2):182–5. https://doi.org/10.4274/ejbh.galenos.2025.2024-12-10. PMID: 40066723 PMCID: PMC11934831.

10. Abdennour RB, Derbel F, Mokni M, Ben Mabrouk M, Jemni H, Hamida MBH, Mazhoud J, Youssef S, Ali AB, Mama N, Ibtissem H, Nadia A, Ouni CE, Naija W. Hydatid cysts of the liver—diagnosis, complications and treatment. In: Derbel F, editor. Abdominal surgery. Rijeka: IntechOpen; 2012.
11. Menghi C, Gatta C, Arias L. Human cystic and alveolar echinococcosis. Curr Treat Options Infect Dis. 2017;9(2):210–22. https://doi.org/10.1007/s40506-017-0121-0.
12. Aydin F, Yalcin A, Karaman A, Sade R, Ozturk G, Alper F. Diagnostic and management perspectives in alveolar echinococcosis: review of literature. Eurasian J Med. 2022;54(Suppl1):10–5. https://doi.org/10.5152/eurasianjmed.2022.22308. PMID: 36655439 PMCID: PMC11163348.
13. D'Alessandro A, Rausch RL. New aspects of neotropical polycystic (Echinococcus vogeli) and unicystic (Echinococcus oligarthrus) echinococcosis. Clin Microbiol Rev. 2008;21(2):380–401, table of contents. https://doi.org/10.1128/cmr.00050-07. PMID: 18400802 PMCID: PMC2292577.
14. Borhani M, Fathi S, Lahmar S, Ahmed H, Abdulhameed MF, Fasihi HM. Cystic echinococcosis in the Eastern Mediterranean region: neglected and prevailing! PLoS Negl Trop Dis. 2020;14(5):e0008114. https://doi.org/10.1371/journal.pntd.0008114. PMID: 32379760 PMCID: PMC7205190.
15. Hogea MO, Ciomaga BF, Muntean MM, Muntean AA, Popa MI, Popa GL. Cystic echinococcosis in the early 2020s: a review. Trop Med Infect Dis. 2024;9(2) https://doi.org/10.3390/tropicalmed9020036. PMID: 38393125 PMCID: PMC10891927.
16. McManus DP, Zhang W. Helminthic diseases: echinococcosis. 2024; https://doi.org/10.1016/B978-0-12-801238-3.03103-2.
17. McManus DP, Zhang W, Li J, Bartley PB. Echinococcosis. Lancet. 2003;362(9392):1295–304. https://doi.org/10.1016/s0140-6736(03)14573-4. PMID: 14575976.
18. Tuergan T, Abulaiti A, Tulahong A, Zhang R, Shao Y, Aji T. Global, regional and national burden of echinococcosis in 204 countries and territories from 1990 to 2021: a systematic analysis based on the Global Burden of Disease Study 2021. J Infect Public Health. 2025;8:102810. https://doi.org/10.1016/j.jiph.2025.102810.
19. Wen H, Vuitton L, Tuxun T, Li J, Vuitton DA, Zhang W, McManus DP. Echinococcosis: advances in the 21st century. Clin Microbiol Rev. 2019;32(2) https://doi.org/10.1128/cmr.00075-18. PMID: 30760475 PMCID: PMC6431127.
20. WHO Informal Working Group. International classification of ultrasound images in cystic echinococcosis for application in clinical and field epidemiological settings. Acta Trop. 2003;85(2):253–61. https://doi.org/10.1016/s0001-706x(02)00223-1. PMID: 12606104.
21. Moro P, Schantz PM. Echinococcosis: a review. Int J Infect Dis. 2009;13(2):125–33. https://doi.org/10.1016/j.ijid.2008.03.037. PMID: 18938096.
22. Govindasamy A, Bhattarai PR, John J. Liver cystic echinococcosis: a parasitic review. Ther Adv Infect Dis. 2023;10:20499361231171478. https://doi.org/10.1177/20499361231171478. PMID: 37197609 PMCID: PMC10184195.
23. Rostami A, Lundström-Stadelmann B, Frey CF, Beldi G, Lachenmayer A, Chang BCH, Norouzian MM, Hemphill A, Gasser RB. Human alveolar echinococcosis-A neglected zoonotic disease requiring urgent attention. Int J Mol Sci. 2025;26(6) https://doi.org/10.3390/ijms26062784. PMID: 40141427 PMCID: PMC11943292.
24. Gessese AT. Review on epidemiology and public health significance of hydatidosis. Vet Med Int. 2020;2020:8859116. https://doi.org/10.1155/2020/8859116. PMID: 33354312 PMCID: PMC7735834.
25. Grosso G, Gruttadauria S, Biondi A, Marventano S, Mistretta A. Worldwide epidemiology of liver hydatidosis including the Mediterranean area. World J Gastroenterol. 2012;18(13):1425–37. https://doi.org/10.3748/wjg.v18.i13.1425. PMID: 22509074 PMCID: PMC3319938.
26. Larrieu E, Zanini F. Critical analysis of cystic echinococcosis control programs and praziquantel use in South America, 1974–2010. Rev Panam Salud Publica. 2012;31(1):81–7. https://doi.org/10.1590/s1020-49892012000100012. PMID: 22427169.
27. Deplazes P, Rinaldi L, Alvarez Rojas CA, Torgerson PR, Harandi MF, Romig T, Antolova D, Schurer JM, Lahmar S, Cringoli G, Magambo J, Thompson RC, Jenkins EJ. Global distribution of alveolar and cystic echinococcosis. Adv Parasitol. 2017;95:315–493. https://doi.org/10.1016/bs.apar.2016.11.001. PMID: 28131365.

28. Eckert J, Gemmell M, Meslin F-X, Pawlowski Z. WHO/OIE manual on echinococcosis in humans and animals: a public health problem of global concern. France: World Organisation for Animal Health Paris; 2001.
29. Massolo A, Liccioli S, Budke C, Klein C. Echinococcus multilocularis in North America: the great unknown. Parasite. 2014;21:73. https://doi.org/10.1051/parasite/2014069. PMID: 25531581 PMCID: PMC4273702.
30. Fallahizadeh S, Arjmand R, Jelowdar A, Rafiei A, Kazemi F. Determination of Echinococcus granulosus genotypes in livestock slaughtered in Shush County, Southwest Iran using PCR-RFLP. Helminthologia. 2019;56(3):196–201. https://doi.org/10.2478/helm-2019-0023. PMID: 31662691 PMCID: PMC6799581.
31. Zhou H, Wang X, Han S, Xiao N. Advances and challenges in the prevention, control and research of echinococcosis in China. Decod Infect Transm. 2025;3:100041. https://doi.org/10.1016/j.dcit.2025.100041.
32. Cassini R, Canali M, Tamarozzi F, Angheben A, Capelli G, Gobbi F, Legnardi M, Brichese M, Napoletano G, Cestaro F, Casulli A, Drigo M, Aragrande M. A One-Health evaluation of the burden of cystic echinococcosis and its prevention costs: case study from a hypo-endemic area in Italy. One Health. 2021;13:100320. https://doi.org/10.1016/j.onehlt.2021.100320. PMID: 34504939 PMCID: PMC8413884.
33. Salman H, Salman Z, Kart Y, Akçam M. Cystic echinococcosis in children: ten years of experience and which laboratory result in evaluating ruptured cases? J Contemp Med. 2023;13(1):54–9. https://doi.org/10.16899/jcm.1207494.
34. Erfani A, Shahriarirad R, Eskandarisani M, Rastegarian M, Sarkari B. Management of liver hydatid cysts: a retrospective analysis of 293 surgical cases from Southern Iran. J Trop Med. 2023;2023:9998739. https://doi.org/10.1155/2023/9998739. PMID: 37377601 PMCID: PMC10292944.
35. Aydin Y, Ulas AB, Ahmed AG, Eroglu A. Pulmonary hydatid cyst in children and adults: diagnosis and management. Eurasian J Med. 2022;54(Suppl1):133–40. https://doi.org/10.5152/eurasianjmed.2022.22289. PMID: 36655457 PMCID: PMC11163342.
36. Doganay Z, Guven H, Aygun D, Altintop L, Yerliyurt M, Deniz T. Blunt abdominal trauma with unexpected anaphylactic shock due to rupture of hepatic hydatid cysts. Emerg Med Grand Rounds. 2002;2:17–20.
37. Yilmaz M, Akbulut S, Kahraman A, Yilmaz S. Liver hydatid cyst rupture into the peritoneal cavity after abdominal trauma: case report and literature review. Int Surg. 2012;97(3):239–44. https://doi.org/10.9738/cc116.1. PMID: 23113853 PMCID: PMC3723224.
38. Karakas E, Tuna Y, Basar O, Koklu S. Primary pancreatic hydatid disease associated with acute pancreatitis. Hepatobiliary Pancreat Dis Int. 2010;9(4):441–2. PMID: 20688612.
39. Gharde P, Wagh D, Patil A. Left renal hydatid cyst presenting as hematuria and macroscopic hydatiduria since last ten years. Trop Parasitol. 2012;2(1):58–60. https://doi.org/10.4103/2229-5070.97242. PMID: 23508941 PMCID: PMC3593509.
40. Schantz PM, Kern P, Brunetti E. Chapter 114—Echinococcosis. In: Guerrant RL, Walker DH, Weller PF, editors. Tropical infectious diseases. 2nd ed. Philadelphia: Churchill Livingstone; 2006. p. 1304–26.
41. Sen P, Demirdal T, Nemli SA. Evaluation of clinical, diagnostic and treatment aspects in hydatid disease: analysis of an 8-year experience. Afr Health Sci. 2019;19(3):2431–8. https://doi.org/10.4314/ahs.v19i3.17. PMID: 32127814 PMCID: PMC7040281.
42. Öztürk G, Uzun MA, Özkan ÖF, Kayaalp C, Tatlı F, Eren S, Aksungur N, Çoker A, Bostancı EB, Öter V, Kaya E, Taşar P. Turkish HPB Surgery Association consensus report on hepatic cystic echinococcosis (HCE). Turk J Surg. 2022;38(2):101–20. https://doi.org/10.47717/turkjsurg.2022.5757. PMID: 36483170 PMCID: PMC9714645.
43. Brunetti E, Tamarozzi F, Macpherson C, Filice C, Piontek MS, Kabaalioglu A, Dong Y, Atkinson N, Richter J, Schreiber-Dietrich D, Dietrich CF. Ultrasound and cystic echinococcosis. Ultrasound Int Open. 2018;4(3):E70–e78. https://doi.org/10.1055/a--0650-3807. PMID: 30364890 PMCID: PMC6199172.
44. Gharbi HA, Hassine W, Brauner MW, Dupuch K. Ultrasound examination of the hydatic liver. Radiology. 1981;139(2):459–63. https://doi.org/10.1148/radiology.139.2.7220891. PMID: 7220891.

45. Kern P, Wen H, Sato N, Vuitton DA, Gruener B, Shao Y, Delabrousse E, Kratzer W, Bresson-Hadni S. WHO classification of alveolar echinococcosis: principles and application. Parasitol Int. 2006;55(Suppl):S283–7. https://doi.org/10.1016/j.parint.2005.11.041. PMID: 16343985.
46. Anand S, Rajagopalan S, Mohan R. Management of liver hydatid cysts—current perspectives. Med J Armed Forces India. 2012;68(3):304–9. https://doi.org/10.1016/j.mjafi.2012.04.010. PMID: 24532894 PMCID: PMC3862559.
47. Alshoabi SA, Alkalady AH, Almas KM, Magram AO, Algaberi AK, Alareqi AA, Hamid AM, Alhazmi FH, Qurashi AA, Abdulaal OM, Aloufi KM, Alsharif WM, Alsultan KD, Omer AM, Gareeballah A. Hydatid disease: A radiological pictorial review of a great neoplasms mimicker. Diagnostics (Basel). 2023;13(6) https://doi.org/10.3390/diagnostics13061127. PMID: 36980435 PMCID: PMC10047450.
48. Yesilyurt M, Esdur V. Anatomical-based imaging of cystic echinococcosis and review of the current literature. Eurasian J Med. 2022;54(Suppl1):1–9. https://doi.org/10.5152/eurasianjmed.2022.22309. PMID: 36655438 PMCID: PMC11163361.
49. Erden A, Ormeci N, Fitoz S, Erden I, Tanju S, Genç Y. Intrabiliary rupture of hepatic hydatid cysts: diagnostic accuracy of MR cholangiopancreatography. AJR Am J Roentgenol. 2007;189(2):W84–9. https://doi.org/10.2214/ajr.07.2068. PMID: 17646444.
50. Zaharie F, Valean D, Zaharie R, Popa C, Mois E, Schlanger D, Fetti A, Zdrehus C, Ciocan A, Al-Hajjar N. Surgical management of hydatid cyst disease of the liver: an improvement from our previous experience? World J Gastrointest Surg. 2023;15(5):847–58. https://doi.org/10.4240/wjgs.v15.i5.847. PMID: 37342852 PMCID: PMC10277938.
51. Dolay K, Akbulut S. Role of endoscopic retrograde cholangiopancreatography in the management of hepatic hydatid disease. World J Gastroenterol. 2014;20(41):15253–61. https://doi.org/10.3748/wjg.v20.i41.15253. PMID: 25386073 PMCID: PMC4223258.
52. Gümüşoğlu AY, Kabuli HA, Değerli MS, Atar B, Köneş O, Kocataş A, Bingül ND, Dolay K. Endoscopic definitive treatment of liver hydatid cysts ruptured into bile duct. Surg Endosc. 2024;38(12):7187–94. https://doi.org/10.1007/s00464-024-11305-9. PMID: 39367134.
53. Husmann L, Deibel A, Skawran S, Ledergerber B, Muehlematter UJ, Hasse B, Huellner MW, Reiner CS, Muellhaupt B. Follow-up PET/CT of alveolar echinococcosis: comparison of metabolic activity and immunodiagnostic testing. PLoS One. 2022;17(6):e0270695. https://doi.org/10.1371/journal.pone.0270695. PMID: 35767557 PMCID: PMC9242476.
54. Husmann L, Muehlematter UJ, Grimm F, Ledergerber B, Messerli M, Kudura K, Gruenig H, Muellhaupt B, Hasse B, Huellner MW. PET/CT helps to determine treatment duration in patients with resected as well as inoperable alveolar echinococcosis. Parasitol Int. 2021;83:102356. https://doi.org/10.1016/j.parint.2021.102356. PMID: 33872794.
55. Graeter T, Eberhardt N, Shi R, Schmidberger J, Beer AJ, Beer M, Henne-Bruns D, Hillenbrand A, Barth TFE, Grimm J, Kratzer W, Gruener B. Hepatic alveolar echinococcosis: correlation between computed tomography morphology and inflammatory activity in positron emission tomography. Sci Rep. 2020;10(1):11808. https://doi.org/10.1038/s41598-020-68624-9. PMID: 32678174 PMCID: PMC7366930.
56. Hadipour M, Nazari M, Sanei B, Ghayour Z, Sharafi SM, Yazdani H, Darani HY. Immunological diagnosis of human hydatid cyst using Western immunoblotting technique. J Res Med Sci. 2016;21:130. https://doi.org/10.4103/1735-1995.196612. PMID: 28331516 PMCID: PMC5348824.
57. Demirkazık M, Koltas İS, İnceboz T, Korkmaz M, Gümürdülü D. Use of the ELISA (Em2-Em18) and Western blotting methods on diagnosis of alveolar echinococcosis. Turkiye Parazitol Derg. 2019;43(Suppl 1):13–7. https://doi.org/10.4274/tpd.galenos.2019.6447. PMID: 31591873.
58. Ito A, Sako Y, Yamasaki H, Mamuti W, Nakaya K, Nakao M, Ishikawa Y. Development of Em18-immunoblot and Em18-ELISA for specific diagnosis of alveolar echinococcosis. Acta Trop. 2003;85(2):173–82. https://doi.org/10.1016/s0001-706x(02)00221-8. PMID: 12606094.
59. Gottstein B, Lachenmayer A, Beldi G, Wang J, Merkle B, Vu XL, Kurath U, Müller N. Diagnostic and follow-up performance of serological tests for different forms/courses of alveolar echinococcosis. Food Waterborne Parasitol. 2019;16:e00055. https://doi.org/10.1016/j.fawpar.2019.e00055. PMID: 32095626 PMCID: PMC7034017.

60. Ormeci N. PAIR vs Örmeci technique for the treatment of hydatid cyst. Turk J Gastroenterol. 2014;25(4):358–64. https://doi.org/10.5152/tjg.2014.13018. PMID: 25254515.
61. Nazligul Y, Kucukazman M, Akbulut S. Role of chemotherapeutic agents in the management of cystic echinococcosis. Int Surg. 2015;100(1):112–4. https://doi.org/10.9738/intsurg-d-14-00068.1. PMID: 25594649 PMCID: PMC4301274.
62. Suntur BM, Çavuş G. Pooled analysis of 86 published spinal hydatid cyst cases from Turkey. Mediterr J Infect Microb Antimicrob. 2019;8(1):7. https://doi.org/10.4274/mjima.galenos.2019.2019.7.
63. Alvela-Suárez L, Velasco-Tirado V, Belhassen-Garcia M, Novo-Veleiro I, Pardo-Lledías J, Romero-Alegría A, Pérez del Villar L, Valverde-Merino MP, Cordero-Sánchez M. Safety of the combined use of praziquantel and albendazole in the treatment of human hydatid disease. Am J Trop Med Hyg. 2014;90(5):819–22. https://doi.org/10.4269/ajtmh.13-0059. PMID: 24615131 PMCID: PMC4015571.
64. Najdaghi S, Alizadehasl A, Abbaszade N, Jebelli SFH, Davani DN, Aliabadi AY, Haghazali M, Gloverdi AY. Combination therapy of albendazole and praziquantel for treatment of cardiac and multiorgan hydatid cyst: a case report of novel treatment and literature review. Clin Case Rep. 2024;12(12):e9610. https://doi.org/10.1002/ccr3.9610. PMID: 39619313 PMCID: PMC11605159.
65. Akhan O, Erdoğan E, Ciftci TT, Unal E, Karaağaoğlu E, Akinci D. Comparison of the long-term results of Puncture, Aspiration, Injection and Re-aspiration (PAIR) and catheterization techniques for the percutaneous treatment of CE1 and CE3a liver hydatid cysts: a prospective randomized trial. Cardiovasc Intervent Radiol. 2020;43(7):1034–40. https://doi.org/10.1007/s00270-020-02477-7. PMID: 32382859.
66. Akhan O, Yildiz AE, Akinci D, Yildiz BD, Ciftci T. Is the adjuvant albendazole treatment really needed with PAIR in the management of liver hydatid cysts? A prospective, randomized trial with short-term follow-up results. Cardiovasc Intervent Radiol. 2014;37(6):1568–74. https://doi.org/10.1007/s00270-014-0840-2. PMID: 24464258.
67. Ferrer Inaebnit E, Molina Romero FX, Segura Sampedro JJ, González Argenté X, Morón Canis JM. A review of the diagnosis and management of liver hydatid cyst. Rev Esp Enferm Dig. 2022;114(1):35–41. https://doi.org/10.17235/reed.2021.7896/2021. PMID: 34034501.
68. Bektasoglu HK, Hasbahceci M, Tasci Y, Aydogdu I, Malya FU, Kunduz E, Dolay K. Comparison of laparoscopic and conventional cystotomy/partial cystectomy in treatment of liver hydatidosis. Biomed Res Int. 2019;2019:1212404. https://doi.org/10.1155/2019/1212404. PMID: 30868067 PMCID: PMC6379859.
69. Nunnari G, Pinzone MR, Gruttadauria S, Celesia BM, Madeddu G, Malaguarnera G, Pavone P, Cappellani A, Cacopardo B. Hepatic echinococcosis: clinical and therapeutic aspects. World J Gastroenterol. 2012;18(13):1448–58. https://doi.org/10.3748/wjg.v18.i13.1448. PMID: 22509076 PMCID: PMC3319940.
70. Aydinli B, Ozturk G, Arslan S, Kantarci M, Tan O, Ahıskalioglu A, Özden K, Colak A. Liver transplantation for alveolar echinococcosis in an endemic region. Liver Transpl. 2015;21(8):1096–102. https://doi.org/10.1002/lt.24195. PMID: 26074280.
71. Bresson-Hadni S, Koch S, Miguet JP, Gillet M, Mantion GA, Heyd B, Vuitton DA. Indications and results of liver transplantation for Echinococcus alveolar infection: an overview. Langenbeck's Arch Surg. 2003;388(4):231–8. https://doi.org/10.1007/s00423-003-0394-2. PMID: 12905036.
72. Ozdemir F, Ince V, Barut B, Onur A, Kayaalp C, Yilmaz S. Living donor liver transplantation for Echinococcus Alveolaris: single-center experience. Liver Transpl. 2015;21(8):1091–5. https://doi.org/10.1002/lt.24170. PMID: 25981332.
73. Zavoikin VD, Zelya OP, Tumolskaya NI. The importance of uninterrupted albendazole treatment in patients with unresectable alveolar echinococcosis undergoing liver transplantation. Transpl Infect Dis. 2020;22(4):e13291. https://doi.org/10.1111/tid.13291. PMID: 32277859.
74. Jing QD, Ji-De A, Liu LX, Fan HN. Current status of drug therapy for alveolar echinococcosis. World J Hepatol. 2024;16(11):1243–54. https://doi.org/10.4254/wjh.v16.i11.1243. PMID: 39606163 PMCID: PMC11586754.
75. Sadjjadi FS, Ahmadi N, Rezaie-Tavirani M, Zali H. Following up of surgical treated human liver cystic echinococcosis: a proteomics approach. Iran J Parasitol. 2021;16(1):11–22. https://doi.org/10.18502/ijpa.v16i1.5507. PMID: 33786043 PMCID: PMC7988678.

76. Rosa DD, Figueredo E, Rosas M, Goñi F. Follow up study of symptomatic human cystic echinococcosis treatment with albendazole and praziquantel, in Uruguay. BMC Infect Dis. 2024;24(1):737. https://doi.org/10.1186/s12879-024-09539-y. PMID: 39060987 PMCID: PMC11282602.
77. Abbasi B, Akhavan R, Ghamari Khameneh A, Darban Hosseini Amirkhiz G, Rezaei-Dalouei H, Tayebi S, Hashemi J, Aminizadeh B, Darban Hosseini Amirkhiz S. Computed tomography and magnetic resonance imaging of hydatid disease: a pictorial review of uncommon imaging presentations. Heliyon. 2021;7(5):e07086. https://doi.org/10.1016/j.heliyon.2021.e07086. PMID: 34095581 PMCID: PMC8166760.
78. Abdukeyim N, Liu Z, Yan C. CT images classification of hepatic hydatid disease by different methods. In: Proceedings of the 2019 6th International Conference on Systems and Informatics (ICSAI); 2019, 2–4 November 2019. p. 398–403.
79. Abdulla SH, Soran HM. Comparing and evaluating machine learning techniques for liver-hydatid-cyst classification 2024: 307–321.
80. Akbari MN, Azizi A. Enhancing hydatid cyst classification with deep learning and convolutional neural networks using CT scans. Afghan J Infect Dis. 2024;2(1):9–16. https://doi.org/10.60141/AJID/V.2.I.1.2.
81. Chen J, Li K, Wang Z, Wang Z, Huang J. HES-UNet: A U-net for hepatic echinococcosis lesion segmentation. arXiv preprint arXiv:241206530 (2024).
82. Gul Y, Muezzinoglu T, Kilicarslan G, Dogan S, Tuncer T. Application of the deep transfer learning framework for hydatid cyst classification using CT images. Soft Comput. 2023;27(11):7179–89.
83. Lissandrin R, Cicerone O, Vola A, D'Alessandro G, Frassini S, Manciulli T, Famularo S, De Silvestri A, Viganò J, Quaretti P, Ansaloni L, Brunetti E, Maestri M. Machine learning to predict the decision to perform surgery in hepatic echinococcosis. HPB (Oxford). 2025;27(4):480–8. https://doi.org/10.1016/j.hpb.2024.12.014. PMID: 39755481.
84. Manciulli T, Brunetti E. Stepping stones and challenges in the use of artificial intelligence in the diagnosis of echinococcosis. Lancet Digit Health. 2023;5(11):e750–1. https://doi.org/10.1016/s2589-7500(23)00183-8. PMID: 37770336.
85. Nijiati M, Tuerdi M, Damola M, Yimit Y, Yang J, Abulaiti A, Mutailifu A, Aihait D, Wang Y, Zou X. A deep learning radiomics model based on CT images for predicting the biological activity of hepatic cystic echinococcosis. Front Physiol. 2024;15:1426468. https://doi.org/10.3389/fphys.2024.1426468. PMID: 39175611 PMCID: PMC11338923.
86. Peng J, Duan Z, Guo Y, Li X, Luo X, Han X, Luo J. Identification of candidate biomarkers of liver hydatid disease via microarray profiling, bioinformatics analysis, and machine learning. J Int Med Res. 2021;49(3):300060521993980. https://doi.org/10.1177/0300060521993980. PMID: 33787392 PMCID: PMC8020228.
87. Shang F, Song T, Wang Z, Wu M, Yan C, Wang X. Clinical application of artificial intelligence in the ultrasound classification of hepatic cystic echinococcosis. Am J Trop Med Hyg 2024; 111(1): 93–101. DOI: https://doi.org/10.4269/ajtmh.23-0519. PMID: 38806047 PMCID: PMC11229658.
88. Tegshee T, Dorjsuren T, Lee S, Batjargal D. A study on staging cystic echinococcosis using machine learning methods. Bioengineering (Basel). 2025;12(2) https://doi.org/10.3390/bioengineering12020181. PMID: 40001700 PMCID: PMC11852189.
89. Wang Z, Bian H, Li J, Xu J, Fan H, Wu X, Cao Y, Guo B, Xu X, Wang H, Zhang L, Zhou H, Fan J, Ren Y, Geng Y, Feng X, Li L, Wei L, Zhang X. Detection and subtyping of hepatic echinococcosis from plain CT images with deep learning: a retrospective, multicentre study. Lancet Digit Health. 2023;5(11):e754–62. https://doi.org/10.1016/s2589-7500(23)00136-x. PMID: 37770335.
90. Wu M, Yan C, Sen G. Computer-aided diagnosis of hepatic cystic echinococcosis based on deep transfer learning features from ultrasound images. Sci Rep. 2025;15(1):607. https://doi.org/10.1038/s41598-024-85004-9. PMID: 39753933 PMCID: PMC11698856.
91. Xin S, Shi H, Jide A, Zhu M, Ma C, Liao H. Automatic lesion segmentation and classification of hepatic echinococcosis using a multiscale-feature convolutional neural network. Med Biol Eng Comput. 2020;58(3):659–68. https://doi.org/10.1007/s11517-020-02126-8. PMID: 31950330.

92. Yang Y, Cairang Y, Jiang T, Zhou J, Zhang L, Qi B, Ma S, Tang L, Xu D, Bu L, Bu R, Jing X, Wang H, Zhou Z, Zhao C, Luo B, Liu L, Guo J, Nima Y, Hua G, Wa Z, Zhang Y, Zhou G, Jiang W, Wang C, De Y, Yu X, Cheng Z, Han Z, Liu F, Dou J, Feng H, Wu C, Wang R, Hu J, Yang Q, Luo Y, Wu J, Fan H, Liang P, Yu J. Ultrasound identification of hepatic echinococcosis using a deep convolutional neural network model in China: a retrospective, large-scale, multicentre, diagnostic accuracy study. Lancet Digit Health. 2023;5(8):e503–14. https://doi.org/10.1016/s2589-7500(23)00091-2. PMID: 37507196.
93. Yildirim M. Image visualization and classification using hydatid cyst images with an explainable hybrid model. Appl Sci. 2023;13(17):9926. https://doi.org/10.3390/app13179926.

Contents

Contributors

Ez El Din Abu Zeid Department of Surgery, Soroka Medical Center, Beer Sheva, Israel

Akande Oyebanji Azeez Department of Medical Microbiology and Parasitology, Bayero University, Kano, Nigeria

Sami Akbulut Department of Surgery and Liver Transplantation Institute, Inonu University Faculty of Medicine, Malatya, Türkiye
Department Biostatistics and Medical Informatics, Inonu University Faculty of Medicine, Malatya, Türkiye

Ali Aloun Department of General Surgery, Jordanian Royal Medical Services, Amman, Jordan
Department of Hepatobiliary Surgery, Jordanian Royal Medical Services, Amman, Jordan

Aliyu Aminu Department of Medical Microbiology and Parasitology, Bayero University, Kano, Nigeria
Aminu Kano Teaching Hospital, Kano, Nigeria
Department of Medical Microbiology and Parasitology, Faculty of Basic Clinical Sciences, Bayero University, Kano, Nigeria

Kemal Dolay Hepatopancreatobiliary Surgery, Istanbul, Türkiye

Ahmed Elsarawy Department of Surgery and Organ Transplantation, Gaziosmanpasa Hospital, Istanbul Yeni Yuzyil University Faculty of Medicine, Istanbul, Türkiye

Ibrahim Umar Garzali Department of Surgery, Aminu Kano Teaching Hospital, Kano, Nigeria

Muhammad Ibrahim Getso Department of Medical Microbiology and Parasitology, Faculty of Basic Clinical Sciences, Bayero University, Kano, Nigeria

Abdirahman Sakulen Hargura Department of Surgery, Kenyatta University Teaching, Research and Referral Hospital, Nairobi, Kenya

Mucahit Karaduman Department of Software Engineering, Malatya Turgut Ozal University, Malatya, Türkiye

Merve Gulbiz Dagoglu Kartal Department of Radiology, Istanbul University, Istanbul Faculty of Medicine, Istanbul, Türkiye

Tevfik Tolga Sahin Department of Surgery and Liver Transplant Institute, Inonu University Faculty of Medicine, Malatya, Türkiye

Artür Salmaslioglu Department of Radiology, Istanbul University, Istanbul Faculty of Medicine, Istanbul, Türkiye

Abdurrahman Abba Sheshe Department of Surgery, Aminu Kano Teaching Hospital, Kano, Nigeria

Adem Ucar Department of Radiology, Istinye University, Faculty of Medicine, Istanbul, Türkiye

Veysel Umman Department of Surgery, Division of Organ Transplantation and HPB Surgery, Yeditepe University Faculty of Medicine, Istanbul, Türkiye

Muhammed Yildirim Department of Computer Engineering, Malatya Turgut Ozal University, Malatya, Türkiye

Mustapha Ahmed Yusuf Department of Medical Microbiology and Parasitology, Faculty of Clinical Sciences, Bayero University, Kano, Nigeria
Aminu Kano Teaching Hospital, Kano, Nigeria

Terminology Used for Hydatid Disease: Parasitological and Surgical Aspects

1

Ahmed Elsarawy, Sami Akbulut, and Abdirahman Sakulen Hargura

1.1 Introduction

Echinococcus species are parasites categorized under the class of Cestoda, which cause a variety of clinical diseases in humans. Humans are intermediate hosts to this parasite, and infestation results in occult or clinically relevant, single or multi-organ as well as indolent or rather aggressive diseases, under the broad term of Echinococcosis. The medical literature has been predominantly heterogeneous when describing the epidemiology, clinical course, and treatment approaches of the clinicopathological entities secondary to Echinococcosis. Efforts have been made by local and international clinical authorities and scientific societies in order to firstly unify the nomenclature of the disease epidemiology and natural history, secondly to address the burden of the disease on international scale, and lastly to standardize the management approaches. In this chapter, a unified approach to the parasitic/clinical and surgical nomenclature is pursued, in order to further guide clinicians and surgeons in this domain. Since the liver is affected in almost three quarters of patients with Echinococcosis, this chapter mainly focuses on the liver and the terminology is especially liver oriented.

1.1.1 Surgical Anatomy of the Liver

Anatomically, the liver occupies the right hypochondrium, the epigastric area, and partially extends to the left hypochondrium. The liver is completely covered by a fibrous capsule and almost entirely encapsulated by the peritoneum, except for a small subdiaphragmatic bare area. The liver can be described as having a convex diaphragmatic surface that lies anterosuperiorly and a relatively concave visceral surface that is inferoposterior. There is centrally located H-shaped impression formed by two sagittal fissures and one transversely oriented fissure. The transversely located fissure is the impression of the porta hepatis, whereas the sagittal fissures are formed by the gall bladder and the inferior vena cava (IVC) on the right and the ligamentum teres hepatis and the ligamentum venosus on the left.

Except for the posterosuperiorly located small bare area, the liver is covered entirely by visceral

A. Elsarawy
Department of Surgery and Organ Transplantation, Gaziosmanpasa Hospital, Istanbul Yeni Yuzyil University Faculty of Medicine, Istanbul, Türkiye

S. Akbulut (✉)
Department of Surgery and Liver Transplantation Institutes, Inonu University Faculty of Medicine, Malatya, Türkiye

Department Biostatistics and Medical Informatics, Inonu University Faculty of Medicine, Malatya, Türkiye

A. S. Hargura
Department of Surgery, Kenyatta University Teaching, Research and Referral Hospital, Nairobi, Kenya

S. Akbulut (ed.), *Hydatid Disease*, https://doi.org/10.1007/978-3-031-97277-5_1

peritoneum that forms ligaments as it reflects onto the diaphragm to continue with the parietal peritoneum covering the anterior abdominal wall. The right and left triangular peritoneal reflections from the right triangular and the left triangular ligaments, respectively. The peritoneal reflection around the bare area forms the coronary ligaments posteriorly, while the falciform ligament is an anterior peritoneal reflection extending to the umbilicus with ligamentum teres hepatis running along its free border.

Owing to the advancements in expertise as well as the equipment in the domain of hepatobiliary and liver transplant surgery, the description of the liver anatomy over the last 2–3 decades has been surgically oriented to address all forms of minor and major liver resections. Successive systems had been adopted for this purpose and were dictated by insights from hepatobiliary and transplant surgeons. Several controversial systems, namely Couinaud's system, Bismuth, International Hepato-Pancreato-Biliary Association (IHPBA) terminology, and Brisbane terminology among others, have been proposed but not universally adopted. However, in most of these nomenclature systems, the segmental anatomy of the liver is described along the lines of the original Couinaud's system.

Generally, the plane of the middle hepatic vein, running from the IVC to the left of the gallbladder fossa, dividing the liver into the right and left lobes or hemilivers. This plane runs along the Cantlie's line. This subdivision is based on the hepatic portal and venous vasculature, hence surgically significant.

The two hemilivers are further subdivided into sections/sectors formed by a total of 8 segments. The falciform ligament subdivides the left hemiliver into lateral and medial sections, with segments II and III being on its left side, and segments IV to VIII on its ride side. This subdivision is along an oblique plane that runs between the falciform ligament and the left hepatic vein.

The right hemiliver is further subdivided into anterior and posterior sections (sectors) by the plane of the middle hepatic vein. The portal and hepatic arterial supply give anterior and posterior sectoral branches hence asserting the surgical relevance of this subclassification.

Each of the eight segments has its distinct arterial and portal blood supply, as well as venous and biliary drainage. The caudate lobe, regarded as segment I of the liver, has its own portal and arterial blood supply, and independent venous and biliary drainage.

The following are the eight Couinaud's segments: (commonly mentioned with Roman letters)

Segment I: Caudate lobe
Segment II: Superior left lateral
Segment III: Inferior left lateral
Segment IV: Left medial, subdivided into IVa superiorly and IVb inferiorly
Segment V: Right anterior, located inferiorly
Segment VI: Right posterior, located inferiorly
Segment VII: Right posterior, located superiorly
Segment VIII: Right anterior, located superiorly

Most recently, with the wide application of indocyanine green (ICG) dye in minimally invasive liver surgery, better visualization of the territories of liver parenchyma has been achieved. Based on that, a consensus conference re-visited this surgical-anatomical nomenclature and came up with further subdivision, namely the term subsegment or cone units. It is specifically applied to segment VIII (dorsal and ventral subsegments), segment IV (apical and basal subsegments) and segment I (Spiegel, caudate process and paracaval subsegments) [1].

1.1.2 Hilar (Portal) Anatomy

The Glisson's capsule, which is a thin fibrous layer covering the liver, envelops the parenchyma and reflects onto the porta hepatis where the triad of portal veins, hepatic arteries, and bile ducts enters into the various hepatic territories. So the Glisson's capsule forms a sheath that envelops the hilar structures. At the level of the liver hilum, this sheath is rather thickened into a fibrous plate and hence named the hilar plate. This plate needs to be divided and taken down in order to dissect and isolate portal pedicles during liver resection for either lobes, a maneuver known as *lowering of hilar plate*. Of note, the Glissonian sheaths continue investing the portal pedicles in their intrahepatic parts.

1.1.3 Hepatic Vasculature

The liver blood inflow comes from the portal vein and the hepatic artery. The portal vein supplies approximately 75% of inflow, whereas the hepatic artery supplies the remainder inflow [2]. The portal vein provides the majority of the liver's nutrient-rich blood flow, while the hepatic artery supplies oxygenated blood, delivering approximately 50% of the liver's total oxygen requirements [3]. At the porta hepatis is a triad of crucial structures: the portal vein formed by the fusion of mesenteric veins and the splenic vein dividing into the left and the right portal veins, the hepatic artery originating from the celiac axis and giving off left and right hepatic arteries, and the common hepatic duct formed by fusion of the left and right hepatic ducts.

At the level of the hilum, the proper hepatic artery usually divides outside the parenchyma at a lower level than does the hepatic duct or portal vein. Noteworthy the segment IV artery (in some texts referred to as middle hepatic artery) originates either from left hepatic artery or the right hepatic artery. An accessory or replaced right hepatic artery from SMA is found most posterolaterally in the porta hepatis region, while an accessory or replaced left hepatic artery from LGA is found in the middle of the gastrohepatic ligament. Precise mapping of hepatic arterial anatomy via CT angiography is mandatory during planning of liver resectional surgery. It is of utmost importance to preserve the integrity of the hepatic arterial supply being the exclusive blood supply to the biliary tree.

The portal vein anatomy is the least variable among all. The main portal vein is the most posterior portal structure, making the anterior margin of the foramen of Winslow. The right hepatic vein is shorter and divides immediately into sectoral veins. Like the left hepatic duct, the left portal vein is longer and has a horizontal course within the hilar plate before entering the left lobe.

The venous drainage of the liver (also referred to as hepatic outflow) is via the three major branches of the hepatic veins: the right hepatic vein, the middle hepatic vein, and the left hepatic vein. These veins run in a posteroinferior direction and drain into the inferior vena cava (IVC) just inferior to the diaphragm. The middle hepatic vein generally joins the left hepatic vein and drain into the IVC as a single short common trunk. As mentioned earlier, the hepatic veins divide the liver into sectors, hence it is surgically important to guide resections. The right hepatic vein divides the right hemiliver into anterior and posterior sectors. The right hepatic vein divides the right hemiliver into anterior and posterior sectors. In approximately 60% of cases, it has no tributaries within the first 1 cm from its confluence with the inferior vena cava (IVC). In the remaining cases, the presence of an early tributary may increase the risk of bleeding during isolation of the right hepatic vein stump in right hepatectomy procedures [4]. An accessory right hepatic vein, also known as the inferior right hepatic vein, is present in approximately 60% of individuals and drains into the inferior vena cava (IVC) at a more distal location [5].

The middle hepatic vein and left hepatic vein make a common trunk before joining the IVC extrahepatically. The extrahepatic portion of the left hepatic vein is quiet long and could be dissected and isolated, in contrary to the stump of middle hepatic vein which is almost completely surrounded by liver parenchyma. One of the most important aspects of preoperative surgical anatomy planning is identification of the vast variations of the middle hepatic vein tributaries, coming from the anterior sector (i.e., segments V and VIII).

1.1.4 Anatomy of the Biliary System

The common hepatic duct divides at the level of the liver hilum into main right and left hepatic ducts. This division is more to the right and overlies the portal vein bifurcation. An important posterior relation of the biliary confluence in the vast majority of subjects is the right hepatic artery coming from left to right. In rare instances, the right hepatic artery crosses anteriorly, carrying the risk of inadvertent bleeding in hilar biliary reconstructive surgeries. The right hepatic duct is short and immediately divides into anterior and posterior sectoral ducts. The left hepatic duct is longer (approximately 3 cm) and has a more horizontal extrahepatic course in the hilar plate, passing from right to left before dividing into medial and lateral sectoral ducts. In 28% of subjects, the main left hepatic duct received one of the right sectoral ducts before joining the main right hepatic duct.

1.1.5 Surgical Topography of Hepatic Echinococcal Disease

The blood in the portal vein flows at a sluggish rate causing a streaming pattern where the blood from the splenic vein remains on the left of the portal vein whereas the blood from the mesenteric veins flows on the right side within the portal vein. Consequently, infectious diseases from the bowels tend to affect the right lobe of the liver. As elucidated in other chapters of this book, the Echinococcal disease also tends to affect the right lobe more often than the left lobe.

Histopathologically, a mature cyst consists of the following layers from outside inward: The pericyst (also known as ectocyst) is the outermost layer formed by the host tissue reaction, then the endocyst which in turn consists of a laminated membrane fused with the germinal living layer which is the innermost layer surrounding the Echinococcal fluid filled cavity [6]. Frequently, the ectocyst becomes calcified. Hence, the extensions, adhesions, and ramifications of the lesion must be taken into account to undergo a sound surgical procedure with diminished possibility of recurrence. A recent description from an otherwise surgical perspective defined the same three layers: adventitial layer from host reaction, laminated layer, and germinal layer to the innermost. Generally speaking, complete removal of an echinococcal cyst mandates preservation of the whole three-layer envelope to guard against spillage and recurrence.

1.1.6 Parasitological Terminology

Currently, nine species have been precisely described. From a clinical standpoint, three important clinical diseases are considered: *Cystic echinococcosis* caused by the *E. granulosus* species' cluster, which is the most commonly encountered worldwide, *Alveolar echinococcosis* caused by *E. multilocularis*, and notorious of being of rather aggressive cancer-like clinico-pathological behavior, and less commonly *Neutropical Echinococcosis* has been reported, which is caused by *E. vogeli* or *E. oligoarthra* and presents with multiple cysts.

To overcome the confusion and heterogeneity of Echinococcosis literature, a consensus report has been recently published with the following main clinical recommendations: The broad term Echinococcosis preceded by the aforementioned terminology (cystic-alveolar-neutropical) should be predominantly adopted. The term hydatid disease should be exclusively applied to *E. granulosus* species and its geno-subtypes only. Finally, it would be clinically sound to describe the above diseases as follows: Cystic echinococcosis (CE) cysts, alveolar echinococcosis (AE) lesions, and neutropical echinococcosis (NE) multi-cystic disease [7]. Herein after the term Echinococcosis will be employed.

Another important updates relevant to the microbiological and parasitological terminologies of Echinococcal disease are summarized below [7]:

Echinococcus: It is a genus within cestoda, a parasitic class of the platyhelminthes phylum (Kingdom: Animalia, Phylum: Platyhelminthes, Class: Cestoda, Subclass: Eucestoda, Order: Cyclophylidea, Familiy: Taeniidae, Genus: Echinococcus).

***Echinococcus* sp.**: One species within the genus *Echinococcus.*

***Echinococcus* spp.**: More than one (or all) species within the genus *Echinococcus.*

Echinococcosis: Disease related to infection with parasites of the *Echinococcus* genus.

Hydatid disease: Disease related to infection with *Echinococcus* spp.

Hydatidosis: Disease related to infection with *Echinococcus* spp.

***E. granulosus*:** One species cluster that contains all agents of cystic echinococcosis.

Cystic echinococcosis: The disease caused by *E. granulosus*, which belongs to the class Cestoda and family Taeniidae.

E. multilocularis: A species in the genus *Echinococcus.*

Alveolar echinococcosis: The disease caused by *E. multilocularis*, which belongs to the class Cestoda and family Taeniidae. Disease due to *E. multilocularis*, with multiple microcysts giving an "alveolar" aspect on sectioning of the infected organ.

Hydatid cyst: Anatomical lesion due to infection with *Echinococcus granulosus*, excluding all other species of *Echinococcus.*

Aborted cyst: Nonviable parasitic structure, as evidenced by imaging (complete calcification in alveolar echinococcosis, cystic echinococcosis type 5 cyst), or histological examination (absence of viable parasitic cells).

Oncosphere: Infectious form of *Echinococcus* spp. parasites resulting from the release from the egg through the action of proteolytic enzymes in the digestive system of intermediate hosts.

Metacestode: Larval parasitic structure that contains viable protoscoleces and thus allows infection of the definitive hosts through scoleces or in some pathological or experimental circumstances production of newly developed metacestodes in the intermediate host. In other words metacestode is a fluid-filled, spherical, unilocular cyst that consists of an inner germinal layer of cells supported by a characteristic acidophilic-staining, acellular, laminated membrane of variable thickness.

Proglottis: Part of the adult form of *Echinococcus* spp. parasites resulting from the segmentation of the scolex in the intestine of definitive hosts.

Protoscolex: Prefiguration of the scolex, produced by the brood capsules budding to the outside of the germinal layer of *Echinococcus* spp. metacestode and released in the cyst (cystic echinococcosis) or micro-cyst (alveolar echinococcosis) fluid.

Scolex: First segment ("head") of the adult form of cestodes.

Segment: Part of the adult form of *Echinococcus* spp. parasites resulting from segmentation of the scolex in the intestine of definitive hosts.

Strobilar stage: Sexual reproduction stage (also called adult form) of *Echinococcus* spp. parasites in their definitive hosts.

Worm: Strobilar stage of *Echinococcus* spp. parasite in its definitive hosts.

Adult form: Sexual reproduction stage of *Echinococcus* spp. parasites in their definitive hosts.

Larval form: Asexual reproduction form of *Echinococcus* spp. parasites in their intermediate hosts.

Structure of hydatid cyst: The correct description of the parasitic structures in the intermediate host includes three similar expressions, from inside to outside: (i) nucleated germinal layer; (ii) acellular laminated layer; and (iii) adventitial layer (outer pericyst, ectocyst).

Adventitial layer: Partly cellular and fibrous layer of host origin surrounding the inner two layers of parasite origin (germinal and laminated layers) of the metacestode in the intermediate host of *E. granulosus.*

Germinal layer: Inner cellular part of *Echinococcus* spp. metacestode, which includes several types of cells and produces several types of biological components of the metacestode (cyst fluid, brood capsules, and protoscoleces).

Laminated layer: Peripheral acellular outer part of *Echinococcus* spp. metacestode, mostly composed of mucopolysaccharides.

Daughter cyst: Newly formed hydatids inside the cystic echinococcosis cyst in the development of *Echinococcus granulosus*. Not applicable to alveolar echinococcosis.

Hydatid fluid: Liquid secreted by the germinal layer of *Echinococcus* spp. metacestode.

Anti-parasitic treatment: Pharmacological or biological treatment of echinococcosis, able to kill *Echinococcus* spp. or stop or delay their development at the various stages of the parasitic cycle.

Protoscolecide: Compound (natural or chemical) which is able to kill the protoscoleces.

1.1.7 Surgical Terminology

The most frequently used surgical and interventional radiological terminology in echinococcal disease are summarized below:

Subtotal cystectomy defined as nearly total cystectomy with incomplete removal of the adventitial layer of an echinococcal cyst. That is, only parts of the adventitial layer are preserved because of surgical safety. In this procedure, the cyst cavity is opened and the cyst contents are evacuated before the cyst is removed [7].

Partial cystectomy refers to the incomplete resection of any of the three layers (usually of the adventitial layer) due to technical and safety issue. In this procedure, the cyst cavity is opened and the cyst contents are evacuated before the cyst is removed [7].

Total cystectomy defined as complete removal of an echinococcal cyst, including the content (fluid and protoscoleces) and resection of all layers of the cyst (germinal, laminated, and adventitial). In this procedure, it is mostly performed without opening the cyst cavity [7].

Peri-adventitial cystectomy defined as total cystectomy performed without opening the cyst cavity, and which uses the dissection space between the adventitial layer and the healthy liver parenchyma to remove the cyst totally [7].

Hepatectomy requires the en bloc resection of a part of liver parenchyma, following the rules of hepatic resection [7].

Ex vivo liver resection with autotransplantation defined as surgical operation which includes: total hepatectomy followed by ex vivo resection of the pathological parts of the liver associated with various types of reconstruction of the bile ducts and/or vessels then by reimplantation (autotransplantation) of the liver [7].

Liver transplantation for hydatid disease defined as removal of the recipient's liver followed by the transplantation of the donor's liver (or part of liver) at the same anatomical location [7].

Puncture-aspiration-injection-reaspiration (PAIR) defined as percutaneous treatment of echinococcal cyst using needle puncture and use of protoscolecide agents [7].

Standard catheterization technique (S-CAT) defined as Modification of the PAIR technique, which includes the insertion of a catheter left or not in the cyst temporarily for the treatment of selected echinococcal cysts [7].

Modified catheterization technique (Mo-CAT) percutaneous procedure for treatment of echinococcal cysts that results in removal of the cyst layers, possibly including daughter cysts, in addition to the cyst fluid and protoscoleces [7].

Reactivation: Appearance of daughter cyst(s) in the solid matrix of spontaneously inactivated CE4 cysts that shows an evolution toward a CE3b stage.

Recurrence: Appearance of an active cystic echinococcosis cyst (types CE1-3, usually CE3b) in the same location where a treated cyst was located, independently of the type of previous treatment.

Secondary cyst: Newly formed cysts from the dissemination of protoscoleces or germinal layer fragments, in the serosal cavities, spontaneously, accidentally, or after surgery or cyst puncture.

1.1.8 Surgical Approaches

To date there is no clear consensus or standardized approaches for surgical treatment of hepatic echinococcal disease. Surgical procedures are extrapolated from the cumulative experience coming from liver surgeries. The nomenclature of surgical approaches for hepatic echinococcal disease is largely dictated by the surgical terminology applied for general liver resectional surgeries. However, owing to the unique pathological behavior of echinococcal disease, different nomenclature was used. Most recently, a consensus report from an international panel in the domain of Echinococcosis has proposed a standard approach for describing any surgical intervention under the acronym of (**AORC**): **A**pproach (laparatomy, laparoscopy, or robotic); **O**pening (opened cyst or non-opened cyst); **R**esection (cystectomy, hepatectomy, or liver transplantation); **C**ompleteness (total cystectomy, subtotal cystectomy, or partial cystectomy). Though not universally adopted, this AORC recording system could provide a good basis for the related literature [7]. One could argue that there might be some overlap between R and C items.

In this section, the nomenclature of surgical treatment of Echinococcal disease is presented aiming at unification of terminology among liver surgeons. Generally, surgical approaches and procedures are either conservative or radical. Conservative approaches are in favor of limited surgical resection but are notorious of high rate of recurrence [8]. They revolve around emptying the cyst content and managing the remaining cavity. On the other hand, radical approaches entail resection of a significant volume of liver tissue. Though associated with lower recurrence rate, it is a more demanding procedure that needs specialized surgical expertise and set up [9, 10].

Liver resection in the setting of hepatic echinococcal disease is a more radical surgical intervention and preferably to be done within a specialized hepatobiliary surgery center. It entails removal of the cyst with adjacent volume of liver parenchyma and includes the following: segmentectomy (resection of a single liver segment), sectionectomy (resection of two adjoining liver segments, e.g., right posterior sectionectomy of segments VI and VII), lobectomy (resection of a whole right or left hemiliver), or more extended resections depending on the anatomical confinements of the cyst within the liver.

1.1.9 Conservative Approaches

1. **Marsupialization**: Also interchangeably referred to as de-roofing, entails removal of a larger part of the cyst wall closer to the peritoneal cavity, and evacuation of the cyst. This can be followed by suturing of the cyst edge circumferentially to the adjacent liver parenchyma, and hence the term marsupialization applied.
2. **Partial Cystectomy**: Partial removal of the cyst wall with subsequent debulking and evacuation of cyst content. These approaches might be suitable in elderly or frail patients presenting with otherwise superficial lesions and should be discouraged due to the high rate of recurrence.

1.1.10 Radical Approaches

1. **Total (Radical) Cystectomy**: Also referred to as pericystectomy. It implies circumferential dissection of the plane between the liver parenchyma and the outer cyst wall (pericyst) with removal of the cyst in toto without violation of its outermost wall. This is accomplished either via an open or minimally invasive approach. Violation of the cyst wall will end up with spillage, incomplete removal, or recurrence. The cyst can be intentionally punctured in advance, aspirated, and lavaged before total dissection. It is mostly adopted in single lesions close to the liver surface and away from portal pedicles. In this approach, the adjoining liver parenchyma is neither severed nor resected. Developing the potential space between the pericyst and the surrounding liver parenchyma is sometimes challenging and needs vigilant dissection [11]. With the advent of minimally invasive approaches, either laparoscopic or robotic, identification of this liver/pericyst plane could be accomplished more precisely, especially with the application of ICG dye [12].
2. **Cone Unit Resection (or Subsegmentectomy)**: According to the aforementioned most recent liver resectional nomenclature, a subsegmental resection is a valid option and spares liver parenchyma in smaller scattered lesions, namely carried out for lesions in segments I, IV, and VIII.
3. **Segmentectomy**: Isolated resection of a Couinaud's segment is a demanding procedure and aims at parenchymal sparing. Segment I (caudate lobe) is the most demand-

ing due to the high variability and bilaterality of blood and biliary supply. In addition, an intense reaction from the Echinococcal cyst might mandate caval resection as well. Isolated segmentectomy on the right and left lobe are less commonly practiced in real surgical scenarios, though theoretically each segmental inflow and outflow can be isolated and transected. Otherwise, sectoral resections are more practiced.

4. **Sectionectomy (or Sectoral resection):** On the right side includes right anterior (segments V and VIII) and posterior (segments VI and VII) sectionectomies. The right portal pedicle is dissected and isolated after lowering the hilar plate. Further higher dissection along the Glisson's sheath will identify the right posterior pedicle at a more posterior location than the right anterior one, and hence isolated from each other before starting parenchymal resection. Noteworthy that significant bleeding could occur during parenchymal transection from communicating venous channels between right and middle hepatic veins. Left lateral sectionectomy (Segment II and III) necessitates isolation of the portal pedicle at the hilum and this is achieved more easily than the right side owing to the relatively longer left portal pedicle and horizontal passage. A very useful landmark is to follow the round ligament stump toward its joining to the left portal pedicle. The line of transection is immediately to the left of the falciform ligament. Left medial sectionectomy is a more demanding procedure and usually a left lobectomy is formed for Echinococcal lesions in this rather centrally located area. The lines of parenchymal transection are just to the right of falciform ligament and to the right of the Cantlie's line. Significant bleeding can originate from connecting channels between middle and left hepatic veins.
5. **Major hepatectomy**: Deeply seated cysts larger than 12 cm or cysts in close proximity to portal pedicles or major veins are amenable for major hepatectomy. This term applies when more than three Couinaud segments are resected. Variations include: Right hepatectomy (segments V, VI, VII, and VIII), left lobectomy (segments I, II, III, and IV). Trisegmentectomy is a term used when more than anatomical full lobe resected, including right hepatectomy + segment IV or left hepatectomy + segment V and or VIII. Major hepatectomies for hepatic Echinococcal disease had been practiced on a regular basis in many hepatobiliary and tertiary centers [13]. Another form of major hepatectomies is the central hepatectomy, that's applied for massive centrally located Echinococcal cyst(s). It entails removal of varying portions of segments IV, V, and VIII.
6. **Liver transplantation**: Liver transplantation is a more radical treatment modality for hepatic Echinococcal disease and comes after extensive interventions or resections, with persistent voluminous disease or resultant hepatic insufficiency [14]. However, in the context of alveolar echinococcosis, liver transplantation could be the only curative option in view of its cancer-like pathological behavior, namely infiltrative nature and metastatic potential [15–17]. Liver transplantation for alveolar echinococcosis had been reported to be associated with en block resection of diaphragm and IVC, exactly like hepatic oncological surgical procedures [17].

References

1. Wakabayashi G, Cherqui D, Geller DA, Abu Hilal M, Berardi G, Ciria R, Abe Y, Aoki T, Asbun HJ, Chan ACY, Chanwat R, Chen KH, Chen Y, Cheung TT, Fuks D, Gotohda N, Han HS, Hasegawa K, Hatano E, Honda G, Itano O, Iwashita Y, Kaneko H, Kato Y, Kim JH, Liu R, López-Ben S, Morimoto M, Monden K, Rotellar F, Sakamoto Y, Sugioka A, Yoshiizumi T, Akahoshi K, Alconchel F, Ariizumi S, Benedetti Cacciaguerra A, Durán M, Garcia Vazquez A, Golse N, Miyasaka Y, Mori Y, Ogiso S, Shirata C, Tomassini F, Urade T, Wakabayashi T, Nishino H, Hibi T, Kokudo N, Ohtsuka M, Ban D, Nagakawa Y, Ohtsuka T, Tanabe M, Nakamura M, Tsuchida A, Yamamoto M. The Tokyo 2020 terminology of liver anatomy and resections: updates of the Brisbane 2000 system. J Hepatobiliary Pancreat Sci. 2022;29(1):6–15. https://doi.org/10.1002/jhbp.1091. PMID: 34866349.
2. Carneiro C, Brito J, Bilreiro C, Barros M, Bahia C, Santiago I, Caseiro-Alves F. All about portal vein: a pictorial display to anatomy, variants and physiopathology. Insights Imaging. 2019;10(1):38. https://doi.org/10.1186/s13244-019-0716-8).

3. Chouillard EK, Gumbs AA, Cherqui D. Vascular clamping in liver surgery: physiology, indications and techniques. Ann Surg Innov Res. 2010;4:2. https://doi.org/10.1186/1750-1164-4-2.
4. Cawich SO, Naraynsingh V, Pearce NW, Deshpande RR, Rampersad R, Gardner MT, Mohammed F, Dindial R, Barrow TA. Surgical relevance of anatomic variations of the right hepatic vein. World J Transplant. 2021;11(6):231–43.
5. Gündoğdu E. Relationship of the Presence of the Inferior Right Hepatic Vein with the Right Hepatic Vein Diameter and CT Liver Volumetry. Indian J Radiol Imaging. 2023;33(3):332–37.
6. Bhutani N, Kajal P. Hepatic echinococcosis: a review. Ann Med Surg (Lond). 2018;36:99–105. https://doi.org/10.1016/j.amsu.2018.10.032. PMID: 30450204 PMCID: PMC6226561.
7. Vuitton DA, DP MM, Rogan MT, Romig T, Gottstein B, Naidich A, Tuxun T, Wen H, Menezes da Silva A, World Association of E. International consensus on terminology to be used in the field of echinococcoses. Parasite. 2020;27:41. https://doi.org/10.1051/parasite/2020024. PMID: 32500855 PMCID: PMC7273836.
8. Deo KB, Kumar R, Tiwari G, Kumar H, Verma GR, Singh H. Surgical management of hepatic hydatid cysts—conservative versus radical surgery. HPB (Oxford). 2020;22(10):1457–62. https://doi.org/10.1016/j.hpb.2020.03.003. PMID: 32229090.
9. Akbulut S, Senol A, Sezgin A, Cakabay B, Dursun M, Satici O. Radical vs conservative surgery for hydatid liver cysts: experience from single center. World J Gastroenterol. 2010;16(8):953–9. https://doi.org/10.3748/wjg.v16.i8.953. PMID: 20180233 PMCID: PMC2828599.
10. Sozuer E, Akyuz M, Akbulut S. Open surgery for hepatic hydatid disease. Int Surg. 2014;99(6):764–9. https://doi.org/10.9738/intsurg-d-14-00069.1. PMID: 25437585 PMCID: PMC4254238.
11. Öztürk G, Uzun MA, Özkan ÖF, Kayaalp C, Tatlı F, Eren S, Aksungur N, Çoker A, Bostancı EB, Öter V, Kaya E, Taşar P. Turkish HPB Surgery Association consensus report on hepatic cystic echinococcosis (HCE). Turk J Surg. 2022;38(2):101–20. https://doi.org/10.47717/turkjsurg.2022.5757. PMID: 36483170 PMCID: PMC9714645.
12. Goja S, Saha SK, Yadav SK, Tiwari A, Soin AS. Surgical approaches to hepatic hydatidosis ranging from partial cystectomy to liver transplantation. Ann Hepatobiliary Pancreat Surg. 2018;22(3):208–15. https://doi.org/10.14701/ahbps.2018.22.3.208. PMID: 30215042 PMCID: PMC6125266.
13. Ramia JM, Serrablo A, Serradilla M, Lopez-Marcano A, de la Plaza R, Palomares A. Major hepatectomies in liver cystic echinococcosis: a bi-centric experience. Retrospective cohort study. Int J Surg. 2018;54(Pt A):182–6. https://doi.org/10.1016/j.ijsu.2018.04.049. PMID: 29733994.
14. Moreno-Gonzalez E, Loinaz Segurola C, Garcia Urena MA, Garcia Garcia I, Gomez Sanz R, Jimenez Romero C, Gonzalez Pinto I, Corral Sanchez MA, Palma CF. Liver transplantation for Echinococcus granulosus hydatid disease. Transplantation. 1994;58(7):797–800. PMID: 7940713.
15. Akdur A, Kirnap M, Ayvazoglu Soy EH, Ozcay F, Moray G, Arslan G, Haberal M. Unusual indications for a liver transplant: a single-center experience. Exp Clin Transplant. 2017;15(Suppl 1):128–32. https://doi.org/10.6002/ect.mesot2016.P11. PMID: 28260452.
16. Aydinli B, Ozturk G, Arslan S, Kantarci M, Tan O, Ahiskalioglu A, Ozden K, Colak A. Liver transplantation for alveolar echinococcosis in an endemic region. Liver Transpl. 2015;21(8):1096–102. https://doi.org/10.1002/lt.24195. PMID: 26074280.
17. Ozdemir F, Ince V, Barut B, Onur A, Kayaalp C, Yilmaz S. Living donor liver transplantation for Echinococcus alveolaris: single-center experience. Liver Transpl. 2015;21(8):1091–5. https://doi.org/10.1002/lt.24170. PMID: 25981332.

2 Epidemiology, Life Cycle, and Prevention of Hydatid Disease

Tevfik Tolga Sahin and Sami Akbulut

2.1 Introduction to Hydatid Disease

Hydatid disease is caused by the larval stage of tapeworms belonging to the genus *Echinococcus*. Echinococcosis has a global distribution, and *Echinococcus granulosus* is found worldwide, causing diseases in humans, livestock, and other intermediate hosts. The life-threatening complications occur when hydatid cysts swell and apply pressure on vital organs such as the brain, heart, and large blood vessels; rupture of a cyst results in life-threatening anaphylaxis [1]. Transmission is associated with risk factors that increase the likelihood of people coming into contact with food or water contaminated with *Echinococcus granulosus* eggs from the feces of infected canines. Hydatid disease is an important public health issue globally. Knowledge about epidemiology, clinical features, and prevention is paramount for its control [2].

Although known for many years, this infection of domestic animals (including pigs, sheep, and cattle) can also cause considerable disease when acquired by humans. At the adult stage, the female worm releases eggs to the environment via bile, and subsequently, eggs are swallowed by the dogs and other canids that serve as definitive hosts. The next stage, oncosphere, penetrates the wall of the small intestine and then circulates through the systemic circulation to the liver and other organs. In these organs, oncosphere "evolves" into the first larvae called coenurus. The dog can become infected by consuming coenurus-infected offal of slaughtered sheep, goats, camels, etc., which were infected with *Echinococcus* species [3]. There is a predator and pray relationship between the definitive and intermediate host. Recent epidemiologic study from China has stated that the complex life cycle of the parasite is dependent on the food chain involving many different carnivores and herbivores [4].

Echinococcus granulosus will not cause any disease in dogs, but the adult worms can produce eggs, which are then passed out through feces to the environment. In a study, disease in sheep and dogs did not occur above this latitude, indicating that sheep and dogs were unable to transmit this disease higher, due principally to low temperatures, but wide variations in a variety of hosts and the cycle were shown to still exist. To prevent this disease from occurring, it is necessary for those working in agriculture and health to know about *Echinococcus*. It is the aim of this chapter to provide a detailed and well-illustrated review of this infection prevention capability [3].

T. T. Sahin
Department of Surgery and Liver Transplant Institute, Inonu University Faculty of Medicine, Malatya, Türkiye

S. Akbulut (✉)
Department of Surgery and Liver Transplantation Institute, Inonu University Faculty of Medicine, Malatya, Türkiye

Department of Biostatistics and Medical Informatics, Inonu University Faculty of Medicine, Malatya, Türkiye

S. Akbulut (ed.), *Hydatid Disease*, https://doi.org/10.1007/978-3-031-97277-5_2

Hydatid disease (or hydatid disease) is caused by the larval form of *Echinococcus*, a cestode belonging to the Taeniidae family. The infection begins when oncospheres grow in various tissues. There is a cyclic infection between the intermediate and definitive hosts, involving different stages of the parasite's life cycle, including metacestodes and adult worm forms, respectively [5]. Echinococcosis is endemic in sheep and cattle-raising areas. It poses a significant public health challenge in endemic regions and leads to substantial workforce and economic losses [6].

It is a predator-prey relationship. The definitive hosts are dogs, cats, and wild carnivores. The intermediate hosts are marsupials, rodents, and ruminants. The adult stage of the parasite is found in the carnivores and the larval stage develops in a wide variety of intermediate hosts including humans [5]. There is a wide range of intermediate hosts, and the parasite can infect any prey if it is a mammal. There are three types of hosts in the life cycle of the parasite [7]. The definitive host is where the full life cycle of the parasite is completed. On the other hand, the intermediary hosts are infected accidentally, and the parasite invades the venous system and implants into different organs (especially the lung and the liver). The "dead-end hosts" are intermediary hosts, and the parasite does not complete the full life cycle but metacestodes can develop and cause the spread of infection [7]. The subtype of *Echinococcus* determines the geographic distribution. Specifically, *E. granulosus* has a wide geographic distribution including Eurasia, Africa, Australia, and South America. *E. multilocularis* can also be seen in the northern hemisphere including central Europe, northern and central Eurasia, regions of North America, and Northern Africa [8]. The geographic distribution and the propensity of the infestation in the region depend on the species of the *Echinococcus*, the seasonal changes in that region, and the ability of the metacestodes to develop in the specific type of the intermediate host (domestic versus sylvatic) in that region. Therefore, determination of the distribution of the parasite within a particular region is challenging [9].

Four species are recognized to cause infections in humans. These include *Echinococcus granulosus* sensu lato, *Echinococcus multilocularis*, *Echinococcus oligarthra*, and *Echinococcus vogeli* [10]. These pathogens are responsible for three distinct types of infections. Cystic echinococcosis (CE) is caused by *Echinococcus granulosus* sensu lato and its genotypic subtypes (G1-G10). Polycystic echinococcosis (PE) is attributed to *Echinococcus oligarthra* and *Echinococcus vogeli*. Alveolar echinococcosis (AE) is associated with *Echinococcus multilocularis* [10, 11]. The disease in both intermediate and definitive hosts typically remains asymptomatic, even in the later stages. This leads to high transmission rates of the disease [5]. According to data from the World Health Organization (WHO), in untreated cases, the annual mortality rate for cystic echinococcosis is typically between 2% and 4%. In contrast, the mortality rate for untreated alveolar echinococcosis surpasses 90% [4, 12]. Two epidemiologic studies in lietarure have reported that disability-adjusted life years (DALYs) range between 666,434 to 871,000 attributed to cystic echinococcosis. The economic burden amounts to nearly three billion dollars due to losses in the livestock industry. Despite these alarming statistics, echinococcosis remains a neglected disease [4, 12]. In this chapter, our objective is to assess the parasite's biology, life cycle, and epidemiology to guide physicians and researchers involved in echinococcosis. We aimed to give the epidemiologic and structural details of the *Echinococcus granulosus* sensu lato, *Echinococcus multilocularis*, *Echinococcus oligarthra*, and *Echinococcus vogeli* in this chapter. In the treatment of hydatid cysts, albendazole and mebendazole, in addition to surgery, are important. Therapeutic education is important as it can result in economic loss and low quality of life if not well managed. In control, all preventive measures are paramount; events of ivermectin salt licks in addition to dog control are suggested [1, 5].

This chapter addresses these issues by providing an update on the transmission of the parasite, pathogenetic aspects, prevention, and treatment options. Hydatid disease is caused by the metacestode larval stage of the cestode *Echinococcus* species. The objectives of this chapter are to demonstrate the epidemiological and clinical aspects associated with hydatid disease and thereby propose an integrated application of both traditional and modern techniques in its control, hence depicting the effects to the end user.

2.2 Epidemiology

Hydatid disease is a globally distributed zoonosis caused when a human or an intermediate host ingests the eggs of *Echinococcus granulosus* [13]. Between three and eight million people may be infected worldwide at any one time. Echinococcosis is widespread, particularly in regions where livestock are farmed, including areas of China, the Mediterranean region, the Middle East, and parts of the Southern continent [13, 14]. Surveillance for *Echinococcus* eggs is crucial in endemic regions, particularly in rural areas and slaughter facilities [13, 14]. Transmission depends, among other factors, on the number of dogs, foxes, and other carnivores in the area. Dogs are the primary cause of environmental contamination, which varies significantly with local disease burden and environmental conditions [15]. In most endemic areas, sheep are the intermediate host, although the subsequent definitive host and organism that larvae are destined to adhere to also vary [15]. Regions of New Zealand, Australia, and Sudan, middle and eastern Turkey have cattle as the main intermediate hosts [15, 16]. Numerous studies have examined the relationship between prevalence, risk factors, and environmental fecal contamination in several regions. Though variable by country or region, *E. granulosus* disease is found to be positively correlated to age, and this is suggested to be indicative of long-term exposure patterns [17]. There is some evidence to suggest that the relative risk of infection increases more rapidly with the age of exposure in females than in males [18]. Multiple studies discussing subpopulations have found that higher seroprevalence is associated with subjects who have been in contact with dogs or with infected livestock. Data have been gathered in ranges of both urban and rural populations [17, 18]. In the majority of studies, this is regarded as being due to the migration of dogs and contamination of the area with eggs of the parasite. Livestock (mainly sheep) are the main reservoir of human hydatid disease [17, 18].

A study by Kumaresan et al. [19] have shown that seroprevalance rate of the disease is high reaching 40% in general and even higher in females reaching 60%. The results of the study emphasizes the necessity of surveillance programs in endemic areas [19]. When available, surveillance of the subjects was performed with serological data and backed up by ultrasonographic and/or computerized tomography (CT) imaging to determine the number, size, and location of hydatid cysts in affected subjects [20]. Reports that increased eating of mutton in developed countries have given rise to a control strategy involving animal reservoirs. Rodents are also prevalent definitive hosts in areas where human behavior has led to an excess of rodents [21]. There have been reviews and commentaries on the prevention of *E. granulosus* transmission in the wild. Variability of environmental conditions is an important factor for disease transmission [21, 22]. The prevalence of infection in humans, the number of tapeworm carriers, the ratio of pigs to humans in the area, and the improvements required for effective control of transmission [22].

2.2.1 Global Burden

Hydatid disease is a worldwide zoonoses present on every continent except Antarctica [23]. It is a comparatively neglected disease that is frequently observed in areas with poverty and many developing countries. Large gaps exist in knowledge of the true global burden and world distribution of the parasite [23]. The number of new cases of hydatid disease in the world every year is predicted to be millions of patients. A global estimate of disability-adjusted life years is reported to be nearly 900,000 [24]. The average annual mortality rate ranges from 0.5% to 1% [25, 26].

Hydatid disease burden is disproportionately concentrated where livestock are kept and slaughtered. Hydatid disease has been estimated to be responsible for the loss of economic resources in many countries, one-third of which are in Africa south of the Sahara [27]. In Turkey, the reported annual economic loss due to hydatid disease is nearly $600,000 [28]. The parasite is thought to have existed for over 3000 years [29]. The extended latent period of this disease complicates accurate prevalence estimates [30]. Industrialized countries have significantly decreased their domestic disease burden through measures such as routine deworming programs, the culling of infected livestock, and thorough meat inspection [31]. However, migration patterns from high- to low-prevalence areas are influencing prevalence estimates in various regions [20].

2.2.2 Distribution and Prevalence

Hydatid disease has a global distribution and is endemic in many countries, particularly those with livestock farming, suitable environmental and cultural conditions, and accessible abattoir facilities [20]. Consumption of raw or undercooked animal products, often habitual or culturally significant in some regions, increases the risk of infection [20, 31, 32]. Transmission is further facilitated by close contact between dogs/foxes and livestock, poor or free-range animal husbandry practices, the presence of suitable wild and semidomestic definitive hosts, and socioeconomic conditions prevalent in rural and isolated populations [20, 31, 32]. While hydatid disease is typically associated with rural areas and, to a lesser extent, isolated urban settings, two geographical surveys have identified exceptions to this pattern. Specifically, the absence of the disease in much of the central highlands has been attributed to the geohistorical evolution of domestic dogs [20, 31, 32].

In endemic areas, children are most frequently infected, typically through contact with dogs that have access to infected livestock viscera or by consuming raw, infected offal. Sixteen to 30% of the newly diagnosed cases are seen in pediatric patients [33, 34]. Cancer awareness programs have incidentally led to the detection of hydatid cysts in many individuals by using CT surveillance scans, with an average age of 40 years and a male predominance [34]. Furthermore, about 15% of the patients are incidentally diagnosed, which emphasizes the importance of surveillance [34].

2.3 Structure, Etiology, and Life Cycle

The adult worm consists of a scolex, neck region, and 3–4 segments. Its length varies between 2 and 9 mm, depending on the species. Both the structure of the scolex and the number of segments vary among different *Echinococcus* species. The characteristic morphology of the *Echinococcus* tapeworm is illustrated in Fig. 2.1.

2.3.1 *Echinococcus granulosus*

The length of this species ranges from 3 to 6 mm. The scolex region features a rostellum at the top, which contains several rows of hooks that aid in attaching to the crypts of Lieberkühn [35]. The scolex region is followed by a narrow neck and three to four segments. The smallest segment, called the immature segment, is closest to the neck region. It is followed by the mature segment, which contains the uterus and testes. The ovaries are kidney-shaped [35, 36].

2.3.2 *Echinococcus multilocularis*

Similarly, the tapeworm measures 2 to 4 mm in length. The scolex region features a rostellum at the top, equipped with a few sets of hooks, followed by a row of suckers. There are three to five segments, with the terminal segment being the gravid segment. In the taxonomy of *Echinococcus multilocularis*, it is classified under the phylum Platyhelminthes and the order Cyclophyllidea. Morphologically, the adult tapeworm consists of a scolex with suckers and a chain of proglottids, each containing reproductive organs. The larval stage, on the other hand, forms a multilocular structure in the host's liver, resembling a tumor. Understanding its taxonomy and morphology is crucial for the accurate identification and study of this parasitic organism [5, 37].

2.3.3 *Echinococcus vogeli*

This tapeworm is relatively short, measuring between 4 and 6 mm. However, the terminal gravid segment is notably long and sac-like, which is highly pathognomonic. Additionally, the scolex contains a higher number of hooks (up to 36) within the rostellum [5].

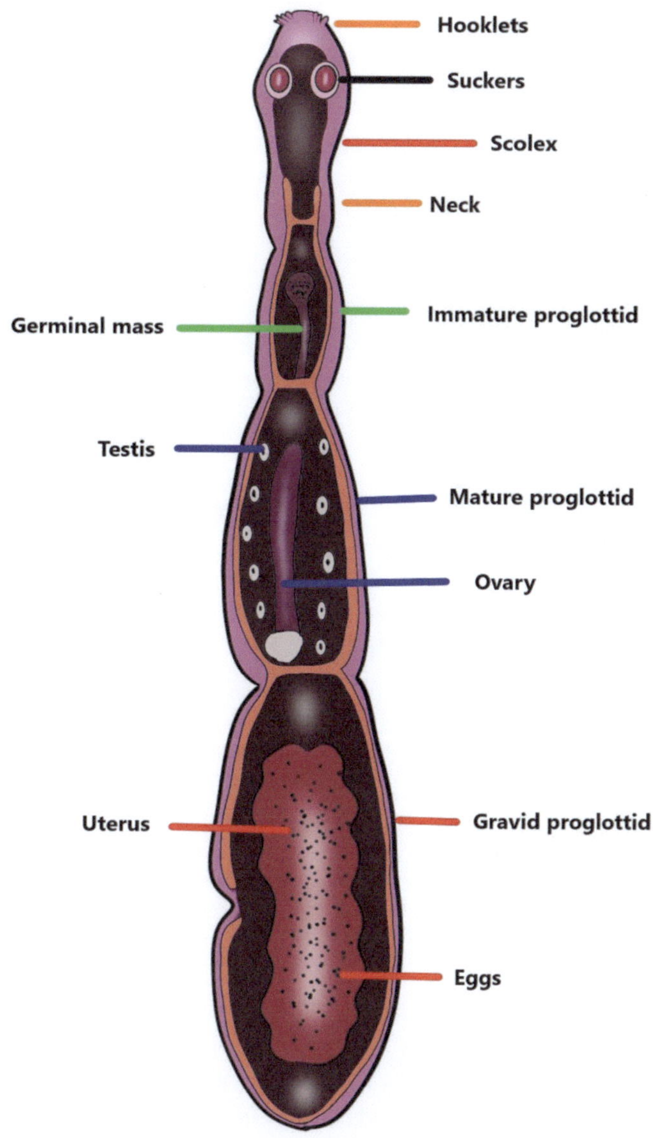

Fig. 2.1 The characteristic morphology of the *Echinococcus* tapeworm. (Image courtesy of DPDx, Centers for Disease Control and Prevention (https://www.cdc.gov/dpdx))

2.3.4 *Echinococus oligarthus*

The adult worm is typically 2–3 mm long and usually has three segments. The terminal segment is saccular and elongated [5].

2.3.5 The Life Cycle of *Echinococcus*

The parasite has two types of life cycles. The sylvatic cycle involves wild animals, with definitive hosts like wolves and wild dogs, and intermediate

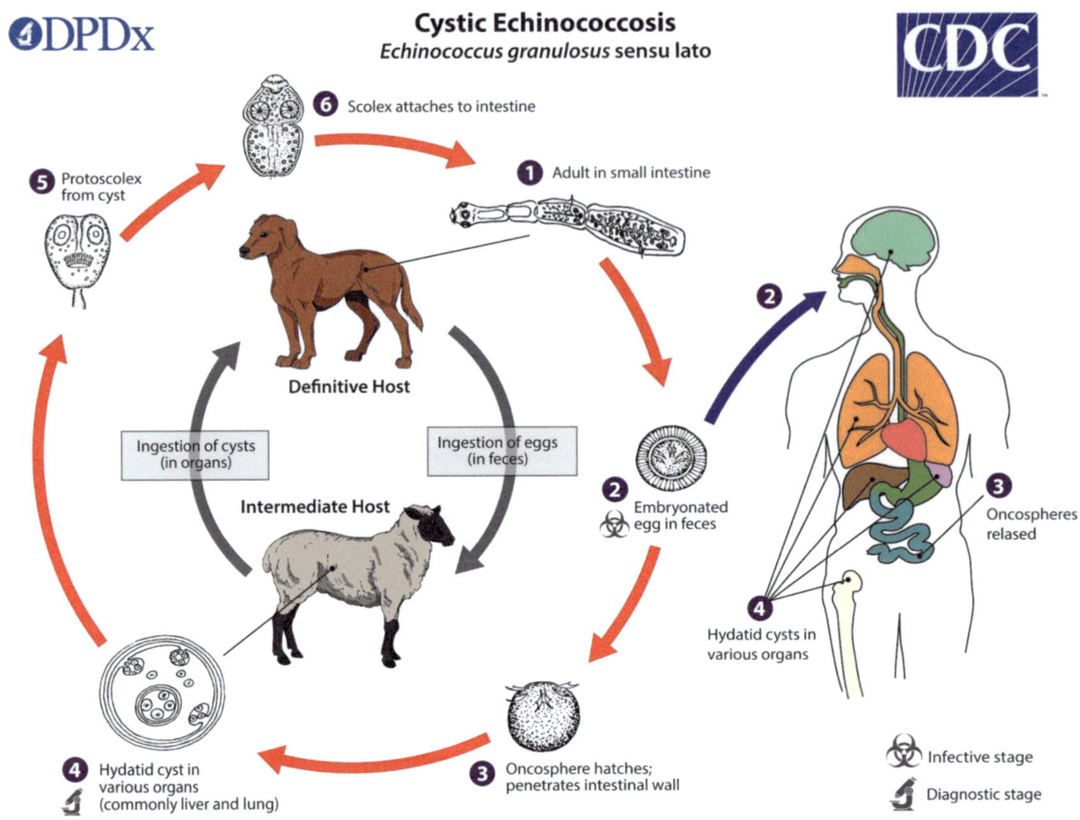

Fig. 2.2 Cystic echinococcosis. Parasitic life cycle of *Echinococcus multilocularis*. (Source: Centers for Disease Control and Prevention (CDC): https://www.cdc.gov/dpdx/echinococcosis/index.html. Reference to specific commercial products, manufacturers, companies, or trademarks does not constitute its endorsement or recommendation by the U.S. Government, Department of Health and Human Services, or Centers for Disease Control and Prevention)

hosts such as cervids [5, 38]. The domestic cycle includes domestic dogs as the definitive hosts and herd animals like sheep and cattle as the intermediate hosts [38]. The parasite undergoes the same transitions in both cycles, summarized in Figs. 2.2 and 2.3.

The cycle starts in the definitive host. The terminal segment of the adult worm, known as the gravid segment or *proglottid*, carries fertilized eggs [16, 39–41]. This segment ruptures, releasing the eggs, which then pass out in the feces of the host. The larval stage of the parasite is known as the metacestode. The life cycle of *Echinococcus* begins when intermediate hosts ingest food or water contaminated with the eggs [35]. These eggs develop into oncospheres, which penetrate the intestinal wall and travel to the liver, lungs, and lymph nodes [42]. Two-thirds of the echinococcal lesions occur in the liver due to the vast portal blood supply [40, 43].

In the end organ, metacestodes develop into fluid-filled, bladder-like cysts. In CE, it forms a unilocular cystic lesion, and in PE, a multilocular cystic lesion [12, 44, 45]. In AE, it presents as a tumor-like metastasizing mass that invades surrounding structures, containing multiple small lacunar and fluid-filled cystic lesions as components of the greater mass [21].

Generally, the cysts are surrounded by two types of capsules. The outer capsule consists of fibrous tissue from the host, forming as a reaction to the metacestodes. This capsule contains dead parenchymal tissue, inflammatory cells, and giant multinucleated cells. The inner membrane,

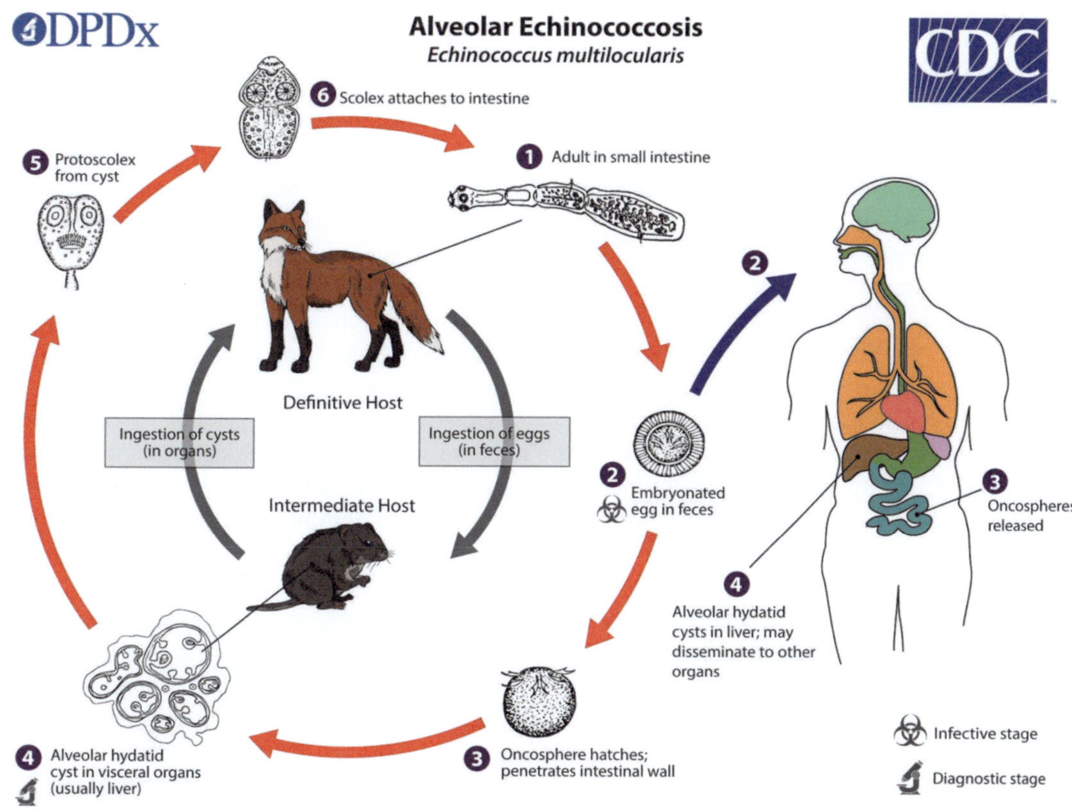

Fig. 2.3 Alveolar echinococcosis. Parasitic life cycle of *Echinococcus multilocularis*. (Source: Centers for Disease Control and Prevention (CDC): https://www.cdc.gov/dpdx/echinococcosis/index.html. Reference to specific commercial products, manufacturers, companies, or trademarks does not constitute its endorsement or recommendation by the U.S. Government, Department of Health and Human Services, or Centers for Disease Control and Prevention)

known as the germinative membrane, consists of brood capsules that contain invaginated protoscolices. Additionally, free protoscolices are present in the cystic fluid. The detached brood capsules and free protoscolices are collectively referred to as hydatid sand [5, 46].

Once the definitive host ingests the meat of an intermediate host infested with metacestodes, the protoscolices evaginate into their scolices form in the intestinal tract [46]. These scolices penetrate the crypts of Lieberkühn and attach to the mucosae using their hooks and sucker apparatus [46]. Humans act as accidental hosts for the parasite in its life cycle. Infection occurs when humans ingest food contaminated with parasite eggs and oncospheres, often due to poor hygiene. Additionally, eggs and oncospheres can be carried to food and water sources by wind and vectors like bugs and flies [5, 21].

2.4 *Echinococcus granulosus*

Echinococcus granulosus is the etiologic agent of hydatid disease in both livestock and humans. It is a small, hermaphroditic tapeworm that belongs to the cestode family Taeniidae. The adult parasites inhabit the small intestines of carnivorous definitive hosts, and the larval forms are in various internal organs of herbivorous intermediate host species including humans [47]. *E. granulosus* displays considerable genetic diversity and comprises up to ten different genotypes, of which G1–G3 appear to cover most parts of the world.

E. granulosus is a significant zoonotic pathogen found in developed and, particularly, developing countries [48]. The presence of domestic dogs infested with the parasite and the contamination of soil with their feces, which contain parasite eggs, contribute significantly to the propagation of hydatid disease [49].

In the intermediate host, the life cycle of *E. granulosus* begins with the ingestion of eggs containing oncospheres [47]. These oncospheres penetrate the intestinal wall and migrate to organs, primarily the liver and lungs, where they develop into hydatid cysts. The larval parasites within these cysts cause pathological changes in both livestock and humans [50]. When a definitive host, such as a dog, ingests these cysts, the protoscoleces develop into adult tapeworms in the intestine, and eggs are released in the feces, completing the cycle. Despite control efforts and widespread knowledge of its epidemiology, *E. granulosus* persists due to its effective transmission strategies and complex host-parasite interactions [50].

2.4.1 Intermediate and Definitive Hosts

Intermediate and definitive hosts play a central role in transmitting hydatid disease and *Echinococcus* cycle dynamics. More than 40 intermediate hosts and more than 30 definitive hosts have been reported worldwide, but only some of these combinations are considered epidemiologically relevant [47]. Intermediate hosts are mainly domesticated animals, particularly livestock, where the larval stage of the parasite causes major economic losses. The relationship between the intermediate and definitive hosts is a predator-prey relationship [47]. The primary mode of transmission is fecal-oral. Infection occurs when an intermediate host ingests food contaminated with the eggs of the parasite, which originate in the feces of the definitive host. Waterborne transmission can also occur. Furthermore, insects can serve as vectors, facilitating the dispersal of eggs to distant areas [51]. These animals are necessary for the successful completion of the life cycle of this parasite. After decades of adaptation, the interactions between *Echinococcus* and these hosts have selected the involved species and established the genetic basis for the observed parasite [52–54]. Host susceptibility and immune response, parasite damage type, and cyst fertility influence transmission and infection dynamics. Experimental and field data show parasite-induced susceptibility in intermediate hosts, which may increase the likelihood of successful infection depending on the *Echinococcus* species and genotype [52–54].

Definitive hosts are generally carnivorous mammals, particularly canids. Adult tapeworms develop in the intestines of these animals, releasing eggs into the environment where they are infective for intermediate hosts [55]. Dogs, in particular, are the most appropriate hosts for this parasite, based on their dietary habits and the morphological characteristics of the worms [55]. Therefore, the genus *Echinococcus* has a remarkable ability to develop and maintain its life cycle using the domestic dog as a definitive host, which is why domestic dogs are considered *Echinococcus* maintenance hosts [47–51]. Adult segments containing eggs are released with the feces, contaminating the environment. Currently, it is accepted that people are usually infected accidentally. Over the long-term course of the infection, continuous exposure from egg-tainted dogs closely related to people was a necessary condition for the hyperendemic focus [47–55].

Current data on cystic echinococcosis transmission in humans, linked to domestic animal sources, emphasize the role of dogs as definitive hosts and environmental contamination as a major risk factor. Beyond their economic importance, dogs, due to their free-roaming habits in urban and suburban areas, can act as carriers and vectors of both human and animal echinococcosis [55]. This increases the risk of infection for both immunologically naive and immunosuppressed individuals exposed to environments contaminated by definitive hosts [56]. However, further research is needed to fully elucidate the complex relationship between the parasite and its various intermediate and definitive hosts.

2.4.2 Clinical Presentation and Diagnosis

Hydatid cysts can cause a range of symptoms, from mild or nonspecific to potentially life-threatening [13]. The clinical manifestations of echinococcosis are highly variable, depending on the location, size, and number of cysts. This variability is often preceded by a latent asymptomatic period during which the cyst develops slowly [13]. While hydatid cysts can develop in any part of the body, the liver and lungs are the most common sites [57]. Extrapulmonary hydatid cyst localization is less frequent but has been documented in a wide range of sites. These include the brain, heart, spleen, bones and joints, kidneys, muscles and soft tissues, pancreas, pelvis, peritoneum, retroperitoneum, diaphragm, skin, parotid gland, and mediastinum [57, 58].

The clinical presentation of hydatid cysts can vary widely, encompassing asymptomatic or atypical cases, symptoms directly related to the cyst itself, and severe complications [13]. Although often asymptomatic in early stages, cyst growth eventually leads to the development of symptoms. Hydatid cyst disease frequently presents incidentally on imaging studies obtained for other indications. A prolonged asymptomatic period is characteristic, with cysts potentially growing for years before causing clinical manifestations [59]. Symptomatic disease can manifest with a range of presentations, including pressure effects on adjacent structures, biliary obstruction and associated jaundice, sepsis (often related to cyst rupture or infection), and anaphylactic reactions (typically triggered by cyst rupture and release of hydatid fluid) [59]. Rupture is a recognized complication that can lead to peritoneal dissemination of the parasite and systemic anaphylaxis [57–59]. Figure 2.4 illustrates a complex case of a ruptured hydatid cyst with peritoneal seeding. Preoperative images as well as postoperative first year images are presented.

The spectrum of clinical symptoms reflects the localization of the cyst. An important example is a cyst in the liver leading to jaundice, and a bronchial cyst in the lungs can cause a cough and other respiratory symptoms [60]. Cystic lesions are initially identified through imaging studies, with definitive diagnosis confirmed by clinical and serologic findings [61]. Ultrasonography is the most commonly used imaging study for the diagnosis and classification of hydatid cysts [62]. However, ultrasonography can miss microcystic forms in the early stages because of several reasons such as a lack of reliable visualization of the pericyst [63]. While computed tomography (CT) and magnetic resonance imaging (MRI) are

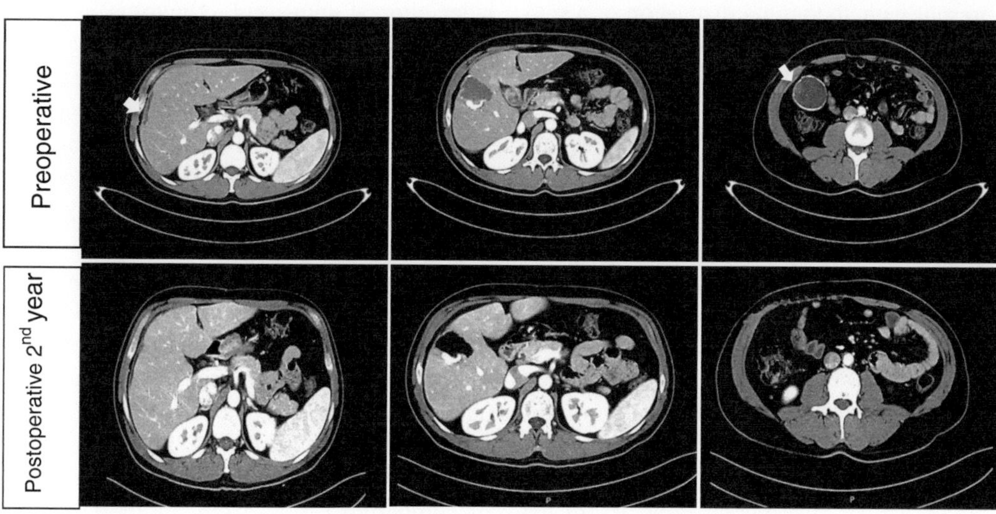

Fig. 2.4 Ruptured hydatid cysts in segment 5 with and ectopic peritoneal spread (White arrows). Both preoperative and postoperative second year imaging of the liver and peritoneal lesion is shown

valuable for diagnosing and characterizing hydatid cysts, they may miss microcystic forms in the early stages [62, 63]. Diagnosis is often incidental due to the insidious and frequently asymptomatic nature of the disease. An additional point to be considered for diagnosis is the patient's travel history and exposure to endemic areas [44, 46]. The diagnosis of hydatid cysts is often challenging due to vague symptoms and reliance on clinical suspicion. Furthermore, diagnostic workup varies widely across institutions.

2.4.3 Symptoms and Complications

Many patients with hydatid cysts remain asymptomatic throughout their lives. Even in cases with multiple cysts, symptoms may be vague, even at advanced stages [64]. When present, pain is often caused by cyst enlargement and compression of surrounding tissues. Biliary communication can lead to suppuration and cholangitis, further contributing to pain. The rupture of cysts into the peritoneal cavity can cause severe anaphylaxis [64].

The location of the cyst, particularly its proximity to vital organs such as the liver, lungs, brain, and vasculature, and the size of the cysts, determines the spectrum of symptoms and potential complications [65]. Occasionally, patients may present with abdominal distension or a sensation of fluid movement, typically due to the mass effect of the cysts. In the liver, hydatid cysts can alter the texture of the organ and mimic a liver mass. Extensive cyst growth or chronic compression of vascular and biliary structures may lead to cirrhosis, potentially resulting in portal hypertension, ascites, and the development of umbilical varices [65].

Echinococcal cyst rupture is a serious complication. Rupture involving major vascular structures can lead to severe intracystic hemorrhage and anaphylactic shock [66]. Additionally, vasculobiliary connections may result in vessel thrombosis and embolism to distant sites, including the lungs [64]. It may take a long time before the complications develop. There are extreme reports of hydatid liver disease associated with Budd-Chiari and acute decompensation. In addition, the patients may suffer from hepatorenal syndrome as the decompensation progresses, leading to acute kidney injury [66–69]. In patients with an active liver or extrahepatic cysts, diagnosis can be confirmed by PCR testing of samples obtained via puncture, aspiration, injection, and respiration (PAIR), paracentesis, or thoracocentesis, along with cytological or serological assays [64].

In cases of abdominal cystic echinococcosis (CE), it is essential to use chest X-ray imaging for diagnosing potential pulmonary dissemination complications [64]. Long-term surveillance with CT scans is recommended for patients with abdominal CE, including presumed inactive forms, as they may develop disseminated cysts originating from early or inactive lesions identified during initial diagnosis and evaluation. Patients who have been treated and are no longer exposed to the pathogen require 3 years surveillance before any follow-up procedure is discontinued [50]. Follow-up protocols should be based on pre-treatment assessments and can be streamlined if the patient can access healthcare.

2.4.4 Diagnostic Methods

To diagnose echinococcosis, clinicians must consider clinical signs and symptoms, patient history, epidemiological risks, serological test results, and imaging findings concurrently. Pathological analysis of surgical or biopsy samples may be required in certain cases [70]. However, clinical signs can be nonspecific and misleading to a considerable degree. Imaging methods can help to clarify, even in part, different morphologies of the cyst, but cannot be used solely in establishing the diagnosis [71]. Despite having a specificity of 77–90% and a sensitivity of 88–90%, specific serological tests remain crucial tools in the diagnosis of echinococcosis. However, their availability can be inconsistent, even within endemic regions [72, 73]. The value of comparing results from traditional serologic methods has been limited by significant interlaboratory heterogeneity in the preparation and quality of the antigens used [72]. Consequently, diagnosing hydatid cysts requires an integrated evaluation of clinical and epidemiologic factors together with radiologic and serologic findings.

2.4.5 Preventive Measures

Preventive measures could also significantly help to reduce or even eliminate the incidence of hydatid disease in populations at risk. How to control and eliminate hydatid disease or echinococcosis is now broadly well recognized worldwide. This mainly involves conducting extensive public health education and promoting prevention to raise awareness and address the pernicious impact these diseases have. This can be supported by implementing regular community mobilization and health education programs, which remind target communities that prevention is critical, with all components of effective control depending upon these preventive activities. Thus, all populations at risk should attain a clear understanding of these varying transmission patterns and critical disease-related control and preventive aspects. Ultimately, it is important for departments responsible for health or hospital management to contribute toward the elimination and control of such diseases.

Sanitation and hygiene could interrupt the transmission cycle if consistently maintained; thus, the control of eco-health or environmental health plays a key role in controlling and decreasing such parasitic diseases over time. Sanitation and hygiene are also key factors underlined in the new program. In addition, vaccines must and can be delivered to increase the immunity of animals and, if possible, prevent the re-entry and re-establishment of *E. granulosus* in situ. If we can conduct echinococcosis control programs effectively and work within the field using best practices concurrently in a One Health approach, we can establish clear control priorities and procedures, and can implement field control, overall decreasing and impacting echinococcosis infection rates within reasonable control targets. The primary emphasis must be to regularly monitor and control livestock and canine populations, and especially infected dogs. Additionally, approval of hydatid disease vaccination projects using best practices for dogs and vulnerable animal-intermediate hosts should be pursued in the future.

2.4.6 Public Health Strategies

Countries such as Australia, New Zealand, and Cyprus have eradicated hydatid disease through comprehensive and sustained health education programs. These programs educate local populations on: (a) the conditions leading to human hydatid disease, (b) the epidemiology of the disease, including the behavior of intermediate and definitive hosts in the local environment, (c) simple and practical preventative measures, (d) the importance of regular dog deworming protocols, and (e) livestock vaccination [74]. Traditional media and social media both play an essential role in providing information and education to their intended audience. Vaccination of cattle and sheep with the E95 vaccine significantly reduces the risk of *Echinococcus* infestation [75, 76]. A combination of 50 μg of antigen and 5 mg of the adjuvant QuilA® is effective in controlling the disease. QuilA®, a plant-based extract from *Quillaja saponaria*, is used in veterinary vaccines as an adjuvant to stimulate CD4+ helper and cytotoxic CD8+ T cell functions, and to provide a sustained antibody response [77].

Preventing the disease requires a nationwide effort involving regulations, substantial funding for public education, surveillance of both humans and animals, deworming hosts, and vaccination programs for livestock. Healthcare professionals, including veterinarians, must identify and report the disease to ensure effective control [78]. With the implementation of appropriate measures, a decline in disease prevalence is anticipated within 10 years. Additionally, as part of nationwide regulations, officials should evaluate the efficacy of preventive measures every 5 years [78].

2.4.7 Prevention

Preventive measures for echinococcosis include: (a) improved slaughterhouse inspection and livestock management practices, (b) vaccination of livestock, and (c) regular deworming of definitive hosts, particularly domestic dogs. The control of slaughterhouses and prevention of illegal or house slaughtering is paramount. Slaughterhouses must undergo regular inspections conducted by veterinarians under regulatory standards [75]. Whenever the inspections fall short, the infected organs can be accessed by the dogs in the region causing infection and production of the eggs thus contaminating the soil and causing infestation in the intermediary host [75, 79]. Therefore, the carcasses in the slaughterhouses should be inspected for the control of hydatid disease. Prevention of illegal slaughtering and home slaughter requires continuing education of the public and legislation [79].

The E95 anti-oncosphere vaccine has proven effective in controlling *E. granulosus* infestation in intermediate hosts [75]. When administered twice a year, the vaccine reduces parasite transmission in sheep and provides protection for nearly 5 years after the third dose [80, 81]. In cattle, immunity lasts for 12 months after the second dose [82]. The vaccine is being developed by modifying the protein to make it more soluble and easier to administer [83].

The first step in preventing and controlling cystic echinococcosis is managing dog populations. Strategies for dog control include praziquantel deworming, arecoline purgation, and population management through capturing, neutering, deworming, and releasing [84]. Deworming with praziquantel is the most commonly used strategy, as it effectively kills the adult parasite in dogs. However, dogs can become reinfected over time, so the dosing intervals of praziquantel depend on reinfection rates [85]. Typically, administering praziquantel four times a year is recommended in endemic areas. The primary challenge is treating animals in nomadic and seminomadic populations living in rural areas, where governmental funding is necessary for mass deworming of canines [86].

Mass education programs are essential for controlling the disease. Collaboration between veterinarians and public health officials is crucial for sustainably managing hydatid infection and training pet owners about the risk factors for echinococcosis [86]. Fertile eggs excreted by untreated stray dogs can remain viable for many months in surface soil, as they are resistant to desiccation, freezing, and UV light exposure [86]. Pet owners should be informed that close contact with infected dogs and consumption of infected offal can cause the disease, especially in domestic dogs [86]. They can bring their animals to municipal kennels for 2 to 3 days, where the cost of deworming is supported by the local government. Regular deworming with praziquantel by a veterinarian and feeding dogs appropriately cooked commercial food are important preventive measures [85, 86]. Currently, there is no commercially available echinococcus vaccine for definitive hosts, particularly dogs [75].

Regular praziquantel treatment of carnivores eliminates adult cestodes, thereby preventing parasite transmission to intermediate hosts and the production and excretion of eggs by infected dogs. This approach is effective for controlling the disease in canines, particularly in rural enzootic (endemic) areas [85]. Community education, through methods like didactic seminars, is a crucial component of these control strategies. This education should inform livestock owners about hydatid infection, and improved veterinary services should provide tailored deworming recommendations for individual herds [86, 87]. In addition to public education programs, school education initiatives are essential. Control of reservoir hosts offers the benefit of reducing or eliminating animal-to-human transmission from enzootic areas, thereby limiting the dissemination of infections to previously unaffected regions [87].

2.4.8 Treatment Options

The pillars for treatment of cystic echinococcosis are: (a) Medical therapy, (b) Watch and wait, (c) Percutaneous treatment options, and (d) Surgical therapy. The type of treatment is chosen according to the symptoms of the patients, the presence of complications, and the characteristics of the cysts such as location, size, and stage [46, 88–90].

The World Health Organization Informal Working Group on Echinococcosis (WHO-IWGE) classification is the standard for determining the stage of the disease. The WHO-IWGE classification includes the following categories [91]:

CE1 Uniocular cysts characterized by a "double line" sign and fine internal echoes or granulations (hydatid sand), which contain protoscoleces and daughter cysts.

CE2 Multilocular or multiseptated lesions showing daughter cysts within the main unilocular cystic lesion, resulting in a "honeycomb" or "rosette" appearance.

CE3a Characterized by the germinative membrane detaching from the pericyst, a pathognomonic finding known as the "water lily" sign.

CE3b Daughter cysts are present within a solid or cystic matrix.

CE4 A heterogeneous solid lesion with internal canalicular formations and no daughter cysts, identified by the pathognomonic "ball of wool" sign.

CE5 A unilocular, degenerated solid lesion with diffuse calcification of its wall.

CE1 and CE2 lesions are considered active and capable of causing infestation. CE4 and CE5 are degenerative and inactive, with no potential for infestation. CE3 lesions represent a transitional stage with a high risk of causing infestation [91, 92].

Albendazole is the preferred medical treatment for CE1, CE2, CE3a, and CE3b cysts measuring less than 5 cm [93]. It is also the primary therapy for inoperable cysts, including multiple cysts of the liver that are not amenable to surgery and cases with multi-organ involvement [94]. The total daily dosage is 10–15 mg/kg, administered in two divided doses. Albendazole is also indicated in the neoadjuvant setting before percutaneous or surgical intervention for liver cysts. Medical therapy should be initiated 3 days to 3 months pre-procedure and continued post-procedure [95]. Treatment duration following percutaneous or surgical intervention should be 4 to 8 weeks for uncomplicated cysts and 3 to 6 months for complicated cysts [96].

The watch-and-wait strategy is applied for CE4 and CE5 cysts. They can be followed by ultrasound without any pharmacologic or surgical intervention [89].

Percutaneous therapy for hydatid cysts includes PAIR, standard catheterization, and modified catheterization (MoCATH) [97–102]. PAIR is recommended for CE1 and CE3a cysts smaller than 10 cm in diameter but is contraindicated in complicated cysts, particularly those with cystobiliary communication [102]. Additionally, PAIR is contraindicated for CE2 and CE3b cysts smaller than 10 cm that cannot be safely localized by ultrasound [102]. For CE1 and CE3a cysts larger than 10 cm or with a volume over 1000 cc, standard catheterization is the preferred approach [103–106]. This technique is also indicated when cystobiliary communication is present. However, standard catheterization is often associated with a longer hospital stay [106]. In contrast, MoCATH is typically employed for uncomplicated CE2 and CE3b cysts, as well as for those larger than 10 cm. Following either procedure, if significant fluid drainage persists, the catheter may be left in place, potentially prolonging the length of hospitalization [103–106].

The surgical indications for cystic echinococcosis include: (1) CE2 and CE3b cysts with multiple lesions compressing adjacent structures; (2) solitary superficial cysts at high risk of rupture; (3) absence of percutaneous drainage; and (4) complicated cysts, such as those with secondary infections (suppurative lesions) or cystobiliary communication [107]. The success rate of PAIR in CE1 and CE3a cysts is 96%, with a recurrence rate of 3%. The major complication rate is less than 10%, while the minor complication rate is 13% [108]. In contrast, the recurrence rate of PAIR in CE2 and CE3b cysts is 61%. MoCATH demonstrates greater success in CE2 and CE3b cysts, with a recurrence rate of 4% and a major complication rate of 10% [107]. The reported mortality rate associated with percutaneous therapies is 0.01% [107]. Major complications include cyst suppuration, cystobiliary fistula, and anaphylactic reactions, while minor complications consist of fever, urticaria, itching, and hypotension [107].

Indications for surgery have been previously stated. Surgical therapy for cystic echinococcosis is categorized into radical and conservative surgical treatments. Conservative surgical options include partial cyst wall resections that involve the removal of the germinative membrane, with or without capitonnage, as well as simple drainage procedures [45]. In contrast, radical surgical resections encompass total cyst excisions, pericystectomies, and major hepatectomies [45]. Radical procedures require expertise in hepatobiliary surgery and transplantation [92]. Furthermore, the complication rate following radical procedures is high [92]. Therefore, conservative surgery is the preferred method. Additionally, conservative surgery allows for the treatment of more cysts than the radical approach, which results in a significant loss of parenchyma [108]. This approach includes the drainage and evacuation of cyst contents, removal of the endocyst, and closure of the cavity using various techniques, such as capitonnage or omentopexy [109]. The conservative technique is associated with complications including intraoperative spillage of the daughter cysts, suppuration of the cyst cavity, and biliary fistula [109–111]. Treatment of the biliary fistula may require ERCP with papillotomy and stent placement. The recurrence rates reach 20–30% following conservative surgery, and intraoperative spillage plays an important role [42, 109]. Conversely, one of the causes of recurrence is the exogenous vesiculation of the cyst. This phenomenon results from increased intracystic pressure, causing small breaks in the pericyst that allow the germinal layer and cystic fluid to herniate into the surrounding parenchyma and form new daughter cysts outside of the mother cysts [42]. Performing intraoperative ultrasound may prevent the surgeon from missing the exogenous vesiculations.

Khuroo et al. [112] reported that percutaneous treatment, compared to surgery, resulted in a lower incidence of procedure-related complications (32% vs. 84%) and a shorter hospital stay (4.2 ± 1.5 days vs. 12.7 ± 6.5 days). Therefore, percutaneous treatment is the preferred approach when clinically indicated.

2.5 Echinococcus Multilocularis

This tapeworm mainly infects the small intestine of canids, including foxes and dogs, but can also cause severe disease in humans as an accidental intermediate host [112]. The larval stage of the parasite causes alveolar echinococcosis. Its growth pattern mimics that of an infiltrative tumor, and it often metastasizes in the progression of the disease. The untreated cases have a 90% mortality rate [112, 113]. Treatment of alveolar echinococcosis is complex and includes treatment modalities such as palliative (reduction) hepatectomy, resection, *including lobar hypertrophy strategies*, ex vivo resection and reimplantation, liver transplantation, and percutaneous treatment. Usually, patients require a combination of these procedures [112].

2.5.1 Life Cycle and Transmission

Echinococcus multilocularis is a small, parasitic tapeworm that belongs to the class Cestoda and the family Taeniidae. Morphologically, it is characterized by a small size, measuring only 1–4 mm in length, and a distinct multilocular structure of its larval stage. The life cycle of *Echinococcus multilocularis* involves complex interactions between definitive hosts, such as foxes and other canids, and intermediate hosts, including rodents and occasionally humans. A parasite with a life cycle primarily in wildlife generally results in a limited rate of human infections. Conversely, the infection rate in humans and livestock becomes higher once domestic dogs are infected and can spread the parasite [114, 115]. The adult tapeworm releases eggs in the feces of the definitive host, which the intermediate host then ingests. The eggs hatch into oncospheres that penetrate the portal vein and migrate to the liver, the primary site of alveolar echinococcosis. The oncospheres transform into metacestodes in the liver, which proliferate and form the characteristic multicystic, invasive lesions observed in alveolar echinococcosis [114]. The peripheral structure of these cysts is formed of three layers, with the outermost layer being the acellular, carbohydrate-rich laminated layer. The larval cyst wall has two main layers. The outermost laminated layer is an acellular, carbohydrate-rich matrix that is secreted by the inner layer. Nutrient uptake for the developing parasite occurs across the tegument to the germinal layer, the innermost, cellular layer [71, 114, 116]. The tegument is responsible for the transport of nutrients from the environment. Furthermore, the surface area of the tegument is increased by digitations into the laminated layer, which are called the microtriches. The germinal layer contains connective tissue, glycogen for energy storage, and cells capable of proliferation and differentiation to form new parasitic structures (metacestodes), which are essential for the parasite's life cycle progression in the definitive host [71, 114, 116].

The metacestode exhibits slow and insidious growth, with an incubation period of 5–15 years. The liver, particularly the right lobe, is the primary site of involvement, while the lung is the second most common site, often due to metastasis. The patients are usually asymptomatic until the disease reaches advanced stages. The morphology of the cysts is classified according to Kodama classification [117–119]. There are five types of cysts: *Type 1*: Multiple cystic lesions without solid component, *Type 2*: Multiple small cysts with solid component, *Type 3*: Large irregular cysts with solid component, *Type 4*: Solid tissue without cystic component, and *Type 5*: Large cystic component [117–119]. Type 2 and 3 cysts are the most frequent forms of the disease, observed in nearly 50% of the cases [120, 121].

2.5.2 Epidemiology of *Echinococcus multilocularis*

Determining the global burden of this disease is very difficult, primarily due to underreporting [122]. As previously stated, liver lesions mimic malignancy, and definitive diagnostic tests suitable for field surveillance are lacking, hindering diagnosis confirmation and underestimating the disease's prevalence and annual incidence rate (AIR) [122, 123]. Consequently, prevalence and incidence data are largely derived from studies in endemic areas. Globally, approximately 10,000 new cases are reported annually, with over 90% (nearly 10,150 cases) originating from China [124].

In the Asian continent, the annual incidence rate (AIR) in China (highest rates in Qinghai and Sichuan provinces) is 0.7 cases per 100,000 population, and the current prevalence is reported to be 130,000 cases [124]. In eastern Asia, Japan has one of the lowest incidences, with an AIR reported to be nearly 0.02 new annual cases per 100,000 population [114]. The AIR in Kyrgyzstan is reported to be 2.6 new cases per 100,000 population, equating to nearly 180 new cases per year [114, 124]. Kazakhstan, however, reports an annual incidence rate (AIR) of 0.04 new cases per 100,000 population, which equates to 7 new cases annually [114]. The annual incidence rate (AIR) of alveolar echinococcosis in Uzbekistan (0.008), Iran (0.002), Armenia (0.04), and Tajikistan (0.1) per 100,000 population is relatively low among Asian countries [114]. Similarly, the AIR in Turkey is reported to be 0.02 new cases annually per 100,000 population, which equates to nearly 20 new cases per year [114, 125].

The Netherlands in Western Europe reported the lowest AIR of alveolar echinococcosis, with 0.005 new cases per 100,000 population, representing one new case per year. The AIR ranges from 0.05 to 0.08 new cases annually per 100,000 population in the remaining Western European countries, including Switzerland, Austria, Belgium, France, and Germany [125]. In Northern Europe, the highest incidence of the disease is 0.52 new annual cases per 100,000 population, with 15 new annual cases. Unfortunately, the AIR is rising in Northern and Eastern Europe [114]. While the annual incidence rate (AIR) of alveolar echinococcosis in Canada is reported as 0.007 new cases per 100,000 population, the incidence of the disease is reportedly on the rise in North America [23, 126, 127].

2.5.3 Clinical Manifestations and Diagnosis

The extent of disease is determined by the PNM staging system proposed by the WHO-IWGE. In this system, P stands for the localization of the parasite, N stands for the extrahepatic extension of the mass lesion, and M stands for the presence of metastasis [127, 128]. The staging system is summarized in Tables 2.1 and 2.2 [128].

Upon infection with *Echinococcus multilocularis*, individuals may remain asymptomatic for a prolonged period, with the disease often presenting with nonspecific symptoms such as abdominal pain, malaise, and weight loss. Abdominal pain, hepatomegaly, weight loss, fatigue, and jaundice can be observed in the advanced stages of the disease. Distant organ metastasis is very rare, but once it develops the patient mortality reached 90% in 10 years [114, 129–131]. The diagnosis of this parasitic infection usually involves visualizing the characteristic multivesicular cysts in the liver and other affected organs through imaging techniques like ultrasound, CT, and MRI. Serologic tests targeting the Em16 and Em18 antigens are useful for monitoring the response to therapy. However, definitive diagnosis ultimately requires histopathologic analysis [132, 133, 141].

Table 2.1 PNM staging of hepatic alveolar echinococcosis

Stage	Definition
P	Location of the lesion in the liver
P X	Primary lesion cannot be assessed
P 0	Hepatic lesion not determined
P 1	Primary lesion in either the right or the left lobe without any vascular or biliary involvement
P 2	Central location of the lesion with vascular and/or biliary involvement of one lobe
P 3	Central localization of the primary lesion with invasion of the portal vessels and biliary duct at the hilar level and/or invasion of at least two hepatic veins
P4	Regardless of the location of the lesion, there is extension along the vasculature and bile ducts
N	Extrahepatic extension to the adjacent organs, including the diaphragm, lungs, pleura, pericardium, heart, gastric and duodenal wall, adrenal glands, peritoneum, retroperitoneum, parietal wall (muscles, skin, bone), pancreas, regional lymph nodes, liver ligaments, and kidneys
NX	Extrahepatic extension cannot be assessed
N0	There is no invasion of the adjacent structures
N1	Extrahepatic extension to the neighboring organs and structures is present
M	Presence of distant (hematogenous) metastasis to lungs, distant lymph nodes, spleen, CNS, orbital, bone, skin, muscle, kidney, distant peritoneum, and retroperitoneum
MX	Metastasis status cannot be assessed
M0	Absence of distant metastasis
M1	Presence of distant metastasis

Ultrasound is the initial imaging method for determining and evaluating hepatic alveolar echinococcal lesions [134]. It allows visualization of the lesion's borders, vascularization, solid and cystic components, and localization. Furthermore, ultrasound aids in detecting any vascular and biliary invasion [135, 136]. CT provides greater detail regarding the characteristics of hepatic lesions, including invasion of major vascular and biliary structures, and extension to adjacent vital organs [137]. Importantly, contrast-enhanced CT helps differentiate alveolar echinococcosis lesions from hepatic tumors, as alveolar lesions typically do not show contrast enhancement. CT is a crucial tool for identifying vascular involvement, invasion of neighboring vital organs, and distant metastasis [137]. MRI can delineate the cystic and solid components of hepatic parasitic lesions. Alveolar echinococcosis lesions typically show heterogeneous signal intensity on T2-weighted images and are hypointense on T1-weighted images [138]. The cystic component is usually hyperintense on T2-weighted images, while fibrotic lesions exhibit attenuated signals [138, 139]. Due to its effectiveness in

Table 2.2 The PNM staging of the disease according to the extension of the hepatic lesion

PNM stage of alveolar echinococcosis			
Stage I	P1	N0	M0
Stage II	P2	N0	M0
Stage IIIa	P3	N0	M0
Stage IIIb	P1–3	N1	M0
	P4	N0	M0
Stage 4	P4	N1	M0
	Any P	Any N	M1

visualizing soft tissues, MRI is also valuable for detecting invasion of neighboring structures. Furthermore, MRI-guided cholangiography is highly effective in identifying lesions invading or compressing the biliary system and the resulting biliary complications [140–142].

2.5.4 Treatment and Management

Management of *Echinococcus multilocularis* infection necessitates a multidisciplinary strategy encompassing pharmacological and surgical options [112]. These approaches aim to eradicate the parasitic infestation and address associated complications. Treatment regimens are tailored to the disease burden and the overall health status of the patient. Close surveillance and long-term follow-up are essential for ensuring treatment success and preventing disease recurrence [143, 144].

2.5.5 Surgical Options

Various surgical options exist for the treatment of alveolar echinococcosis. Curative resection is the preferred surgical option, and similar principles applied during radical resection for hepatic tumors are also used for resections in treating alveolar echinococcosis [145–148]. The resectability of alveolar echinococcosis lesions depends on their size and the extent of disease with invasion of the inferior vena cava (IVC), hilar structures, diaphragm, and retroperitoneal space. The main determinant of the safety of curative resection is the adequacy of the future liver remnant volume following the planned resection [145–148]. When the future liver remnant is inadequate, procedures to induce hypertrophy before resection, such as portal vein embolization and two-stage hepatectomy, may be necessary [145–148]. Although the number of patients and reports is scarce, these techniques have shown effectiveness in achieving satisfactory hypertrophy in patients with insufficient future remnant liver volume [149]. Inevitably, patients who cannot meet these criteria may have to undergo liver transplantation, but unfortunately, the long-term results of this treatment are very poor, with a high rate of locoregional and distant recurrences [150]. Overall mortality following radical liver resection ranges between 0.1% and 3.5% [151–153]. WHO-IWGE stage I (P1N0M0) and II (P2N0M0) lesions are amenable to radical resection. Unfortunately, 70% of the patients with alveolar echinococcosis do not meet the criteria for curative resection [112, 137]. Radical resection in alveolar echinococcosis aims to remove all macroscopic and microscopic parasitic tissue, necessitating a clear residue classification. This classification defines: R0 as no residual parasite tissue with a safety margin of at least 2 cm; R1 as microscopic residual parasite tissue, including cases with a surgical margin less than 1 cm; and R2 as macroscopic residual disease [151]. Available data indicate comparable overall survival outcomes for patients achieving R0 resection and those with R1 resection followed by albendazole-based antiparasitic therapy [151]. R0 resection was associated with no local recurrence and a low pulmonary recurrence rate of approximately 2%. In contrast, R1 and R2 resections showed a significantly higher intrahepatic recurrence rate, approaching 40% [148]. Joliat et al. [148] reported progression-free survival rates of 97% after R0 resection, 64% after R1 resection, and 33% after R2 resection. Ex vivo liver resection and autotransplantation (ERAT) requires careful consideration due to limited data and reported mortality rates approaching 10%. However, it represents a radical, effective, and novel surgical method for end-stage alveolar echinococcosis. Notably, ERAT avoids the need for organ donation and long-term immunosuppressive therapy, potentially offering preferable overall survival rates compared to liver transplantation [154].

Palliative (debulking) resection (all R2 resections) may be considered in operable cases, particularly with parasitic invasion of the hepatic hilum, IVC, or adjacent vital organs like the lungs. While early data suggested palliative resection might be superior to albendazole alone, propensity score matching by Kadry et al. [159] indicated higher progression rates compared to radical resections, with outcomes similar to benzimidazole therapy alone. Furthermore, palliative resection has been associated with increased complication rates, such as biliary fistula and abscess [155–157]. However, Kawamura et al. [152] suggest that if over 90% of the parasitic bulk can be removed, the efficacy of subsequent albendazole therapy improves and complication rates decrease. Therefore, palliative surgery might be justified when a significant debulking (>90%) is achievable, although its overall benefit over medical therapy alone remains questionable based on propensity-matched data [151, 155–161].

2.5.6 Pharmacological and Other Medical Interventions

The mainstay of pharmacological intervention for *Echinococcus multilocularis* is continuous treatment with benzimidazole anthelmintics, primarily albendazole and mebendazole [63]. Following radical resection, a 2-year course is typically advised, whereas lifelong therapy is necessary in other situations [63]. These medications target parasite viability and lesion reduction. The treatment duration, which can extend from months to years, is guided by the patient's response and the presence of residual disease [161]. Careful monitoring for potential adverse effects and regular imaging are crucial components of this pharmacological management. On the other hand, there are some advancements in the medical treatment of alveolar echinococcosis because the efficacy of benzimidazoles has varying degrees of effects related to the size and blood supply to the cysts. For this reason, there are some preclinical studies regarding the anthelminthic capacity of various drugs such as antibiotics (mefloquine, nitazoxanide, and amphotericin B), protease inhibitor (bortezomib, sorafenib, and imatinib), and immune checkpoint inhibitor (anti-PD-1/PD-L1 drugs) [162].

2.5.7 Prevention Strategies

Preventing *Echinococcus multilocularis* infection relies on a multifaceted approach encompassing public health measures and educational campaigns. These strategies target definitive hosts (foxes and rodents) to reduce parasite prevalence and minimize environmental contamination through measures like effective waste management and controlling host populations in high-risk areas. Early detection and intervention are vital for preventing disease progression and further transmission.

The substantial burden of alveolar echinococcosis, estimated at 700,000 disability-adjusted life years (DALYs), is compounded by its increasing incidence, strongly indicating the influence of environmental contamination [163]. The eradication of rabies in Switzerland caused an increase in the fox population, which, in turn, led to a higher incidence of *Echinococcus multilocularis*. This rise is attributed to environmental contamination with feces from these canids [164–166], serving as a good example of how environmental factors influence disease spread. Significantly, dogs are responsible for 20% of environmental contamination and parasite spread [167]. In Switzerland, the growing number of alveolar echinococcosis cases in humans and domestic dogs indicates a substantial level of environmental contamination [168]. Surveillance programs are crucial, and deworming domestic dogs with praziquantel is essential, especially considering the stark contrast in *Echinococcus multilocularis* incidence: up to 20% in endemic regions versus less than 1% in low-incidence areas like central Europe [169–171].

2.5.8 Public Health Measures

Controlling the spread of *Echinococcus multilocularis* requires essential public health measures. These involve implementing surveillance and monitoring programs to pinpoint high-risk areas and populations, establishing protocols for managing infected individuals and preventing secondary transmission [114], and promoting safe food-handling practices to reduce infection from contaminated food. Effective implementation of these strategies demands collaboration between public health authorities and medical professionals [73].

2.5.9 Educational Campaigns

Educational campaigns are vital for raising awareness and knowledge about *Echinococcus multilocularis* among at-risk populations. These campaigns should focus on the life cycle of the parasite, preventive measures, and symptom recognition for early diagnosis [73, 114]. Targeted initiatives can address specific risk factors like contact with infected animals and unsafe environmental practices, and emphasize the importance of seeking medical attention for suspected cases and adhering to preventive measures to reduce exposure and transmission [73, 114].

2.6 Polycystic Neotropical Echinococcus: *Echinococcus vogeli* and *Echinococcus oligathrus*

Polycystic neotropical Echinococcosis is a zoonotic infection found in tropical Central and South America, with *Echinococcus vogeli* as the primary causative agent and *E. oligarthrus* playing a lesser role [172, 173]. Similar to other *Echinococcus* species, its infectious cycle involves humans as accidental intermediate hosts, with the oral-fecal route being the mode of contamination. Infection with *E. vogeli* is fatal without treatment and its progression resembles alveolar echinococcosis [173], whereas the very rare *E. oligarthrus* infection causes unicystic neotropical Echinococcosis [174].

2.6.1 Taxonomy and Classification

Echinococcus vogeli and *E. oligarthrus* are cestodes within the Taeniidae family and *Echinococcus* genus. *E. vogeli*, often called the Brazilian strain of *E. granulosus*, uses rodents, rabbits, and pacas as intermediate hosts. Usually, the definitive host is the bush dog and they prey on pacas. In addition, humans hunt pacas for meat as well [6, 175]. In contrast, *E. oligarthrus*, the causative agent of multiple unicystic neotropical echinococcosis, primarily infects the agouti as its intermediate host. Interestingly the definitive hosts are felids such as the jaguar, puma, or bobcat [174, 176]. Both species share a complex life cycle involving canids and felids in the wild as definitive hosts and small mammals as intermediate hosts, where the larval stages develop cystic lesions in the liver [176]. *E. vogeli* is distinguished by its rostellar hook morphology, featuring a notably long blade, and the distinct size and shape of its larval stages. *E. oligarthrus* shares taxonomic similarities with *E. vogeli*, including the presence of rostellar hooks and unique larval characteristics [6].

2.6.2 Morphology and Life Cycle

While *Echinococcus vogeli* and *E. oligarthrus* share a life cycle featuring a small, white, ribbon-like adult tapeworm in the small intestine of canids and a larval metacestode stage that forms hepatic hydatid cysts in intermediate hosts [6], notable differences exist. The adult *E. vogeli* tapeworm, measuring approximately 3–7 cm, is significantly larger than *E. oligarthrus*, which is only around 7 mm long. Furthermore, their intermediate host ranges differ: *E. vogeli* utilizes rodents, monkeys, and humans, whereas *E. oligarthrus* also employs intermediate hosts for liver cyst development [6].

Echinococcus vogeli exhibits a typical cyclophyllidean cestode life cycle involving multiple intermediate and definitive hosts. The adult tapeworm, residing in the canid small intestine, comprises a scolex, neck, and segmented body, releasing eggs into the environment via the definitive host's feces [172–176]. Upon ingestion by intermediate hosts, the eggs hatch, penetrate the intestinal wall, and migrate to various organs, primarily the liver, where they develop into hydatid cysts containing protoscoleces—the infective stage for definitive hosts, thus completing the cycle [172–176].

In contrast, *Echinococcus oligarthrus* involves felids such as jaguars, pumas, or bobcats as definitive hosts and various small mammals, like the agouti, as intermediate hosts [6]. The small adult tapeworm resides in the definitive host's small intestine, while the metacestode develops as multiple unicystic lesions in the liver of the intermediate host [172]. Human infections with *E. oligarthrus* are rare due to minimal contact between the hosts and humans. The life cycle concludes when the felid host consumes infected intermediate hosts, allowing the adult tapeworm to mature and produce eggs. Morphologically, *E. oligarthrus* also differs from *E. vogeli* by possessing smaller and more delicate hooks on the rostellum of its scolex and fewer proglottids in the adult worm [172].

2.6.3 Epidemiology and Geographic Distribution

The epidemiology of *Echinococcus vogeli* is largely confined to the rainforest areas of Central and South America, with the highest prevalence recorded in Brazil, Venezuela, and Colombia [174]. This species is notably distributed throughout the Amazon basin and other rainforest regions of Central and South America [172–174], with a higher prevalence observed in indigenous communities. Its life cycle involves wild carnivores as definitive hosts and small rodents as intermediate hosts. Human infections are mainly associated with hunting and consumption of bushmeat, particularly pacas, which serve as intermediate hosts for the bush dog, the definitive host. Control measures in these communities focus on education, awareness, and the implementation of strict hygiene practices [172–174].

Echinococcus oligarthrus, on the other hand, is primarily found in the Andean regions of South America, Costa Rica, and the La Pampa region of Argentina, with Peru and Bolivia exhibiting higher prevalence [6, 172]. The geographic distribution of this parasite is closely linked to the habitats of its definitive hosts (felids) and intermediate hosts (small mammals like agouti) [172]. The life cycle involves these felids and small mammals, with human infections occurring through accidental ingestion of eggs in contaminated food or water.

2.6.4 Clinical Manifestations and Pathogenesis

Human infection with *Echinococcus vogeli* typically results in severe disease, primarily affecting the liver and manifesting as multiple cystic echinococcosis or polycystic echinococcosis [172]. The larval stage produces numerous large hepatic cysts, leading to abdominal pain, hepatomegaly, obstructive jaundice, and potentially liver failure and ascites. In some instances, the parasite can disseminate to other organs like the lungs, spleen, and brain, causing severe illness and even death. The pathogenesis involves the continuous growth and expansion of these cysts, resulting in compression and destruction of adjacent tissues, thus producing a disease similar to alveolar echinococcosis [173–176].

Human infection with *Echinococcus oligarthrus* can also lead to severe clinical manifestations due to cyst development in the liver and other organs. Hepatic involvement is common, causing symptoms such as abdominal pain, obstructive jaundice, and hepatomegaly. A characteristic feature is the presence of multiple unicystic lesions scattered throughout the liver [172–176]. Fortunately, infestation is very rare.

2.6.5 Diagnosis and Treatment

Diagnosis of *Echinococcus vogeli* infection often relies on imaging techniques such as ultrasonography, computed tomography, and magnetic resonance imaging. Serological tests detecting antibodies against the *E. vogeli* Ev2 antigen are also available [177]. Treatment typically involves surgical cyst removal followed by chemotherapy, particularly with albendazole and other benzimidazoles [178].

In contrast, *Echinococcus oligarthrus* diagnosis primarily depends on imaging studies, as specific serological tests are lacking, although cross-reactivity with *Echinococcus granulosus* antigens has been observed [6, 172]. Similar to *E. vogeli*, treatment for *E. oligarthrus* infection involves surgical cyst removal combined with benzimidazole chemotherapy to prevent recurrence [178].

2.6.6 Prevention and Control

Preventing and controlling *Echinococcus vogeli* infection largely depends on public health education emphasizing proper hygiene, sanitation, and avoiding contact with infected animals. Reducing wild host animal populations and implementing surveillance programs for early detection and treatment are also crucial [172–177]. Similarly, prevention and control of *Echinococcus oligarthrus* infection focus on public education regarding the risks of consuming contaminated food and water, along with proper waste management and controlling animal populations to minimize transmission [172].

2.7 Summary and Key Points

The Echinococcus family has a broad spectrum of microorganisms that result in serious disease in humans. The disease manifests itself in underprivileged regions. Although it is a growing concern due to climate change and mass migration of human and animal populations, unfortunately, it did not receive the attention it deserved. It needs dedicated multidisciplinary team of surgeons, radiologists, epidemiologists, and infectious disease specialists to treat and eradicate the disease caused by this family of parasites. We need studies in the bench as well as clinical trials to develop new drugs and medical treatment regimens to effectively treat humans and deworm domestic animals. Another area that should have advanced is developing a vaccine for both humans and livestock to prevent spread and even development of the disease.

The *Echinococcus* genus encompasses a range of microorganisms causing serious human diseases, disproportionately affecting underprivileged regions. Despite growing concerns due to climate change and increasing human and animal migration, this significant public health issue has unfortunately not received adequate attention. Addressing these parasitic infections requires dedicated multidisciplinary teams, including surgeons, radiologists, epidemiologists, and infectious disease specialists, for effective treatment and potential eradication. Furthermore, robust basic science research and clinical trials are essential for developing novel drugs and treatment regimens for human cases and for deworming domestic animals. Crucially, advancing the development of vaccines for both humans and livestock is vital for preventing disease spread and even its development.

References

1. Kern P, Menezes da Silva A, Akhan O, Müllhaupt B, Vizcaychipi KA, Budke C, Vuitton DA. The echinococcoses: diagnosis, clinical management and burden of disease. Adv Parasitol. 2017;96:259–369. https://doi.org/10.1016/bs.apar.2016.09.006. PMID: 28212790.
2. Torgerson PR, Robertson LJ, Enemark HL, Foehr J, van der Giessen JWB, Kapel CMO, Klun I, Trevisan C. Source attribution of human echinococcosis: a systematic review and meta-analysis. PLoS Negl Trop Dis. 2020;14(6):e0008382. https://doi.org/10.1371/journal.pntd.0008382. PMID: 32569309.
3. Eckert J, Deplazes P. Biological, epidemiological, and clinical aspects of echinococcosis, a zoonosis of increasing concern. Clin Microbiol Rev. 2004;17(1):107–35. https://doi.org/10.1128/CMR.17.1.107-135.2004. PMID: 14726458.
4. Wang X, Kui Y, Xue CZ, Wang Q, Zheng CJ, Zhao JS, Yang YM, Jiang XF, Gong-Sang QZ, Ma X, Feng Y, Wu XL, Chen S, Li FK, Yu WJ, Li BF, Liu BX, Wang Y, Wang LY, Yang SJ, Wang ZH, Hu W, Shen YJ, Zhang WB, Craig PS, Wu WP, Xiao N, Han S, Zhou XN, Li SZ, Cao JP. Past, present and future epidemiology of echinococcosis in China based on nationwide surveillance data 2004–2022. J Infect. 2025;90(3):106445. https://doi.org/10.1016/j.jinf.2025.106445. PMID: 39956527.
5. Gessese AT. Review on epidemiology and public health significance of hydatidosis. Vet Med Int. 2020;2020:8859116. https://doi.org/10.1155/2020/8859116. PMID: 33354312.
6. Ali R, Khan S, Khan M, Adnan M, Ali I, Khan TA, Haleem S, Rooman M, Norin S, Khan SN. A systematic review of medicinal plants used against Echinococcus granulosus. PLoS One. 2020;15(10):e0240456. https://doi.org/10.1371/journal.pone.0240456. PMID: 33048959.
7. Romig T, Deplazes P, Jenkins D, Giraudoux P, Massolo A, Craig PS, Wassermann M, Takahashi K, de la Rue M. Ecology and life cycle pat-

terns of Echinococcus species. Adv Parasitol. 2017;95:213–314. https://doi.org/10.1016/bs.apar.2016.11.002. PMID: 28131364.
8. Tor M, Atasalihi A, Altuntas N, Sulu E, Senol T, Kir A, Baran R. Review of cases with cystic hydatid lung disease in a tertiary referral hospital located in an endemic region: a 10 years' experience. Respiration. 2000;67(5):539–42. https://doi.org/10.1159/000067470. PMID: 11070459.
9. Carmena D, Cardona GA. Canine echinococcosis: global epidemiology and genotypic diversity. Acta Trop. 2013;128(3):441–60. https://doi.org/10.1016/j.actatropica.2013.08.002. PMID: 23954494.
10. Thompson RC. Biology and systematics of Echinococcus. Adv Parasitol. 2017;95:65–109. https://doi.org/10.1016/bs.apar.2016.07.001. Epub 2016 Aug 5. PMID: 28131366.
11. Thompson RCA. The molecular epidemiology of Echinococcus infections. Pathogens. 2020;9(6):453. https://doi.org/10.3390/pathogens9060453. PMID: 32521787.
12. Wen H, Vuitton L, Tuxun T, Li J, Vuitton DA, Zhang W, McManus DP. Echinococcosis: advances in the 21st century. Clin Microbiol Rev. 2019;32(2):e00075–18. https://doi.org/10.1128/CMR.00075-18. PMID: 30760475.
13. Woolsey ID, Miller AL. Echinococcus granulosus sensu lato and Echinococcus multilocularis: a review. Res Vet Sci. 2021;135:517–22. https://doi.org/10.1016/j.rvsc.2020.11.010. PMID: 33246571.
14. Andrabi A, Tak H, Rasool A. Hydatidosis: a review. Int J Sci Technol Res. 2020;9:4664–71.
15. Fong IW. Animals and mechanisms of disease transmission. In: Emerging zoonoses; 2017. p. 15–38. https://doi.org/10.1007/978-3-319-50890-0_2.
16. Aziz H, Seda P, Aswani Y, Gosse MD, Krishnakumari AJ, Pawlik TM. Cystic echinococcosis of the liver. J Gastrointest Surg. 2025;29(3):101974. https://doi.org/10.1016/j.gassur.2025.101974. Epub 2025 Jan 27. PMID: 39864780.
17. Wang Z, Wang X, Liu X. Echinococcosis in China, a review of the epidemiology of Echinococcus spp. EcoHealth. 2008;5(2):115–26. https://doi.org/10.1007/s10393-008-0174-0. PMID: 18787915.
18. Baser S, Macin S, Macin G, Orsten S. An overview of the genetic diversity of Echinococcus granulosus sensu lato in Turkey. J Contemp Med. 2022;12(6):1025–9. https://doi.org/10.16899/jcm.1134747.
19. Kumaresan M, Gunalan A, Pandy G, Agarwal RH, Rajkumari NA. Serological study on hydatid disease from a Tertiary Care Hospital in Southern India: A retrospective observational study. J Clin of Diagn Res. 2024;18(3) https://doi.org/10.7860/JCDR/2024/67156.19107.
20. Fakhri Y, Omar SS, Dadar M, Pilevar Z, Sahlabadi F, Torabbeigi M, Rezaeiarshad N, Abbasi F, Mousavi KA. The prevalence of hydatid cyst in raw meat products: a global systematic review, meta-analysis, and meta-regression. Sci Rep. 2024;14(1):26094. https://doi.org/10.1038/s41598-024-77168-1. PMID: 39478044.
21. Grosso G, Gruttadauria S, Biondi A, Marventano S, Mistretta A. Worldwide epidemiology of liver hydatidosis including the Mediterranean area. World J Gastroenterol. 2012;18(13):1425–37. https://doi.org/10.3748/wjg.v18.i13.1425. PMID: 22509074.
22. Paternoster G, Boo G, Flury R, Raimkulov KM, Minbaeva G, Usubalieva J, Bondarenko M, Müllhaupt B, Deplazes P, Furrer R, Torgerson PR. Association between environmental and climatic risk factors and the spatial distribution of cystic and alveolar echinococcosis in Kyrgyzstan. PLoS Negl Trop Dis. 2021;15(6):e0009498. https://doi.org/10.1371/journal.pntd.0009498. PMID: 34161356.
23. Deplazes P, Rinaldi L, Alvarez Rojas CA, Torgerson PR, Harandi MF, Romig T, Antolova D, Schurer JM, Lahmar S, Cringoli G, Magambo J, Thompson RC, Jenkins EJ. Global distribution of alveolar and cystic echinococcosis. Adv Parasitol. 2017;95:315–493. https://doi.org/10.1016/bs.apar.2016.11.001. PMID: 28131365.
24. Narayan KG, Sinha DK, Singh DK. Hydatidosis. In: Handbook of management of zoonoses. Singapore: Springer; 2024. https://doi.org/10.1007/978-981-99-9885-2_99.
25. Dogru MV, Sezen CB, Aker C, Erdogu V, Erduhan S, Cansever L, Metin M. Evaluating giant hydatid cysts: factors affecting mortality and morbidity. Ann Thorac Cardiovasc Surg. 2021;27(3):164–8. https://doi.org/10.5761/atcs.oa.20-00178. PMID: 33162437.
26. Aghajanzadeh M, Shafaghi A, Ashoobi MT, Mashhor MY, Rad OM, Farzin M. Clinical presentation, diagnosis and surgical treatment of intrabiliary ruptured hydatid disease of the liver. GSC Adv Res Rev. 2023;16(3):162–70. https://doi.org/10.30574/gscarr.2023.16.3.0369.
27. Kere OJ, Joseph E, Jessika BL, Maina KJ. Prevalence and monetary loss due to cystic Echinococcosis in slaughter house livestock: a case study of Migori County, Kenya. Parasite Epidemiol Control. 2019;5:e00105. https://doi.org/10.1016/j.parepi.2019.e00105. PMID: 31193569.
28. Aciöz M, Bozkaya F. The monetary losses associated with hydatidosis in slaughtered ruminants in Turkey. Helminthologia. 2022;59(3):246–52. https://doi.org/10.2478/helm-2022-0028. PMID: 36694834.
29. Eckert J, Thompson RC. Historical aspects of echinococcosis. Adv Parasitol. 2017;95:1–64. https://doi.org/10.1016/bs.apar.2016.07.003. PMID: 28131361.
30. Cappello E, Cacopardo B, Caltabiano E, Li Volsi S, Chiara R, Sapienza M, Nigro L. Epidemiology and clinical features of cystic hydatidosis in Western Sicily: a ten-year review. World J Gastroenterol. 2013;19(48):9351–8. https://doi.org/10.3748/wjg.v19.i48.9351. PMID: 24409062.
31. Jenkins DJ. Hydatid control in Australia: where it began, what we have achieved and where to from

here. Int J Parasitol. 2005;35(7):733–40. https://doi.org/10.1016/j.ijpara.2005.03.001. PMID: 15907851.
32. Brookes VJ, Barnes TS, Jenkins DJ, Van der Saag MR, Dempster R, Wilson CS. The effect of cystic echinococcosis (hydatid disease) on carcase weight in cattle in eastern Australia. Sci Rep. 2024;14(1):7297. https://doi.org/10.1038/s41598-024-57886-2. PMID: 38538664.
33. Alpaca Rodriguez LR, Kirschbaum-Chrem JP, Romero G, Villanueva E, Ugas Charcape CF. The many faces of pediatric hydatid disease: a pictorial review. Pediatr Radiol. 2025;55(1):115–27. https://doi.org/10.1007/s00247-024-06080-7. PMID: 39589488.
34. Hammami F, Gargouri L, Ayed HB, Koubaa M, Rekik K, Jemaa TB, Jemaa MB. Hydatid disease among adults and children: it is time to worry! Electron J Gen Med. 2019;16(6):em151. https://doi.org/10.29333/ejgm/110172.
35. Rahman WA, Elmajdoub LE, Noor SAM, Wajidi MF. Present status on the taxonomy and morphology of Echinococcus granulosus: a review. Austin J Vet Sci Anim Husb. 2015;2(2):1013.
36. Manterola C, Rivadeneira J, Pogue SD, Rojas C. Morphology of Echinococcus granulosus protoscolex. Int J Morphol. 2023;41(2):646–53.
37. Houston S, Belga S, Buttenschoen K, Cooper R, Girgis S, Gottstein B, Low G, Massolo A, MacDonald C, Müller N, Preiksaitis J, Sarlieve P, Vaughan S, Kowalewska-Grochowska K. Epidemiological and clinical characteristics of alveolar echinococcosis: an emerging infectious disease in Alberta, Canada. Am J Trop Med Hyg. 2021;104(5):1863–9. https://doi.org/10.4269/ajtmh.20-1577. PMID: 33755579.
38. Mutwiri T, Muigai AWT, Magambo J, Mulinge E, Gitau L, Muinde P, Bettridge JM, Rogan M, Fèvre EM, Falzon LC. The potential role of roaming dogs in establishing a geographically novel life cycle of taeniids (Echinococcus spp. and Taenia spp.) in a non-endemic area. Vet Parasitol Reg Stud Rep. 2023;38:100829. https://doi.org/10.1016/j.vprsr.2022.100829. PMID: 36725161.
39. Awosanya EJ, Olagbaju A, Peruzzu A, Masu G, Masala G, Bonelli P. Detection of Echinococcus granulosus sensu lato in environmental samples from Ibadan, Oyo State, South West Nigeria. Vet Sci. 2022;9(12):679. https://doi.org/10.3390/vetsci9120679. PMID: 36548840.
40. Fadel SA, Asmar K, Faraj W, Khalife M, Haddad M, El-Merhi F. Clinical review of liver hydatid disease and its unusual presentations in developing countries. Abdom Radiol (NY). 2019;44(4):1331–9. https://doi.org/10.1007/s00261-018-1794-7. PMID: 30306203.
41. Bhutani N, Kajal P. Hepatic echinococcosis: a review. Ann Med Surg (Lond). 2018;36:99–105. https://doi.org/10.1016/j.amsu.2018.10.032. PMID: 30450204.
42. Świderski Z, Miquel J, Conn DB. Functional ultrastructure of eggs and cellular organization of hexacanths of the cyclophyllidean cestode Thysanotaenia congolensis: a phylogenetic implication of obtained results. Parasitology. 2016;143(3):320–33. https://doi.org/10.1017/S0031182015001560. PMID: 26689933.
43. Deo KB, Kumar R, Tiwari G, Kumar H, Verma GR, Singh H. Surgical management of hepatic hydatid cysts—conservative versus radical surgery. HPB (Oxford). 2020;22(10):1457–62. https://doi.org/10.1016/j.hpb.2020.03.003. PMID: 32229090.
44. Frider B, Larrieu E, Odriozola M. Long-term outcome of asymptomatic liver hydatidosis. J Hepatol. 1999;30(2):228–31. https://doi.org/10.1016/s0168-8278(99)80066-x. PMID: 10068100.
45. Wang Y, He T, Wen X, Li T, Waili A, Zhang W, Xu X, Vuitton DA, Rogan MT, Wen H, Craig PS. Post-survey follow-up for human cystic echinococcosis in Northwest China. Acta Trop. 2006;98(1):43–51. https://doi.org/10.1016/j.actatropica.2006.01.009. PMID: 16676417.
46. Govindasamy A, Bhattarai PR, John J. Liver cystic echinococcosis: a parasitic review. Ther Adv Infect Dis. 2023;10:20499361231171478. https://doi.org/10.1177/20499361231171478. PMID: 37197609.
47. Badwaik N, Gharde P, Shinde RK, Tayade H, Navandhar PS, Patil M. Hydatid cyst or echinococcosis: a comprehensive review of transmission, clinical manifestations, diagnosis, and multidisciplinary treatment. Cureus. 2024;16(7):e63713. https://doi.org/10.7759/cureus.63713. PMID: 39099980.
48. Nikmanesh B, Mirhendi H, Ghalavand Z, Alebouyeh M, Sharbatkhori M, Kia E, Mohebali M, Eghbali M, Rokni MB. Genotyping of Echinococcus granulosus isolates from human clinical samples based on sequencing of mitochondrial genes in Iran, Tehran. Iran J Parasitol. 2014;9(1):20–7. PMID: 25642256.
49. Tamarozzi F, Deplazes P, Casulli A. Reinventing the wheel of Echinococcus granulosus sensu lato transmission to humans. Trends Parasitol. 2020;36(5):427–34. https://doi.org/10.1016/j.pt.2020.02.004. PMID: 32298630.
50. Moro P, Schantz PM. Echinococcosis: a review. Int J Infect Dis. 2009;13(2):125–33. https://doi.org/10.1016/j.ijid.2008.03.037. PMID: 18938096.
51. Benelli G, Wassermann M, Brattig NW. Insects dispersing taeniid eggs: who and how? Vet Parasitol. 2021;295:109450. https://doi.org/10.1016/j.vetpar.2021.109450. PMID: 34038808.
52. Manterola C, Totomoch-Serra A, Rojas C, Riffo-Campos ÁL, García-Méndez N. Echinococcus granulosus sensu lato genotypes in different hosts worldwide: a systematic review. Acta Parasitol. 2022;67(1):161–85. https://doi.org/10.1007/s11686-021-00439-8. PMID: 34264444.
53. Daniel Mwambete K, Ponce-Gordo F, Cuesta-Bandera C. Genetic identification and host range of the Spanish strains of Echinococcus granulosus. Acta Trop. 2004;91(2):87–93. https://doi.org/10.1016/j.actatropica.2004.04.001. PMID: 15234657.

54. Örsten S, Çiftçi T, Azizova A, Yüce G, Uysal A, İmamoğlu Ç, Karaağaoğlu E, Akıncı D, Akyön Y, Casulli A, Akhan O. Investigation of the relationship between CE cyst characteristics and genetic diversity of Echinococcus granulosus sensu lato in humans from Turkey. Parasitology. 2020;147(14):1712–7. https://doi.org/10.1017/S0031182020001535. PMID: 32799957.
55. Ali R, Nazeer S, Elahi MMS, Idu EG, Zhang H, Mahmoudvand H, Khan SN, Yang J. Global distribution and definitive host range of Echinococcus species and genotypes: a systematic review. Vet Parasitol. 2024;331:110273. https://doi.org/10.1016/j.vetpar.2024.110273. PMID: 39116549.
56. Ghasemirad H, Bazargan N, Shahesmaeili A, Harandi MF. Echinococcosis in immunocompromised patients: a systematic review. Acta Trop. 2022;232:106490. https://doi.org/10.1016/j.actatropica.2022.106490. PMID: 35490729.
57. Mona C, Meryam M, Nizar K, Yazid B, Amara A, Haithem Z, Dhafer H, Anis BM. Exploring rare locations of hydatid disease: a retrospective case series. BMC Surg. 2024;24(1):160. https://doi.org/10.1186/s12893-024-02443-x. PMID: 38760696.
58. Sachar S, Goyal S, Goyal S, Sangwan S. Uncommon locations and presentations of hydatid cyst. Ann Med Health Sci Res. 2014;4(3):447–52. https://doi.org/10.4103/2141-9248.133476. PMID: 24971224.
59. Ghartimagar D, Ghosh A, Shrestha MK, Talwar OP, Sathian B. 14 years hospital-based study on clinical and morphological spectrum of hydatid disease. JNMA J Nepal Med Assoc. 2013;52(190):349–53. PMID: 24362658.
60. Ahmadi NA, Bodi F. Clinical presentation, localization and morphology of hepato-pulmonary hydatid cysts in patients operated in Tehran. World Appl Sci J. 2011;12(9):1544–8.
61. Yu XK, Zhang L, Ma WJ, Bi WZ, Ju SG. An overview of hepatic echinococcosis and the characteristic CT and MRI imaging manifestations. Infect Drug Resist. 2021;14:4447–55. https://doi.org/10.2147/IDR.S331957. PMID: 34737585.
62. Brunetti E, Tamarozzi F, Macpherson C, Filice C, Piontek MS, Kabaalioglu A, Dong Y, Atkinson N, Richter J, Schreiber-Dietrich D, Dietrich CF. Ultrasound and cystic echinococcosis. Ultrasound Int Open. 2018;4(3):E70–8. https://doi.org/10.1055/a-0650-3807. PMID: 30364890.
63. Tao Y, Wang YF, Wang J, Long S, Seyler BC, Zhong XF, Lu Q. Pictorial review of hepatic echinococcosis: ultrasound imaging and differential diagnosis. World J Gastroenterol. 2024;30(37):4115–31. https://doi.org/10.3748/wjg.v30.i37.4115. PMID: 39474399.
64. Brunetti E, Kern P, Vuitton DA. Writing Panel for the WHO-IWGE. Expert consensus for the diagnosis and treatment of cystic and alveolar echinococcosis in humans. Acta Trop. 2010;114(1):1–16. https://doi.org/10.1016/j.actatropica.2009.11.001. PMID: 19931502.
65. Geramizadeh B. Unusual locations of the hydatid cyst: a review from Iran. Iran J Med Sci. 2013;38(1):2–14. PMID: 23645952.
66. Ben Ismail I, Sghaier M, Boujmil K, Rebii S, Zoghlami A. Hydatid cyst of the liver fistulized into the inferior vena cava. Int J Surg Case Rep. 2022;94:107060. https://doi.org/10.1016/j.ijscr.2022.107060. PMID: 35413675.
67. Narne KG, Vaishnavi KISN, Kakumani J, Vivekanandan T, Gowri Shankar A. Revealing an exceptional case of Budd-Chiari syndrome. Cureus. 2024;16(9):e69949. https://doi.org/10.7759/cureus.69949. PMID: 39445237.
68. Bandyopadhyay C, Biswas S, Paul P, Chattopadhyay S, Bhattacharyya C, Sukul B. Hydatid cyst disease with hepatic encephalopathy and multiorgan failure: a rare case. J Indian Acad Forensic Med. 2014;36(2):209–12. https://doi.org/10.1177/0971097320140.
69. Altay M, Unverdi S, Altay FA, Ceri M, Akay H, Ozer H, Kiraç H, Denizli N, Yilmaz B, Güvence N, Duranay M. Renal injury due to hepatic hydatid disease. Nephrol Dial Transplant. 2010;25(8):2611–5. https://doi.org/10.1093/ndt/gfq092. PMID: 20179010.
70. Lightowlers MW, Gasser RB, Hemphill A, Romig T, Tamarozzi F, Deplazes P, Torgerson PR, Garcia HH, Kern P. Advances in the treatment, diagnosis, control and scientific understanding of taeniid cestode parasite infections over the past 50 years. Int J Parasitol. 2021;51(13–14):1167–92. https://doi.org/10.1016/j.ijpara.2021.10.003. PMID: 34757089.
71. Toews E, Musiani M, Checkley S, Visscher D, Massolo A. A global assessment of Echinococcus multilocularis infections in domestic dogs: proposing a framework to overcome past methodological heterogeneity. Int J Parasitol. 2021;51(5):379–92. https://doi.org/10.1016/j.ijpara.2020.10.008. PMID: 33482171.
72. Maleki F, Akhlaghi L, Tabatabaie F. Evaluation of hydatid cyst antigen for serological diagnosis. Med J Islam Repub Iran. 2023;37:87. https://doi.org/10.47176/mjiri.37.87. PMID: 37655067.
73. Christodoulidis G, Samara AA, Diamantis A, Floros T, Sgantzou IK, Karakantas KS, Zotos PA, Koutras A, Janho MB, Tepetes K. Reaching the challenging diagnosis of complicated liver hydatid disease: a single institution's experience from an endemic area. Medicina (Kaunas). 2021;57(11):1210. https://doi.org/10.3390/medicina57111210. PMID: 34833428.
74. Craig PS, Hegglin D, Lightowlers MW, Torgerson PR, Wang Q. Echinococcosis: control and prevention. Adv Parasitol. 2017;96:55–158. https://doi.org/10.1016/bs.apar.2016.09.002. PMID: 28212791.
75. Borhani M, Fathi S, Harandi MF, Casulli A, Ding J, Liu M, Zhang W, Wen H. Echinococcus granulosus sensu lato control measures: a specific focus on vaccines for both definitive and intermediate hosts. Parasit Vectors. 2024;17(1):533. https://

doi.org/10.1186/s13071-024-06581-2. PMID: 39716337.
76. Heath DD, Robinson C, Shakes T, Huang Y, Gulnur T, Shi B, Zhang Z, Anderson GA, Lightowlers MW. Vaccination of bovines against Echinococcus granulosus (cystic echinococcosis). Vaccine. 2012;30(20):3076–81. https://doi.org/10.1016/j.vaccine.2012.02.073. PMID: 22406459.
77. Tomaiuolo S, Jansen W, Soares Martins S, Devriendt B, Cox E, Mori M. QuilA® adjuvanted Coxevac® sustains Th1-CD8+-type immunity and increases protection in Coxiella burnetii-challenged goats. NPJ Vaccines. 2023;8(1):17. https://doi.org/10.1038/s41541-023-00607-z. PMID: 36788233.
78. Velasco-Tirado V, Alonso-Sardón M, Lopez-Bernus A, Romero-Alegría Á, Burguillo FJ, Muro A, Carpio-Pérez A, Muñoz Bellido JL, Pardo-Lledias J, Cordero M, Belhassen-García M. Medical treatment of cystic echinococcosis: systematic review and meta-analysis. BMC Infect Dis. 2018;18(1):306. https://doi.org/10.1186/s12879-018-3201-y. PMID: 29976137.
79. Khan A, Ahmed H, Simsek S, Afzal MS, Cao J. Spread of cystic echinococcosis in Pakistan due to stray dogs and livestock slaughtering habits: research priorities and public health importance. Front Public Health. 2020;7:412. https://doi.org/10.3389/fpubh.2019.00412. PMID: 32064244.
80. Larrieu E, Mujica G, Araya D, Labanchi JL, Arezo M, Herrero E, Santillán G, Vizcaychipi K, Uchiumi L, Salvitti JC, Grizmado C, Calabro A, Talmon G, Sepulveda L, Galvan JM, Cabrera M, Seleiman M, Crowley P, Cespedes G, García Cachau M, Gino L, Molina L, Daffner J, Gauci CG, Donadeu M, Lightowlers MW. Pilot field trial of the EG95 vaccine against ovine cystic echinococcosis in Rio Negro, Argentina: 8 years of work. Acta Trop. 2019;191:1–7. https://doi.org/10.1016/j.actatropica.2018.12.025. PMID: 30576624.
81. Poggio TV, Jensen O, Mossello M, Iriarte J, Avila HG, Gertiser ML, Serafino JJ, Romero S, Echenique MA, Dominguez DE, Barrios JR, Heath D. Serology and longevity of immunity against Echinococcus granulosus in sheep and llama induced by an oil-based EG95 vaccine. Parasite Immunol. 2016;38(8):496–502. https://doi.org/10.1111/pim.12325. PMID: 27104482.
82. Gauci C, Heath D, Chow C, Lightowlers MW. Hydatid disease: vaccinology and development of the EG95 recombinant vaccine. Expert Rev Vaccines. 2005;4(1):103–12. https://doi.org/10.1586/14760584.4.1.103. PMID: 15757477.
83. Gauci C, Jenkins D, Lightowlers MW. Strategies for optimal expression of vaccine antigens from Taeniid cestode parasites in Escherichia coli. Mol Biotechnol. 2011;48(3):277–89. https://doi.org/10.1007/s12033-010-9368-0. PMID: 21222242.
84. Larrieu E, Zanini F. Critical analysis of cystic echinococcosis control programs and praziquantel use in South America, 1974–2010. Rev Panam Salud Publica. 2012;31(1):81–7. https://doi.org/10.1590/s1020-49892012000100012. PMID: 22427169.
85. Jiang B, Zhou XN, Zhang HB, Tao Y, Huo LL, Liu N. Slow-release praziquantel for dogs: presentation of a new formulation for echinococcosis control. Infect Dis Poverty. 2017;6(1):140. https://doi.org/10.1186/s40249-017-0357-4. PMID: 28911334.
86. Irabedra P, Ferreira C, Sayes J, Elola S, Rodríguez M, Morel N, Segura S, Santos ED, Guisantes JA. Control programme for cystic echinococcosis in Uruguay. Mem Inst Oswaldo Cruz. 2016;111(6):372–7. https://doi.org/10.1590/0074-02760160070. PMID: 27223652.
87. Moss JE, Chen X, Li T, Qiu J, Wang Q, Giraudoux P, Ito A, Torgerson PR, Craig PS. Reinfection studies of canine echinococcosis and role of dogs in transmission of Echinococcus multilocularis in Tibetan communities, Sichuan, China. Parasitology. 2013;140(13):1685–92. https://doi.org/10.1017/S0031182013001200. PMID: 23985352.
88. Inayat F, Azam S, Baig AS, Nawaz G, Ullah W. Presentation patterns, diagnostic modalities, management strategies, and clinical outcomes in patients with hydatid disease of the pelvic bone: a comparative review of 31 cases. Cureus. 2019;11(3):e4178. https://doi.org/10.7759/cureus.4178. PMID: 31106078.
89. Mönnink GLE, Stijnis C, van Delden OM, Spijker R, Grobusch MP. Percutaneous versus surgical interventions for hepatic cystic echinococcosis: a systematic review and meta-analysis. Cardiovasc Intervent Radiol. 2021;44(11):1689–96. https://doi.org/10.1007/s00270-021-02911-4. PMID: 34272589.
90. Popa GL, Popa AC, Mastalier B, Crețu CM, Popa MI. Complicated clinical course of a patient with multivisceral cystic echinococcosis requiring extensive surgical and medical treatment. J Clin Med. 2023;12(17):5596. https://doi.org/10.3390/jcm12175596. PMID: 37685663.
91. Stojkovic M, Rosenberger K, Kauczor HU, Junghanss T, Hosch W. Diagnosing and staging of cystic echinococcosis: how do CT and MRI perform in comparison to ultrasound? PLoS Negl Trop Dis. 2012;6(10):e1880. https://doi.org/10.1371/journal.pntd.0001880. PMID: 23145199.
92. Stojković M, Weber TF, Junghanss T. Clinical management of cystic echinococcosis: state of the art and perspectives. Curr Opin Infect Dis. 2018;31(5):383–92. https://doi.org/10.1097/QCO.0000000000000485. PMID: 30124496.
93. Keong B, Wilkie B, Sutherland T, Fox A. Hepatic cystic echinococcosis in Australia: an update on diagnosis and management. ANZ J Surg. 2018;88(1–2):26–31. https://doi.org/10.1111/ans.14117. PMID: 29024292.
94. Rossi P, Tamarozzi F, Galati F, Akhan O, Cretu CM, Vutova K, Siles-Lucas M, Brunetti E, Casulli A, ERCE network. The European Register of Cystic Echinococcosis, ERCE: state-of-the-art five years

after its launch. Parasit Vectors. 2020;13(1):236. https://doi.org/10.1186/s13071-020-04101-6. PMID: 32381109.
95. Akhan O, Yildiz AE, Akinci D, Yildiz BD, Ciftci T. Is the adjuvant albendazole treatment really needed with PAIR in the management of liver hydatid cysts? A prospective, randomized trial with short-term follow-up results. Cardiovasc Intervent Radiol. 2014;37(6):1568–74. https://doi.org/10.1007/s00270-014-0840-2. PMID: 24464258.
96. Öztürk G, Uzun MA, Özkan ÖF, Kayaalp C, Tatlı F, Eren S, Aksungur N, Çoker A, Bostancı EB, Öter V, Kaya E, Taşar P. Turkish HPB Surgery Association consensus report on hepatic cystic echinococcosis (HCE). Turk J Surg. 2022;38(2):101–20. https://doi.org/10.47717/turkjsurg.2022.5757. PMID: 36483170.
97. Köroğlu M, Erol B, Gürses C, Türkbey B, Baş CY, Alparslan AŞ, Köroğlu BK, Toslak İE, Çekiç B, Akhan O. Hepatic cystic echinococcosis: percutaneous treatment as an outpatient procedure. Asian Pac J Trop Med. 2014;7(3):212–5. https://doi.org/10.1016/S1995-7645(14)60023-7. PMID: 24507642.
98. Ustünsöz B, Akhan O, Kamiloğlu MA, Somuncu I, Uğurel MS, Cetiner S. Percutaneous treatment of hydatid cysts of the liver: long-term results. AJR Am J Roentgenol. 1999;172(1):91–6. https://doi.org/10.2214/ajr.172.1.9888746. PMID: 9888746.
99. Turgut B, Öncü F. The conversion time of cysts to inactive form after percutaneous treatment of hepatic hydatid disease and predisposing factors in this process. Med Bull Haseki. 2020;58(3):286–92. https://doi.org/10.4274/haseki.galenos.2020.5897.
100. Kahriman G, Ozcan N, Dogan S, Karaborklu O. Percutaneous treatment of liver hydatid cysts in 190 patients: a retrospective study. Acta Radiol. 2017;58(6):676–84. https://doi.org/10.1177/0284185116664226. PMID: 27565632.
101. Khuroo MS. Percutaneous drainage in hepatic hydatidosis-the PAIR technique: concept, technique, and results. J Clin Exp Hepatol. 2021;11(5):592–602. https://doi.org/10.1016/j.jceh.2021.05.005. PMID: 34511821.
102. Balli O, Balli G, Cakir V, Gur S, Pekcevik R, Tavusbay C, Akhan O. Percutaneous treatment of giant cystic echinococcosis in liver: catheterization technique in patients with CE1 and CE3a. Cardiovasc Intervent Radiol. 2019;42(8):1153–9. https://doi.org/10.1007/s00270-019-02248-z. PMID: 31119356.
103. Akhan O, Salik AE, Ciftci T, Akinci D, Islim F, Akpinar B. Comparison of Long-term results of percutaneous treatment techniques for hepatic cystic echinococcosis types 2 and 3b. AJR Am J Roentgenol. 2017;208(4):878–84. https://doi.org/10.2214/AJR.16.16131. PMID: 28199132.
104. Akhan O, Erdoğan E, Ciftci TT, Unal E, Karaağaoğlu E, Akinci D. Cystobiliary fistula of liver CE treatment as a major problem. Cardiovasc Intervent Radiol. 2020;43(11):1718–9. https://doi.org/10.1007/s00270-020-02631-1. PMID: 32914369.
105. Akhan O, Erdoğan E, Ciftci TT, Unal E, Karaağaoğlu E, Akinci D. Comparison of the long-term results of Puncture, Aspiration, Injection and Re-aspiration (PAIR) and catheterization techniques for the percutaneous treatment of CE1 and CE3a liver hydatid cysts: a prospective randomized trial. Cardiovasc Intervent Radiol. 2020;43(7):1034–40. https://doi.org/10.1007/s00270-020-02477-7. PMID: 32382859.
106. Akhan O, Gumus B, Akinci D, Karcaaltincaba M, Ozmen M. Diagnosis and percutaneous treatment of soft-tissue hydatid cysts. Cardiovasc Intervent Radiol. 2007;30(3):419–25. https://doi.org/10.1007/s00270-006-0153-1. PMID: 17295079.
107. Smego RA Jr, Bhatti S, Khaliq AA, Beg MA. Percutaneous aspiration-injection-reaspiration drainage plus albendazole or mebendazole for hepatic cystic echinococcosis: a meta-analysis. Clin Infect Dis. 2003;37(8):1073–83. https://doi.org/10.1086/378275. PMID: 14523772.
108. Akhan O. Percutaneous treatment of liver hydatid cysts: to PAIR or not to PAIR. Curr Opin Infect Dis. 2023;36(5):308–17. https://doi.org/10.1097/QCO.0000000000000956. PMID: 37548385.
109. Kayaalp C. Hydatid disease of the liver. J Gastroenterol Hepatol. 2005;20(3):331. https://doi.org/10.1111/j.1440-1746.2005.03741.x. PMID: 15740473.
110. Vagholkar K, Pawanarkar A, Chougle Q, Vagholkar S. Surgery for intra-abdominal hydatid disease: A single Centre experience. Int J Res Med Sci. 2016;4(10):4241–5. https://doi.org/10.18203/2320-6012.ijrms20163106.
111. El Nakeeb A, Salem A, El Sorogy M, Mahdy Y, Ellatif MA, Moneer A, Said R, El Ghawalby A, Ezzat H. Cystobiliary communication in hepatic hydatid cyst: predictors and outcome. Turk J Gastroenterol. 2017;28(2):125–30.
112. Khuroo MS, Wani NA, Javid G, Khan BA, Yattoo GN, Shah AH, Jeelani SG. Percutaneous drainage compared with surgery for hepatic hydatid cysts. N Engl J Med. 1997;337(13):881–7. https://doi.org/10.1056/NEJM199709253371303. PMID: 9302302.
113. Kamiyama T. Recent advances in surgical strategies for alveolar echinococcosis of the liver. Surg Today. 2020;50(11):1360–7. https://doi.org/10.1007/s00595-019-01922-6. PMID: 31768657.
114. Aydin F, Yalcin A, Karaman A, Sade R, Ozturk G, Alper F. Diagnostic and management perspectives in alveolar echinococcosis: review of literature. Eurasian J Med. 2022;54(Suppl. 1):S10–5.
115. Lundström-Stadelmann B, Rostami A, Frey CF, Torgerson PR, Riahi SM, Bagheri K, Kaethner M, Lachenmayer A, Beldi G, Gasser RB, Hemphill A. Human alveolar echinococcosis-global, regional, and national annual incidence and prevalence rates. Clin Microbiol Infect. 2025:S1198-743X(25)00099-0. https://doi.org/10.1016/j.cmi.2025.01.034. PMID: 40054771.

116. Tappe D, Zidowitz S, Demmer P, Kern P, Barth TF, Frosch M. Three-dimensional reconstruction of Echinococcus multilocularis larval growth in human hepatic tissue reveals complex growth patterns. Am J Trop Med Hyg. 2010;82(1):126–7. https://doi.org/10.4269/ajtmh.2010.09-0462. PMID: 20065007.
117. Reinehr M, Micheloud C, Grimm F, Kronenberg PA, Grimm J, Beck A, Nell J, Meyer Zu Schwabedissen C, Furrer E, Müllhaupt B, TFE B, Deplazes P, Weber A. Pathology of echinococcosis: a morphologic and immunohistochemical study on 138 specimens with focus on the differential diagnosis between cystic and alveolar echinococcosis. Am J Surg Pathol. 2020;44(1):43–54. https://doi.org/10.1097/PAS.0000000000001374. PMID: 31567204.
118. Kodama Y, Fujita N, Shimizu T, Endo H, Nambu T, Sato N, Todo S, Miyasaka K. Alveolar echinococcosis: MR findings in the liver. Radiology. 2003;228(1):172–7. https://doi.org/10.1148/radiol.2281020323. PMID: 12750459.
119. Sade R, Kantarci M, Ogul H, Gundogdu B, Aydınlı B. Differentiation between hepatic alveolar echinococcosis and primary hepatic malignancy with diffusion-weighted magnetic resonance imaging. Diagn Interv Imaging. 2018;99(3):169–77. https://doi.org/10.1016/j.diii.2017.09.007. PMID: 29110943.
120. Brumpt É, Liu W, Graeter T, Calame P, Rong S, Jiang Y, Li W, Bao H, Delabrousse É, XUUB consortium. Kodama-XUUB: an informative classification for alveolar echinococcosis hepatic lesions on magnetic resonance imaging. Parasite. 2021;28:66. https://doi.org/10.1051/parasite/2021062. PMID: 34569927.
121. Kantarci M, Bayraktutan U, Karabulut N, Aydinli B, Ogul H, Yuce I, Calik M, Eren S, Atamanalp SS, Oto A. Alveolar echinococcosis: spectrum of findings at cross-sectional imaging. Radiographics. 2012;32(7):2053–70. https://doi.org/10.1148/rg.327125708. PMID: 23150858.
122. Czermak BV, Akhan O, Hiemetzberger R, Zelger B, Vogel W, Jaschke W, Rieger M, Kim SY, Lim JH. Echinococcosis of the liver. Abdom Imaging. 2008;33(2):133–43. https://doi.org/10.1007/s00261-007-9331-0. PMID: 17912581.
123. Paternoster G, Boo G, Wang C, Minbaeva G, Usubalieva J, Raimkulov KM, Zhoroev A, Abdykerimov KK, Kronenberg PA, Müllhaupt B, Furrer R, Deplazes P, Torgerson PR. Epidemic cystic and alveolar echinococcosis in Kyrgyzstan: an analysis of national surveillance data. Lancet Glob Health. 2020;8(4):e603–11. https://doi.org/10.1016/S2214-109X(20)30038-3. PMID: 32199126.
124. Bebezov B, Mamashev N, Umetaliev T, Ziadinov I, Craig PS, Joekel DE, Deplazes P, Grimm F, Torgerson PR. Intense focus of alveolar echinococcosis, South Kyrgyzstan. Emerg Infect Dis. 2018;24(6):1119–22. https://doi.org/10.3201/eid2406.161641. PMID: 29774832.
125. Torgerson PR, Keller K, Magnotta M, Ragland N. The global burden of alveolar echinococcosis. PLoS Negl Trop Dis. 2010;4(6):e722. https://doi.org/10.1371/journal.pntd.0000722. PMID: 20582310.
126. Zewude RT, Corbeil A, Fung S, Moulton CA, Bogoch II. Alveolar Echinococcus in a 70-year-old man in Ontario. J Assoc Med Microbiol Infect Dis Can. 2024;8(4):336–42. https://doi.org/10.3138/jammi-2023-0012. PMID: 38250619.
127. Polish LB, Pritt B, Barth TFE, Gottstein B, O'Connell EM, Gibson PC. First European haplotype of Echinococcus multilocularis identified in the United States: an emerging disease? Clin Infect Dis. 2021;72(7):1117–23. https://doi.org/10.1093/cid/ciaa245. PMID: 32198510.
128. Kern P, Wen H, Sato N, Vuitton DA, Gruener B, Shao Y, Delabrousse E, Kratzer W, Bresson-Hadni S. WHO classification of alveolar echinococcosis: principles and application. Parasitol Int. 2006;55(Suppl):S283–7. https://doi.org/10.1016/j.parint.2005.11.041. PMID: 16343985.
129. Albayrak Y, Kargi A, Albayrak A, Gelincik I, Cakir YB. Liver alveolar echinococcosis metastasized to the breast. Breast Care (Basel). 2011;6(4):289–91. https://doi.org/10.1159/000331314. PMID: 22135627.
130. McManus DP, Zhang W, Li J, Bartley PB. Echinococcosis. Lancet. 2003;362(9392):1295–304. https://doi.org/10.1016/S0140-6736(03)14573-4. PMID: 14575976.
131. Bresson-Hadni S, Spahr L, Chappuis F. Hepatic alveolar echinococcosis. Semin Liver Dis. 2021;41(3):393–408. https://doi.org/10.1055/s--0041-1730925. PMID: 34161992.
132. Kantarci M, Bayraktutan U, Pirimoglu B, Ogul H, Oral A, Eren S, Gundogdu B. Multisystem involvement of alveolar echinococcosis in a child. J Infect Dev Ctries. 2014;8(11):1494–7. https://doi.org/10.3855/jidc.4214. PMID: 25390065.
133. Craig P. Echinococcus multilocularis. Curr Opin Infect Dis. 2003;16(5):437–44. https://doi.org/10.1097/00001432-200310000-00010. PMID: 14501996.
134. Yilmaz-Cankaya B, Pirimoglu B. Hepatic alveolar echinococcosis in a child. Rev Soc Bras Med Trop. 2021;54:e0487–2020. https://doi.org/10.1590/0037-8682-0487-2020. PMID: 33681918.
135. Bresson-Hadni S, Delabrousse E, Blagosklonov O, Bartholomot B, Koch S, Miguet JP, Mantion GA, Vuitton DA. Imaging aspects and non-surgical interventional treatment in human alveolar echinococcosis. Parasitol Int. 2006;55(Suppl):S267–72. https://doi.org/10.1016/j.parint.2005.11.053. PMID: 16403670.
136. Bulakçı M, Kartal MG, Yılmaz S, Yılmaz E, Yılmaz R, Şahin D, Aşık M, Erol OB. Multimodality imaging in diagnosis and management of alveolar echinococcosis: an update. Diagn Interv Radiol.

2016;22(3):247–56. https://doi.org/10.5152/dir.2015.15456. PMID: 27082120.

137. Parry AH, Wani AH, Feroz I. The spectrum of multimodality imaging findings in hepatic alveolar echinococcosis and the potential role of diffusion-weighted imaging in its characterisation. Pol J Radiol. 2020;85:e613–23. https://doi.org/10.5114/pjr.2020.101015. PMID: 33376563.

138. Beaussant-Cohen S, Richou C, Lenoir M, Grenouillet F, Bresson-Hadni S, Delabrousse E. MR imaging features of peritoneal alveolar echinococcosis. Diagn Interv Imaging. 2018;99(7–8):511–2. https://doi.org/10.1016/j.diii.2018.02.009. PMID: 29580747.

139. Okur A, Ogul H, Sengul G, Karaca L, Nalbantoglu NG, Kantarci M. Magnetic resonance spectroscopy and magnetic resonance imaging findings of the intracerebral alveolar echinococcosis. J Craniofac Surg. 2014;25(4):1352–3. https://doi.org/10.1097/SCS.0000000000000815. PMID: 25006915.

140. Reuter S, Nüssle K, Kolokythas O, Haug U, Rieber A, Kern P, Kratzer W. Alveolar liver echinococcosis: a comparative study of three imaging techniques. Infection. 2001;29(3):119–25. https://doi.org/10.1007/s15010-001-1081-2. PMID: 11440381.

141. Becce F, Pomoni A, Uldry E, Halkic N, Yan P, Meuli R, Schmidt S. Alveolar echinococcosis of the liver: diffusion-weighted MRI findings and potential role in lesion characterisation. Eur J Radiol. 2014;83(4):625–31. https://doi.org/10.1016/j.ejrad.2013.12.025. PMID: 24457140.

142. Kantarci M, Pirimoglu B, Aydinli B, Ozturk G. A rare reason for liver transplantation: hepatic alveoloar echinococcosis. Transpl Infect Dis. 2014;16(3):450–2. https://doi.org/10.1111/tid.12222. PMID: 24797290.

143. Tarhan NC, Agildere AM, Gur G, Boyaciğlu S. HASTE MRCP and MRI findings in alveolar echinococcosis of the liver. Australas Radiol. 2001;45(4):496–500. https://doi.org/10.1046/j.1440-1673.2001.00962.x. PMID: 11903184.

144. Torgerson PR, Schweiger A, Deplazes P, Pohar M, Reichen J, Ammann RW, Tarr PE, Halkic N, Müllhaupt B. Alveolar echinococcosis: from a deadly disease to a well-controlled infection. Relative survival and economic analysis in Switzerland over the last 35 years. J Hepatol. 2008;49(1):72–7. https://doi.org/10.1016/j.jhep.2008.03.023. PMID: 18485517.

145. Ishizu H, Uchino J, Sato N, Aoki S, Suzuki K, Kuribayashi H. Effect of albendazole on recurrent and residual alveolar echinococcosis of the liver after surgery. Hepatology. 1997;25(3):528–31. https://doi.org/10.1002/hep.510250305. PMID: 9049192.

146. Ammann RW, Eckert J. Cestodes: Echinococcus. Gastroenterol Clin North Am. 1996;25(3):655–89. https://doi.org/10.1016/s0889-8553(05)70268-5. Erratum in: Gastroenterol Clin North Am 1996;25(4):vii. PMID: 8863045.

147. Bresson-Hadni S, Vuitton DA, Bartholomot B, Heyd B, Godart D, Meyer JP, Hrusovsky S, Becker MC, Mantion G, Lenys D, Miguet JP. A twenty-year history of alveolar echinococcosis: analysis of a series of 117 patients from eastern France. Eur J Gastroenterol Hepatol. 2000;12(3):327–36. https://doi.org/10.1097/00042737-200012030-00011. PMID: 10750654.

148. Joliat GR, Melloul E, Petermann D, Demartines N, Gillet M, Uldry E, Halkic N. Outcomes after liver resection for hepatic alveolar echinococcosis: a single-center cohort study. World J Surg. 2015;39(10):2529–34. https://doi.org/10.1007/s00268-015-3109-2. PMID: 26067633.

149. Kamiyama T, Nakanishi K, Yokoo H, Kamachi H, Tahara M, Yamashita K, Taniguchi M, Shimamura T, Matsushita M, Todo S. Perioperative management of hepatic resection toward zero mortality and morbidity: analysis of 793 consecutive cases in a single institution. J Am Coll Surg. 2010;211(4):443–9. https://doi.org/10.1016/j.jamcollsurg.2010.06.005. PMID: 20822741.

150. Guo Q, Wang M, Zhong K, Li J, Jiang T, Ran B, Shalayiadang P, Zhang R, Tuergan T, Aji T, Shao Y. Application of hepatic lobe hyperplasia techniques in the treatment of advanced hepatic alveolar echinococcosis: a single-centre experience. BMC Surg. 2022;22(1):415. https://doi.org/10.1186/s12893-022-01864-w. PMID: 36474286.

151. Ozdemir F, Ince V, Barut B, Onur A, Kayaalp C, Yilmaz S. Living donor liver transplantation for Echinococcus alveolaris: single-center experience. Liver Transpl. 2015;21(8):1091–5. https://doi.org/10.1002/lt.24170. PMID: 25981332.

152. Kawamura N, Kamiyama T, Sato N, Nakanishi K, Yokoo H, Kamachi H, Tahara M, Yamaga S, Matsushita M, Todo S. Long-term results of hepatectomy for patients with alveolar echinococcosis: a single-center experience. J Am Coll Surg. 2011;212(5):804–12. https://doi.org/10.1016/j.jamcollsurg.2011.02.007. PMID: 21398158.

153. Hillenbrand A, Gruener B, Kratzer W, Kern P, Graeter T, Barth TF, Buttenschoen K, Henne-Bruns D. Impact of safe distance on Long-term outcome after surgical therapy of alveolar echinococcosis. World J Surg. 2017;41(4):1012–8. https://doi.org/10.1007/s00268-016-3813-6. PMID: 27822723.

154. Chen KF, Tang YY, Wang R, Fang D, Chen JH, Zeng Y, Li B, Wen TF, Wang WT, Wu H, Xu MQ, Yang JY, Wei YG, Huang JW, Li JX, Zhang HZ, Feng X, Yan LN, Chen ZY. The choose of different surgical therapies of hepatic alveolar echinococcosis: a single-center retrospective case-control study. Medicine (Baltimore). 2018;97(8):e0033. https://doi.org/10.1097/MD.0000000000010033. PMID: 29465544.

155. Yang C, Yang HJ, Deng SP, Zhang Y. Current status of ex-vivo liver resection and autologous liver transplantation for end-stage hepatic alveolar echinococcosis. Ann Palliat Med. 2020;9(4):2271–8. https://doi.org/10.21037/apm-20-184. PMID: 32576011.

156. Ayifuhan A, Tuerganaili A, Jun C, Ying-Mei S, Xiang-Wei L, Hao W. Surgical treatment for hepatic alveolar echinococcosis: report of 50 cases. Hepato-Gastroenterology. 2012;59(115):790–3. https://doi.org/10.5754/hge10545. PMID: 22469722.
157. Piarroux M, Piarroux R, Giorgi R, Knapp J, Bardonnet K, Sudre B, Watelet J, Dumortier J, Gérard A, Beytout J, Abergel A, Mantion G, Vuitton DA, Bresson-Hadni S. Clinical features and evolution of alveolar echinococcosis in France from 1982 to 2007: results of a survey in 387 patients. J Hepatol. 2011;55(5):1025–33. https://doi.org/10.1016/j.jhep.2011.02.018. PMID: 21354448.
158. Buttenschoen K, Carli Buttenschoen D, Gruener B, Kern P, Beger HG, Henne-Bruns D, Reuter S. Long-term experience on surgical treatment of alveolar echinococcosis. Langenbeck's Arch Surg. 2009;394(4):689–98. https://doi.org/10.1007/s00423-008-0392-5. PMID: 18651165.
159. Kadry Z, Renner EC, Bachmann LM, Attigah N, Renner EL, Ammann RW, Clavien PA. Evaluation of treatment and long-term follow-up in patients with hepatic alveolar echinococcosis. Br J Surg. 2005;92(9):1110–6. https://doi.org/10.1002/bjs.4998. PMID: 16044412.
160. Frei P, Misselwitz B, Prakash MK, Schoepfer AM, Prinz Vavricka BM, Müllhaupt B, Fried M, Lehmann K, Ammann RW, Vavricka SR. Late biliary complications in human alveolar echinococcosis are associated with high mortality. World J Gastroenterol. 2014;20(19):5881–8. https://doi.org/10.3748/wjg.v20.i19.5881. PMID: 24914349.
161. Graeter T, Ehing F, Oeztuerk S, Mason RA, Haenle MM, Kratzer W, Seufferlein T, Gruener B. Hepatobiliary complications of alveolar echinococcosis: a long-term follow-up study. World J Gastroenterol. 2015;21(16):4925–32. https://doi.org/10.3748/wjg.v21.i16.4925. PMID: 25945006.
162. Xu X, Qian X, Gao C, Pang Y, Zhou H, Zhu L, Wang Z, Pang M, Wu D, Yu W, Kong F, Shi D, Guo Y, Su X, Hu W, Yan J, Feng X, Fan H. Advances in the pharmacological treatment of hepatic alveolar echinococcosis: from laboratory to clinic. Front Microbiol. 2022;13:953846. https://doi.org/10.3389/fmicb.2022.953846. PMID: 36003932.
163. Gottstein B, Stojkovic M, Vuitton DA, Millon L, Marcinkute A, Deplazes P. Threat of alveolar echinococcosis to public health—a challenge for Europe. Trends Parasitol. 2015;31(9):407–12. https://doi.org/10.1016/j.pt.2015.06.001. PMID: 26115902.
164. Schweiger A, Ammann RW, Candinas D, Clavien PA, Eckert J, Gottstein B, Halkic N, Muellhaupt B, Prinz BM, Reichen J, Tarr PE, Torgerson PR, Deplazes P. Human alveolar echinococcosis after fox population increase, Switzerland. Emerg Infect Dis. 2007;13(6):878–82. https://doi.org/10.3201/eid1306.061074. PMID: 17553227.
165. Deplazes P, Hegglin D, Gloor S, Romig T. Wilderness in the city: the urbanization of Echinococcus multilocularis. Trends Parasitol. 2004;20(2):77–84. https://doi.org/10.1016/j.pt.2003.11.011. PMID: 14747021.
166. Catalano S, Lejeune M, Liccioli S, Verocai GG, Gesy KM, Jenkins EJ, Kutz SJ, Fuentealba C, Duignan PJ, Massolo A. Echinococcus multilocularis in urban coyotes, Alberta, Canada. Emerg Infect Dis. 2012;18(10):1625–8. https://doi.org/10.3201/eid.1810.120119. PMID: 23017505.
167. Hegglin D, Deplazes P. Control of Echinococcus multilocularis: strategies, feasibility and cost-benefit analyses. Int J Parasitol. 2013;43(5):327–37. https://doi.org/10.1016/j.ijpara.2012.11.013. PMID: 23384814.
168. Meyer A, Olias P, Schüpbach G, Henzi M, Barmettler T, Hentrich B, Gottstein B, Frey CF. Combined cross-sectional and case-control study on Echinococcus multilocularis infection in pigs in Switzerland. Vet Parasitol. 2020;277S:100031. https://doi.org/10.1016/j.vpoa.2020.100031. PMID: 34392944.
169. Budke CM, Jiamin Q, Craig PS, Torgerson PR. Modeling the transmission of Echinococcus granulosus and Echinococcus multilocularis in dogs for a high endemic region of the Tibetan plateau. Int J Parasitol. 2005;35(2):163–70. https://doi.org/10.1016/j.ijpara.2004.10.026. PMID: 15710437.
170. Ziadinov I, Mathis A, Trachsel D, Rysmukhambetova A, Abdyjaparov TA, Kuttubaev OT, Deplazes P, Torgerson PR. Canine echinococcosis in Kyrgyzstan: using prevalence data adjusted for measurement error to develop transmission dynamics models. Int J Parasitol. 2008;38(10):1179–90. https://doi.org/10.1016/j.ijpara.2008.01.009. PMID: 18371969.
171. Dyachenko V, Pantchev N, Gawlowska S, Vrhovec MG, Bauer C. Echinococcus multilocularis infections in domestic dogs and cats from Germany and other European countries. Vet Parasitol. 2008;157(3–4):244–53. https://doi.org/10.1016/j.vetpar.2008.07.030. PMID: 18819752.
172. Coyle CM. Polycystic echinococcosis (polycystic neotropical Echinococcus). In: Hunter's tropical medicine and emerging infectious diseases. Elsevier; 2020. p. 957–9.
173. D'Alessandro A, Rausch RL. New aspects of neotropical polycystic (Echinococcus vogeli) and unicystic (Echinococcus oligarthrus) echinococcosis. Clin Microbiol Rev. 2008;21(2):380–401, table of contents. https://doi.org/10.1128/CMR.00050-07. PMID: 18400802.
174. Soares Mdo C, Rodrigues AL, Moreira Silva CA, Brito EM, Gomes-Gouvêa MS, Corrêa IR, Pinho JR, Malheiros AP, Nunes HM, Póvoa MM. Anatomo-clinical and molecular description of liver neotropical echinococcosis caused by Echinococcus oligarthrus in human host. Acta Trop. 2013;125(1):110–4. https://doi.org/10.1016/j.actatropica.2012.09.004. PMID: 23010160.
175. Vizcaychipi KA, Helou M, Dematteo K, Macchiaroli N, Cucher M, Rosenzvit M, D'Alessandro A. Primera identificación de Echinococcus vogeli en una paca

en la provincia de Misiones, Argentina [First report of Echinococcus vogeli in a paca in Misiones province, Argentina]. Rev Argent Microbiol. 2013;45(3):169–73. https://doi.org/10.1016/s0325--7541(13)70020-8. PMID: 24165140.

176. Tappe D, Stich A, Frosch M. Emergence of polycystic neotropical echinococcosis. Emerg Infect Dis. 2008;14(2):292–7. https://doi.org/10.3201/eid1402.070742. PMID: 18258123.

177. Siles-Lucas M, Casulli A, Conraths FJ, Müller N. Laboratory diagnosis of Echinococcus spp. in human patients and infected animals. Adv Parasitol. 2017;96:159–257. https://doi.org/10.1016/bs.apar.2016.09.003. PMID: 28212789.

178. Siqueira NG, Siqueira CM, Rodrigues-Silva R, Soares Mdo C, Póvoa MM. Polycystic echinococcosis in the state of Acre, Brazil: contribution to patient diagnosis, treatment and prognosis. Mem Inst Oswaldo Cruz. 2013;108(5):533–40. https://doi.org/10.1590/s0074-02762013000500001. PMID: 23903966.

3 Clinical Features of Hydatid Disease

Abdirahman Sakulen Hargura, Ahmed Elsarawy, and Sami Akbulut

3.1 Introduction

The infestation with the metacestode stage of the tapeworm *Echinococcus* of the family Taeniidae is the cause of the echinococcal disease. The life cycle of *Echinococcus* involves a definitive host (usually dogs), an intermediate host (such as sheep, goat, or cattle), and sometimes an incidental host (human). The most common routes of infection in people are ingestion of contaminated water or uncooked food, or direct hand-to-mouth fecal transmission [1]. Once ingested, the eggs develop into oncospheres in the upper gastrointestinal tract that penetrate the intestinal wall, enter the portal vein, and reach the liver [2].

There are four species of the *Echinococcus* that cause infections in humans: *E. granulosus*, *E. multilocularis*, *E. vogelli*, and *E. oligarthus*. The two most common types of echinococcosis seen in humans are cystic echinococcosis and alveolar echinococcosis, caused by *E. granulosus* and *E. multilocularis*, respectively. Both cystic and alveolar echinococcal diseases affect mainly liver, constituting 75% and 90% of disease burden, respectively [3, 4].

Liver hydatid disease caused by *E. granulosus* has been identified as an important health problem worldwide that heavily impacts healthcare systems in view of being mainly a disease of the young active age, protracted clinical course of presentation, diagnosis, management, and follow up as well as utilizing expensive and complex resources during its clinical pathway.

Preliminary diagnosis of liver hydatid disease requires a high index of suspicion to be considered among patients living in—or emigrating from—endemic areas, and presenting with protracted or intermittent course of vague abdominal complaints, not responding completely to over-the-counter medications.

A. S. Hargura
Department of Surgery, Kenyatta University Teaching, Research and Referral Hospital, Nairobi, Kenya

A. Elsarawy
Department of Surgery and Organ Transplantation, Gaziosmanpasa Hospital, Istanbul Yeni Yuzyil University Faculty of Medicine, Istanbul, Türkiye

S. Akbulut (✉)
Department of Surgery and Liver Transplantation Institute, Inonu University Faculty of Medicine, Malatya, Türkiye

Department Biostatistics and Medical Informatics, Inonu University Faculty of Medicine, Malatya, Türkiye

3.1.1 Clinical Features

The disease is generally asymptomatic in its initial stages and therefore patients might be diagnosed on imaging done on routine basis at varying points during the clinical course. On the other hand, many patients might live with the disease in its latent stage for very long periods of time and diagnosed incidentally during autopsies. For

S. Akbulut (ed.), *Hydatid Disease*, https://doi.org/10.1007/978-3-031-97277-5_3

cystic echinococcus, caused by *E. granulosus*, the site and size of the cyst play a role in symptomatic presentations. Though rare, symptoms of mass effect, biliary, blood or lymphatic obstruction, and complications secondary to rupture or bacterial super infection can be encountered. Cyst growth rates over time are highly variable but generally there is 1–5 cm annual increase in diameter [5, 6].

The liver is the most commonly affected organ, reaching up to 50–70% of cases, followed by the pulmonary system which is affected in 20–30% of patients with hydatid disease [7]. Other organs that are affected by the disease, though at considerably lower rates, are the muscles, kidneys, brain, heart, bone, and pancreas. In majority of cases, *E. granulosus* is characterized by a single organ and single cyst involvement. The presence and nature of symptoms vary with the specific disease and the site affected.

3.1.2 Cystic Echinococcosis

The clinical presentation of liver hydatid disease can be categorized as follows: (i) A radiological complaint of an otherwise healthy asymptomatic person who underwent an imaging modality on routine basis or checkup. (ii) Vague, long-standing abdominal pain that waxes and wanes and not relieved completely with over-the-counter medications. (iii) Pressure symptoms caused by larger cysts including nausea, vomiting, or symptoms of reflux. (iv) Less commonly, patients may present with the clinical picture of a complicated cyst including abscess, intraperitoneal rupture, cystobiliary fistula, cystopleural fistula, cystocutaneous fistula, or very rarely intravascular rupture and hematogenous dissemination [8, 9].

The right lobe of the liver is affected in over 60% of cases while the left lobe is affected in around 20% of cases. Caudate lobe cysts had been described in some cases. Symptoms are generally due to mass effect and developed once the lesion grows beyond 10 cm. In very large cysts, patients may manifest with hepatomegaly and associated symptoms such as nausea, vomiting, and epigastric or right upper quadrant pain with palpable mass.

With growth and rise of intracystic pressure, the cyst might rupture resulting in one of the aforementioned complications. Generally, two patterns of cyst rupture had been described, a contained rupture where the endocyst ruptures but the pericyst is intact, hence preventing the contents from spilling, and a communicating rupture where the pericyst is also breached spilling the content freely into the abdominal/pleural cavity or into the biliary tree.

In a considerable percentage of patients, around 6–25%, the cyst may rupture into the biliary tree causing symptoms like obstructive jaundice, biliary colic, cholangitis, and pancreatitis [10]. The rupture can be contained rupture, where the endocyst ruptures but the pericyst is intact hence preventing the contents from spilling; or communicating rupture, where the pericyst is also breached spilling the content freely into the abdominal/pleural cavity or into the biliary tree. This condition is described as cystobiliary fistula and a cut-off cyst diameter between 7.5 and 14 cm had been described to be associated with considerable risk of this complication. The hallmark of this complication is some derangement of liver enzymes, especially chlostatic enzymes, i.e., GGT and ALP [11]. Intrabiliary rupture can be occult (in 10–37% of patients) or frank (in 3–17% of patients). In frank rupture, daughter vesicles and membrane fragments enter the biliary tree causing obstructive jaundice and related complications such as cholangitis, cholecystitis, pancreatitis, and septicemia [12]. However, there are very few reported cases of hydatid material related pancreatitis and cholecystitis [13]. Biliary cirrhosis may also be a rare, delayed complication of intrabiliary rupture of hydatid cyst [14].

Other symptoms associated with cyst rupture are anaphylaxis and multiorgan failure secondary to peritonitis, especially in a free peritoneal spill. Secondary infection of the cyst may also lead to pyogenic abscess of the liver, and this might be the initial presentation. In endemic and underserved areas, careful history and proper investigations should be adopted so as not to miss the diagnosis of hydatid cyst abscess being misdiagnosed as amoebic liver abscess.

In some instances, mass effect of the cyst can result in portal hypertension, cholestasis or Budd-Chiari syndrome depending on the vessel, or structure grossly obstructed by the cyst compression. In some cases, the liver hydatid cyst may also rupture into the lung parenchyma and the pleural space resulting in pulmonary complications.

The hydatid disease less commonly affects other organ systems causing organ specific symptoms with increased morbidity and mortality in some cases and will be mentioned in details.

3.1.3 Alveolar Echinococcosis

Hepatic alveolar echinococcus disease is a rare, slow growing (incubation period of 5–15 years), but potentially life-threatening, parasitic disease that develops because of intrahepatic growth of the *Echinococcus multilocularis* larvae [15]. The disease can also extend to extrahepatic sites such as spleen, lungs, and very rarely pericardium [16]. Epidemiologically, the disease mostly occurs in the northern hemisphere, in parts of Europe, Asia, Alaska, and North America [17].

The final hosts for these parasites are carnivores such as fox, with humans affected as accidental intermediate hosts through indirect contact with natural intermediate hosts like rodents [15].

The commonly reported symptoms of alveolar hydatid disease include right upper quadrant pain, fever, itching, and portal hypertension, especially in patients with history of growing up in rural areas and contact with free-roaming domestic canines. The alveolar echinococcus disease has prognosis similar to that of liver malignancy with reported mortality of over 90% in untreated cases within 10–15 years of diagnosis [18].

3.2 Extrahepatic Localizations

Along with hepatic involvement, other organs and body systems may also be affected simultaneously by echinococcal infection. Less commonly, isolated extrahepatic involvement without liver disease has been reported. Diagnosing extrahepatic hydatid disease is generally easier in patients with a prior history of hepatic echinococcosis compared to those presenting with isolated extrahepatic forms. Physicians practicing in endemic areas should consider this parasitic infection in the differential diagnosis of any persistent or unexplained swelling. Reported extrahepatic sites include the lungs, brain, spleen, parotid gland, bones, and soft tissues.

3.3 Lung

The lungs remain the organ commonly infected by the echinococcal cystic disease after the liver, and its prevalence varies between 20% and 40% [7, 19]. Hydatid disease of the lungs mostly affects the right lung (60%), more specifically the lower lobes (60%) [20]. Lung hydatid disease is a multicystic disease, with around 20% of patients with lung cystic disease also reportedly having liver hydatid disease as well. Hydatid lung disease is more common in children than adults, who have higher liver to lung disease burden [21]. Sign and symptoms depend on cyst size and location. The most common symptoms are chest pain, shortness of breath, cough, fever, hemoptysis, pneumonia, atelectasis, expectoration, anaphylactic reactions (rash, edema, bronchospasm), pulmonary embolism (rarely), and vascular complications [22–24]. Patients may also exhibit nonspecific, generalized symptoms such as malaise and nausea with associated vomiting [23]. Cysts in the lung can also rupture, either into the bronchi or the pleural space. Symptoms associated with cyst rupture into the bronchi are couch, chest pain, hemoptysis while free rupture into the pleural may result in pleural effusion, empyema, or pneumothorax. Bacterial abscesses due to secondary infection may be encountered [25]. Reportedly, children are generally asymptomatic due to weaker immune response and relatively sound lung elasticity that may accommodate the compression or stretching effects of even larger cysts [26].

3.4 Spleen

The worldwide prevalence of splenic hydatid cyst varies between 0.5% and 8% [27, 28]. Echinococcal infestation had been reported in the spleen either primary (isolated form) or in association with hepatic hydatid cyst (secondary form) [27]. Splenic hydatid cysts may remain asymptomatic, slowly growing and diagnosed incidentally in approximately 23.8–50% of cases [28, 29]. An ultrasound done in the setting of blunt upper abdominal trauma led to the diagnosis in some asymptomatic cases. Left upper quadrant abdominal pain, and mass or palpable splenomegaly are the main clinical presentations in symptomatic cases. Constitutional symptoms could ensue if gets infected. Splenic hydatid cyst rupture is a serious complication characterized by peritoneal or pleural dissemination [27, 28, 30].

3.5 Heart

Cardiac hydatid cysts are extremely rare, but they are clinically important because they carry a potential risk of mortality. Cardiac hydatid cysts constitute 0.5% to 2% of all hydatid cyst cases, and most of these cases are secondary involvements [31]. Primary cardiac hydatid disease is very rare and, according to literature data, the prevalence of the disease varies between 0.002% and 2% [32]. The parasite reaches the myocardium of the left ventricle in most of the situations, via coronary arteries or pulmonary veins, originating from a ruptured pulmonary hydatid cyst. The clinical presentation depends on the site and size of the implanted lesion and ranges between asymptomatic to sudden death. Pressure symptoms ensue while the cyst enlarges, leading to dyspnea, chest pain, and palpitation. Stretching of the pericardium and abutting of coronaries can also result in such symptoms. The enlarging cyst could get infected and ruptures resulting in cardiac tamponade and symptoms of heart failure [31–34].

3.6 Pancreas

Although many rates are given regarding the incidence of pancreatic hydatid disease, the actual incidence varies between 0.1% and 2% according to current systematic review studies [35, 36]. The distribution of hydatid cysts in the pancreatic gland varies widely: head (31–58%), body (22–34%), and in the tail (16–47%) [35–37]. The diagnosis should be suspected if a pancreatic cyst is detected by imaging in a known previous case of echinococcosis in another organ system, otherwise it is among the differential diagnosis of pancreatic cyst, including pancreatic pseudocyst, and other cystic benign and malignant pathologies. Patients can remain asymptomatic for a long period and be diagnosed incidentally. It is known that there is an relationship between the localization of pancreatic cysts and clinical symptoms. It is known that cysts located in the head cause more pronounced symptoms even in the early stages due to relationship of the pancreatic head and adjacent organs. On the other hand, cysts located in the body and tail of the pancreas tend to remain asymptomatic unless their size increases significantly. The main symptoms are epigastric pain, left upper quadrant pain, obstructive jaundice, cholangitis, acute pancreatitis, manifestation of duodenal compression (duodenal stenosis, duodenal fistula), manifestation of gastric compression (nausea, vomiting, abdominal pain, early satiety), splenic vein compression (splenic vein thrombosis, left-sided portal hypertension), and secondary infection (abscess, spontaneous rupture) [35, 36, 38, 39].

3.7 Central Nervous System

Hydatid disease of the central nervous system is a serious presentation mainly due to the late diagnosis. Its prevalence ranges between 0.5% and 4%, especially in endemic areas in developing countries [40–42]. A male predominance had been reported. The cyst commonly settles in the supratentorial region and the regions where cysts are most frequently detected are parietal, frontal, temporal, and occipital regions, respectively [41]. Posterior fossa is a rare location for hydatid disease. The patient usually presents with nonspecific symptoms for a protracted period. Clinical signs and symptoms mostly depend on the location and size of the cysts. As cysts enlarge, symptoms may occur of increased intracranial pressure may occur, such as headache, nausea, vomiting, papilledema, altered levels of consciousness, focal neurological deficits, and seizures [40, 42–44].

Spinal cord involvement of hydatid disease is rare, but it is an important clinical entity because it can cause serious complications such as spinal cord compression. It is an important cause of spinal cord compression syndrome in areas where hydatid disease is endemic, and its incidence varies between 1.1% and 14%. The frequency of location of hydatid disease in the spinal cord is thoracic (46–52%), lumbar (20–37%), sacral region (5.5–23%), and cervical (5.5%), respectively [44, 45]. Although there is no specific finding of spinal involvement, the most common symptoms are back pain, paraparesis, radiculopathy, sensory disturbance, sphincter involvement, and rarely paraplegia [44, 45].

3.8 Musculoskeletal and Soft Tissues

Primary (isolated) or secondary (together with other organ involvement) musculoskeletal system involvement accounts for 1–5% of hydatid cyst cases [46, 47]. Most of musculoskeletal involvements are clinically asymptomatic in the early stages of disease and usually detected incidentally during radiological examinations performed for other reasons. Symptoms vary depending on the involved organ, location of the cysts, compression of the adjacent tissue and organs, whether or not ruptured, and secondary infection [46–49]. The most common symptoms related with musculoskeletal located hydatid cysts are slow-growing soft tissue mass with or without pain, myositis, calcified hematoma, motor or sensory neurological deficits, and pathologic fractures [48, 49]. The spine, pelvis, and long bones are commonly affected by the disease. Spinal cord involvement may result in cord compression symptoms. Infection of the ischioanal fossa have been reported as well, and manifested by a painless perianal mass [49, 50].

3.9 Kidney and Retroperitoneum

Although renal involvement is rare, it is accounts for 2–4% of all hydatid cyst cases. When the hydatid lesions are examined in terms of location, 37% of the renal cysts are located in the upper pole, 34% in the lower pole and 16% in the medorenal region. Most patients with renal hydatid disease have nonspecific clinical findings, which are often detected incidentally in radiological images performed for other reasons. Renal involvement of hydatid disease is associated with signs and symptoms such as lumbar or upper abdominal pain, tumor resembling lumbar or abdominal mass, malaise, vomiting, fever, dysuria, eosinophilia, pyuria, hydatiduria, hematuria, and glomerulonephritis related nephrotic syndromes [51–55].

Primary (isolated) retroperitoneal hydatid cyst disease is extremely rare and usually seen secondary to involvement of other organs, which often develops after spontaneous or traumatic cyst rupture. Considering the limited number of case reports published in the literature, it has been reported that the cysts are located in the retroperitoneal space in the majority of cases. The most common symptoms of nonrenal retroperitoneal hydatid cyst disease are palpable mass with or without pain, urinary tract symptoms (urgency, oliguria, dysuria, urinary retention), and rarely dyspareunia. Many of these symptoms depend on the size of the cyst and the compression of adjacent tissues and organs [56, 57].

3.10 Subcutaneous Tissue

Subcutaneous hydatid disease is a rare disease even in endemic areas and constitutes 0.6% to 2.6% of all hydatid cyst cases [58]. The surgical treatment of hydatid cyst of visceral organs can result in cyst spillage and implantation in subcutaneous locations, leading to secondary lesions. However, the mechanism of primary affection of subcutaneous tissue is poorly understood and had been rarely reported, in endemic rural areas. The most frequent locations are thigh, gluteal regions, head and neck, upper extremities, trunk, lower extremities, and pelvic subcutaneous spaces. A slowly growing, painless mobile subcutaneous swelling is mostly the presenting complaint. The overlying skin is otherwise healthy unless secondary infected. The diagnosis in most instances is established after surgical excision and histopathological assessment [58, 59].

3.11 Orbital and Ocular Space

Isolated echinococcus involvement in the periorbital region is a rare entity that accounts for approximately 1% of cases. It had been described in some case reports, especially from endemic regions. Most cysts settle unilaterally within the retrobulbar tissue, in the superomedial or superolateral aspects of the orbit and uncommonly the inferior parts. The symptoms are secondary to pressure exerted from the growing cyst on the globe, including unilateral painless proptosis, downward and angled globe deviation, chemosis, lid edema, pain, and visual impairment. Orbital echinococcosis should be considered in a patient with unilateral proptosis in endemic areas [60–65].

3.12 Head, Neck, Thyroid, and Parotid Glands

Though very rare, isolated involvement of salivary glands has also been reported and should be considered in patients living in endemic areas and presenting with cystic neck swellings. The parotid gland is the main location in the mentioned case reports. A painless slowly growing swelling in the parotid region with intact overlying skin was the main clinical finding. Facial palsy had not been described. Likewise, solitary involvement of the thyroid gland is very rare and the definite diagnosis is almost after surgical excision. Solitary thyroid hydatid cyst presents with a front neck painless swelling in the thyroid anatomic confinements and hence the patient is usually enrolled in the work up of a thyroid swelling unless a clear history of previous hydatid infection is evident. Hoarseness of voice due to recurrent laryngeal nerve affection probably from intense cyst reaction had been described. Bilateral cervical adenopathy was evident, as well. Hydatid affection of other fascial spaces with displacement of thyroid gland, and mimicking branchial cyst had been also described. Another rare and difficult location was hydatid cyst of the infratemporal fossa [66–70].

References

1. Fasihi Harandi M, Budke CM, Rostami S. The monetary burden of cystic echinococcosis in Iran. PLoS Negl Trop Dis. 2012;6(11):e1915. https://doi.org/10.1371/journal.pntd.0001915. PMID: 23209857 PMCID: PMC3510083.
2. Ammann RW, Eckert J. Cestodes. Echinococcus. Gastroenterol Clin North Am. 1996;25(3):655–89. https://doi.org/10.1016/s0889-8553(05)70268-5. PMID: 8863045.
3. Grosso G, Gruttadauria S, Biondi A, Marventano S, Mistretta A. Worldwide epidemiology of liver hydatidosis including the Mediterranean area. World J Gastroenterol. 2012;18(13):1425–37. https://doi.org/10.3748/wjg.v18.i13.1425. PMID: 22509074 PMCID: PMC3319938.
4. Wen H, Vuitton L, Tuxun T, Li J, Vuitton DA, Zhang W, McManus DP. Echinococcosis: advances in the 21st century. Clin Microbiol Rev. 2019;32(2) https://doi.org/10.1128/CMR.00075-18. PMID: 30760475 PMCID: PMC6431127.
5. Frider B, Larrieu E, Odriozola M. Long-term outcome of asymptomatic liver hydatidosis. J Hepatol. 1999;30(2):228–31. https://doi.org/10.1016/s0168-8278(99)80066-x. PMID: 10068100.
6. Moro PL, Gilman RH, Verastegui M, Bern C, Silva B, Bonilla JJ. Human hydatidosis in the Central Andes of Peru: evolution of the disease over 3 years. Clin Infect Dis. 1999;29(4):807–12. https://doi.org/10.1086/520440. PMID: 10589894.
7. Bhutani N, Kajal P. Hepatic echinococcosis: a review. Ann Med Surg (Lond). 2018;36:99–105. https://doi.org/10.1016/j.amsu.2018.10.032. PMID: 30450204 PMCID: PMC6226561.
8. Koc C, Akbulut S, Sahin TT, Tuncer A, Yilmaz S. Intraperitoneal rupture of the hydatid cyst dis-

ease: single-center experience and literature review. Ulus Travma Acil Cerrahi Derg. 2020;26(5):789–97. https://doi.org/10.14744/tjtes.2020.32223. PMID: 32946087.
9. Yilmaz M, Akbulut S, Kahraman A, Yilmaz S. Liver hydatid cyst rupture into the peritoneal cavity after abdominal trauma: case report and literature review. Int Surg. 2012;97(3):239–44. https://doi.org/10.9738/CC116.1. PMID: 23113853 PMCID: PMC3723224.
10. Dadoukis J, Gamvros O, Aletras H. Intrabiliary rupture of the hydatid cyst of the liver. World J Surg. 1984;8(5):786–90. https://doi.org/10.1007/BF01655782. PMID: 6506741.
11. Ozturk G, Uzun MA, Ozkan OF, Kayaalp C, Tatli F, Eren S, Aksungur N, Coker A, Bostanci EB, Oter V, Kaya E, Tasar P. Turkish HPB Surgery Association consensus report on hepatic cystic echinococcosis (HCE). Turk J Surg. 2022;38(2):101–20. https://doi.org/10.47717/turkjsurg.2022.5757. PMID: 36483170 PMCID: PMC9714645.
12. Sparchez Z, Osian G, Onica A, Barbanta C, Tantau M, Pascu O. Ruptured hydatid cyst of the liver with biliary obstruction: presentation of a case and review of the literature. Rom J Gastroenterol. 2004;13(3):245–50. PMID: 15470540.
13. Aydin A, Ersoz G, Tekesin O, Mentes A. Hydatid acute pancreatitis: a rare complication of hydatid liver disease. Report of two cases. Eur J Gastroenterol Hepatol. 1997;9(2):211–4. https://doi.org/10.1097/00042737-199702000-00020. PMID: 9058637.
14. Sikar HE, Kaptanoglu L, Kement M. An unusual appearance of complicated hydatid cyst: necrotizing pancreatitis. Ulus Travma Acil Cerrahi Derg. 2017;23(1):81–3. https://doi.org/10.5505/tjtes.2016.26820. PMID: 28261778.
15. Guidelines for treatment of cystic and alveolar echinococcosis in humans. WHO informal working group on echinococcosis. Bull World Health Organ. 1996;74(3):231–42. PMID: 8789923 PMCID: PMC2486920.
16. Yamamoto Y, Sakamoto Y, Kamiyama T, Nagatsu A, Asahi Y, Orimo T, Kakisaka T, Kamachi H, Otsuka T, Mitsuhashi T, Taketomi A. A case of alveolar echinococcosis in the liver that ruptured into the pericardium treated by a combination of hepatectomy and albendazole. Surg Case Rep. 2022;8(1):63. https://doi.org/10.1186/s40792-022-01417-6. PMID: 35394220 PMCID: PMC8993950.
17. Raether W, Hanel H. Epidemiology, clinical manifestations and diagnosis of zoonotic cestode infections: an update. Parasitol Res. 2003;91(5):412–38. https://doi.org/10.1007/s00436-003-0903-9. PMID: 13680371.
18. Joliat GR, Melloul E, Petermann D, Demartines N, Gillet M, Uldry E, Halkic N. Outcomes after liver resection for hepatic alveolar echinococcosis: a single-center cohort study. World J Surg. 2015;39(10):2529–34. https://doi.org/10.1007/s00268-015-3109-2. PMID: 26067633.
19. Ben Ameur H, Affes N, Abdelhedi C, Kchaou A, Boujelbene S, Beyrouti MI. Hydatid cyst of the spleen: Tunisian series of 21 cases. Indian J Surg. 2015;77(Suppl 2):515–9. https://doi.org/10.1007/s12262-013-0905-5. PMID: 26730056 PMCID: PMC4692885.
20. Morar R, Feldman C. Pulmonary echinococcosis. Eur Respir J. 2003;21(6):1069–77. https://doi.org/10.1183/09031936.03.00108403. PMID: 12797504.
21. Baden LR, Elliott DD. Case records of the Massachusetts General Hospital. Weekly Clinicopathological exercises. Case 4-2003. A 42-year-old woman with cough, fever, and abnormalities on thoracoabdominal computed tomography. N Engl J Med. 2003;348(5):447–55. https://doi.org/10.1056/NEJMcpc020027. PMID: 12556547.
22. Weber TF, Junghanss T, Stojković M. Pulmonary cystic echinococcosis. Curr Opin Infect Dis. 2023;36(5):318–25. https://doi.org/10.1097/qco.0000000000000962. PMID: 37578473 PMCID: PMC10487362.
23. Santivanez S, Garcia HH. Pulmonary cystic echinococcosis. Curr Opin Pulm Med. 2010;16(3):257–61. https://doi.org/10.1097/MCP.0b013e3283386282. PMID: 20216420 PMCID: PMC3362862.
24. Sarkar M, Pathania R, Jhobta A, Thakur BR, Chopra R. Cystic pulmonary hydatidosis. Lung India. 2016;33(2):179–91. https://doi.org/10.4103/0970-2113.177449. PMID: 27051107 PMCID: PMC4797438.
25. Turgut AT, Altinok T, Topcu S, Kosar U. Local complications of hydatid disease involving thoracic cavity: imaging findings. Eur J Radiol. 2009;70(1):49–56. https://doi.org/10.1016/j.ejrad.2008.01.002. PMID: 18291609.
26. Arroud M, Afifi MA, El Ghazi K, Nejjari C, Bouabdallah Y. Lung hydatic cysts in children: comparison study between giant and non-giant cysts. Pediatr Surg Int. 2009;25(1):37–40. https://doi.org/10.1007/s00383-008-2256-z. PMID: 18828025.
27. Eris C, Akbulut S, Yildiz MK, Abuoglu H, Odabasi M, Ozkan E, Atalay S, Gunay E. Surgical approach to splenic hydatid cyst: single center experience. Int Surg. 2013;98(4):346–53. https://doi.org/10.9738/INTSURG-D-13-00138.1. PMID: 24229022 PMCID: PMC3829062.
28. Akbulut S, Sogutcu N, Eris C. Hydatid disease of the spleen: single-center experience and a brief literature review. J Gastrointest Surg. 2013;17(10):1784–95. https://doi.org/10.1007/s11605-013-2303-5. PMID: 23949423.
29. Mejri A, Arfaoui K, Ayadi MF, Aloui B, Yaakoubi J. Primitive isolated hydatid cyst of the spleen: total splenectomy versus spleen saving surgical modalities. BMC Surg. 2021;21(1):46. https://doi.org/10.1186/s12893-020-01036-8. PMID: 33472623 PMCID: PMC7819254.
30. Belli S, Akbulut S, Erbay G, Koçer NE. Spontaneous giant splenic hydatid cyst rupture causing fatal anaphy-

lactic shock: a case report and brief literature review. Turk J Gastroenterol. 2014;25(1):88–91. https://doi.org/10.5152/tjg.2014.3521. PMID: 24918138.
31. Banisefid E, Baghernezhad K, Beheshti R, Hamzehzadeh S, Nemati S, Samadifar Z, Owaysee Osquee H, Javanshir E, Naseri A. Cardiac hydatid disease; a systematic review. BMC Infect Dis. 2023;23(1):600. https://doi.org/10.1186/s12879-023-08576-3. PMID: 37705012 PMCID: PMC10500901.
32. Abhishek V, Avinash V. Cardiac hydatid disease: literature review. Asian Cardiovasc Thorac Ann. 2012;20(6):747–50. https://doi.org/10.1177/0218492312460774. PMID: 23284134.
33. Azari A, Bigdelu L, Maadarani O, Bitar Z. Cardiac hydatid disease with palpitations as the only presenting complaint: the egg in the nest. Eur J Case Rep Intern Med. 2023;10(5):003886. https://doi.org/10.12890/2023_003886. PMID: 37205207 PMCID: PMC10187103.
34. Kahlfuss S, Flieger RR, Roepke TK, Yilmaz K. Diagnosis and treatment of cardiac echinococcosis. Heart. 2016;102(17):1348–53. https://doi.org/10.1136/heartjnl-2016-309350. PMID: 27199228.
35. Akbulut S, Yavuz R, Sogutcu N, Kaya B, Hatipoglu S, Senol A, Demircan F. Hydatid cyst of the pancreas: report of an undiagnosed case of pancreatic hydatid cyst and brief literature review. World J Gastrointest Surg. 2014;6(10):190–200. https://doi.org/10.4240/wjgs.v6.i10.190. PMID: 25346801 PMCID: PMC4208043.
36. Dziri C, Dougaz W, Bouasker I. Surgery of the pancreatic cystic echinococcosis: systematic review. Transl Gastroenterol Hepatol. 2017;2:105. https://doi.org/10.21037/tgh.2017.11.13. PMID: 29354762 PMCID: PMC5763016.
37. Al Laham O, Abdul Khalek G, Alboushi H, Abazid E, Darwish A, Hamza A. An incidentally diagnosed primary pancreatic body hydatid cyst: a case report and literature review. Int J Surg Case Rep. 2024;116:109392. https://doi.org/10.1016/j.ijscr.2024.109392. PMID: 38367420.
38. Afilal I, Abbou W, Oukrid K, Aichouni N, Nasri S, Kamaoui I, Skiker I. Acute pancreatitis revealing an isolated hydatid pancreatic cyst simulating a pseudocyst: a case report. Radiol Case Rep. 2023;18(5):1868–71. https://doi.org/10.1016/j.radcr.2023.01.064. PMID: 36936800 PMCID: PMC10017306.
39. Jajal V, Nag H, K S. A pancreatic hydatid cysts causing recurrent acute pancreatitis mimicking a pancreatic pseudocyst: a case report. Cureus. 2023;15(3):e36402. https://doi.org/10.7759/cureus.36402. PMID: 37090324 PMCID: PMC10115739.
40. Din NU, Raza M, Rahim S, Memon W, Ahmad Z. Hydatid disease of central nervous system, a clinicopathological study of 33 cases. J Neurosci Rural Pract. 2023;14(2):293–7. https://doi.org/10.25259/jnrp_51_2022. PMID: 37181187 PMCID: PMC10174139.
41. Kantzanou M, Karalexi MA, Vassalos CM, Kostare G, Vrioni G, Tsakris A. Central nervous system cystic echinococcosis: a systematic review. Germs. 2022;12(2):283–91. https://doi.org/10.18683/germs.2022.1330. PMID: 36504616 PMCID: PMC9719387.
42. Pour-Rashidi A, Turgut M, Fallahpour M, Mohammadi E, Hanaei S, Rezaei N. Central nervous system hydatidosis around the world: a systematic review. J Neurosurg Sci. 2023;67(5):653–63. https://doi.org/10.23736/s0390-5616.22.05817-9. PMID: 36800683.
43. Nourbakhsh A, Vannemreddy P, Minagar A, Toledo EG, Palacios E, Nanda A. Hydatid disease of the central nervous system: a review of literature with an emphasis on Latin American countries. Neurol Res. 2010;32(3):245–51. https://doi.org/10.1179/016164110X12644252260673. PMID: 20406602.
44. Padayachy LC, Ozek MM. Hydatid disease of the brain and spine. Childs Nerv Syst. 2023;39(3):751–8. https://doi.org/10.1007/s00381-022-05770-7. PMID: 36443475 PMCID: PMC9707099.
45. Pamir MN, Ozduman K, Elmaci I. Spinal hydatid disease. Spinal Cord. 2002;40(4):153–60. https://doi.org/10.1038/sj.sc.3101214. PMID: 11965552.
46. García-Alvarez F, Torcal J, Salinas JC, Navarro A, García-Alvarez I, Navarro-Zorraquino M, Sousa R, Tejero E, Lozano R. Musculoskeletal hydatid disease: a report of 13 cases. Acta Orthop Scand. 2002;73(2):227–3. https://doi.org/10.1080/000164702753671858. PMID: 12079024.
47. Toğral G, Arıkan ŞM, Ekiz T, Kekeç AF, Ekşioğlu MF. Musculoskeletal hydatid cysts resembling tumors: a report of five cases. Orthop Surg. 2016;8(2):246–52. https://doi.org/10.1111/os.12246. PMID: 27384735 PMCID: PMC6584089.
48. Tekin R, Avci A, Tekin RC, Gem M, Cevik R. Hydatid cysts in muscles: clinical manifestations, diagnosis, and management of this atypical presentation. Rev Soc Bras Med Trop. 2015;48(5):594–8. https://doi.org/10.1590/0037-8682-0197-2015. PMID: 26516970.
49. Cattaneo L, Manciulli T, Cretu CM, Giordani MT, Angheben A, Bartoloni A, Zammarchi L, Bartalesi F, Richter J, Chiodini P, Godbole G, Junghanss T, Stojkovic M, Sammarchi L, Dore R, Vercelli A, Benazzo F, Cuzzocrea F, Tamarozzi F, Brunetti E. Cystic echinococcosis of the bone: a European multicenter study. Am J Trop Med Hyg. 2019;100(3):617–21. https://doi.org/10.4269/ajtmh.18-0758. PMID: 30693857 PMCID: PMC6402919.
50. Abdalla KM, Al Sharie AH, ALb YO, Daoud MN, Aleshawi AJ, Gharaibeh KI. Ischio-anal fossa hydatid cyst. Int J Infect Dis. 2020;92:181–3. https://doi.org/10.1016/j.ijid.2020.01.015. PMID: 31981765.
51. Gelman R, Brook G, Green J, Ben-Itzhak O, Nakhoul F. Minimal change glomerulonephritis associated with

hydatid disease. Clin Nephrol. 2000;53(2):152–5. PMID: 10711418.
52. Göger YE, Ozkent MS, Yıldırım MA, Bakdık S, Sönmez MG, Kadiyoran C, Balasar M, Cakir M. Hydatid cyst treatment and management in retroperitoneal organs; is percutaneous drainage an option? Urol J. 2020;17(6):657–63. https://doi.org/10.22037/uj.v16i7.6353. PMID: 33432566.
53. Khalili N, Iranpour P, Khalili N, Haseli S. Hydatid disease: a pictorial review of uncommon locations. Iran J Med Sci. 2023;48(2):118–29. https://doi.org/10.30476/ijms.2022.93123.2442. PMID: 36895460 PMCID: PMC9989246.
54. Yousofi Darani H, Jafari R. Renal echinococcosis; the parasite, host immune response, diagnosis and management. J Infect Dev Ctries. 2020;14(5):420–7. https://doi.org/10.3855/jidc.11496. PMID: 32525824.
55. Turgut AT, Odev K, Kabaalioglu A, Bhatt S, Dogra VS. Multitechnique evaluation of renal hydatid disease. AJR Am J Roentgenol. 2009;192(2):462–7. https://doi.org/10.2214/ajr.08.1129. PMID: 19155411.
56. Akbulut S, Senol A, Ekin A, Bakir S, Bayan K, Dursun M. Primary retroperitoneal hydatid cyst: report of 2 cases and review of 41 published cases. Int Surg. 2010;95(3):189–96. PMID: 21066995.
57. Slavu IM, Gheorghita V, Macovei Oprescu AM, Filipoiu F, Munteanu O, Tulin R, Dogaru IA, Ursuț BM, Tulin A. Primary retroperitoneal hydatid cyst: a diagnostic and treatment conundrum. Cureus. 2024;16(2):e53842. https://doi.org/10.7759/cureus.53842. PMID: 38465152 PMCID: PMC10924441.
58. Kayaalp C, Dirican A, Aydin C. Primary subcutaneous hydatid cysts: a review of 22 cases. Int J Surg. 2011;9(2):117–21. https://doi.org/10.1016/j.ijsu.2010.10.009. PMID: 21029797.
59. Suffee T, Chader H, Foulet F, Herruela C, Djabbari M, Chosidow O, Zehou O. A suspicious subcutaneous tumor. Clin Infect Dis. 2015;61(11):1707, 1759–60. https://doi.org/10.1093/cid/civ662. PMID: 26567173.
60. Al-Muala HD, Sami SM, Shukri MA, Hasson HK, Alaboudy AT. Orbital hydatid cyst. Ann Maxillofac Surg. 2012;2(2):197–9. https://doi.org/10.4103/2231-0746.101365. PMID: 23483684 PMCID: PMC3591069.
61. Debela AS, Abore KW, Worke AB, Wendimagegn ST. Primary intra-orbital hydatid cyst: a case report of a rare cause of exophthalmos. Int Med Case Rep J. 2024;17:89–92. https://doi.org/10.2147/IMCRJ.S454518. PMID: 38344471 PMCID: PMC10854397.
62. Kumar M, Harsh V, Prakash A, Sahay CB, Kumar A. Neglected case of primary intraorbital hydatid cyst. Neurol India. 2022;70(1):337–9. https://doi.org/10.4103/0028-3886.338694. PMID: 35263908.
63. Kumar V, Surve A, Biswal D, Verma N, Kashyap S, Venkatesh P, Mirdha BR. Intra-ocular hydatid cyst in a child: a rare presentation. Eur J Ophthalmol. 2023;33(3):NP70–4. https://doi.org/10.1177/11206721221079481. PMID: 35166612.
64. Lentzsch AM, Gobel H, Heindl LM. Primary orbital hydatid cyst. Ophthalmology. 2016;123(7):1410. https://doi.org/10.1016/j.ophtha.2016.02.042. PMID: 27342324.
65. Chaabouni M, Ben Zina Z, Ben Ayez H, Tounsi R, Trigui A, Ben Mansour H. [Hydatid orbital cyst: a unique intra-orbital locality. A case report]. J Fr Ophtalmol. 1999;22(3):329–334. PMID: 10337589.
66. Boumediene M, Tbini M, Driss S, Bensalah M, Jrad M. Hydatid cyst in the parotid gland, a rare localization of the disease: a case report. Int J Surg Case Rep. 2023;107:108355. https://doi.org/10.1016/j.ijscr.2023.108355. PMID: 37263004 PMCID: PMC10310901.
67. Akbulut S, Demircan F, Sogutcu N. Hydatid cyst disease of the thyroid gland: report of two cases. Int Surg. 2015;100(4):643–7. https://doi.org/10.9738/intsurg-d-14-00124.1. PMID: 25598424 PMCID: PMC4400932.
68. Yilmaz M, Akbulut S, Sogutlu G, Arabaci E, Kayaalp C. Hydatid cyst of the thyroid gland: report of three cases. Surg Today. 2013;43(8):937–41. https://doi.org/10.1007/s00595-012-0269-7. PMID: 22836546.
69. Rasool Hammoodi SA, Aftan KT, Ali MR. Hydatid cysts of parotid glands- diagnosis, treatment and recurrences. Int J Surg Protoc. 2021;25(1):135–40. https://doi.org/10.29337/ijsp.154. PMID: 34395959 PMCID: PMC8323526.
70. Hammoodi SAR, Aftan KT, Ali MR. Management of hydatid cysts of parotid glands. J Stomatol Oral Maxillofac Surg. 2023;124(6):101465. https://doi.org/10.1016/j.jormas.2023.101465. PMID: 37030440.

Radiological Features of Liver Hydatid Disease

4

Artür Salmaslioglu
and Merve Gulbiz Dagoglu Kartal

4.1 Cystic Echinococcosis

4.1.1 Introduction

Cystic echinococcosis, or hydatidosis, is a parasitic disease caused by the larva of the *Echinococcus granulosus* (*E. granulosus*). Dogs are the main carriers of the adult worm and herbivores (and sometimes humans) are intermediate hosts. The eggs hatched in the gastrointestinal system can be transported to various organs [1]. The liver is the most frequent location of echinococcal cysts, representing approximately 70% of cases [2]. The lungs are the second most common location. In the liver, cysts grow to 1 cm during the first 6 months and approximately 2–3 cm annually, thereafter, depending on the host tissue resistance [3].

The microscopic evaluation of the hydatid cyst demonstrates the presence of three layers: (i) The outer layer, also called the pericyst, is a fibrous structure produced by the host cells and is a defense mechanism to limit the growth of the cyst. (ii) The middle layer (often called ectocyst) is an acellular membrane, which acts as a conduit for the transfer of nutrients. (iii) The inner germinal layer is a thin layer which gives rise to scolices, the infectious part of the cyst, in the form of outpouchings [4].

A. Salmaslioglu (✉) · M. G. Dagoglu Kartal
Department of Radiology, Istanbul University, Istanbul Faculty of Medicine, Istanbul, Türkiye

4.1.2 Classification Systems

Various systems (more than 15) have been proposed for the classification of the hydatid disease in the liver [5]. The first and one of the most popular classification systems was developed by Gharbi et al. [6] in 1981. The World Health Organization (WHO) Informal Working Group's classification system proposed on 2003 is also widely used [7]. The importance of the classification systems is that the treatment approach depends on the cysts' type [8].

Ultrasonography is the main radiological modality the classification systems are based on. While the liver may harbor many types of cysts or predominantly cystic lesions, including simple cysts, polycystic disease, cystadenomas, traumatic cysts, bile duct hamartomas, and cystic malignancies [9], there are pathognomonic sonographic findings for hydatid cysts. Those findings include anechoic cysts with visible cyst walls and membranes/hydatid sand, and cysts with multiple vesicles/daughter cysts, mimicking a wheel or honeycomb.

The WHO classification system considers unilocular cystic lesions with no visible walls in Cystic Lesion (CL) category. Type CL cysts, while not being pathognomonic for hydatid disease, may still have the probability to be an early-stage hydatid cyst, so further diagnostic work-up or follow-up may be required. If a CL lesion is ultimately diagnosed as a hydatid cyst, it is considered as an active lesion but is probably not fertile yet.

S. Akbulut (ed.), *Hydatid Disease*, https://doi.org/10.1007/978-3-031-97277-5_4

Other than the CL category, the WHO classification includes five more types of cysts (Cystic Echinococcosis (CE) 1–5). In all categories, cysts smaller than 5 cm are suffixed with the letter (s), cysts measuring between 5 and 10 cm are suffixed with the letter (m), and cysts larger than 10 cm are suffixed with the letter (l).

The CE1 category includes anechoic cysts with a visible wall (Fig. 4.1a). The cysts may include fine internal echoes “hydatid sand”(Fig. 4.1b), representing brood capsules. These echoes are mobile, which mimics falling snowflakes (“snowflake sign”) (Fig. 4.1c), and they can create a layer inside the cyst. CE1 cysts are considered as active and are generally fertile.

The CE2 category is characterized by the presence of daughter cysts, creating multivesicular or sometimes multiseptated cystic lesions (Fig. 4.2). The daughter cysts can be described as having honeycomb-like features or forming rosette-like configurations. CE2 cysts, like CE1 cysts, are also considered as active and fertile.

The CE3 category represents a transitional phase, where the cyst begins to degenerate and has two subcategories, CE3a and CE3b. In CE3a, the typical and pathognomonic finding is the detachment of the endocyst membrane, which floats freely inside the cyst (“the water-lily sign”—Fig. 4.3a). In CE3b, anechoic daughter cysts are observed interspersed with areas of echogenicity due to the presence of degenerated daughter cysts and debris. The anechoic areas in CE3b cysts carry the risk of being alive (Fig. 4.3b) [10].

The CE4 and CE5 categories represent inactive cysts. In CE4, no daughter cyst is visible, and the lesion mostly looks solid, with heterogenous

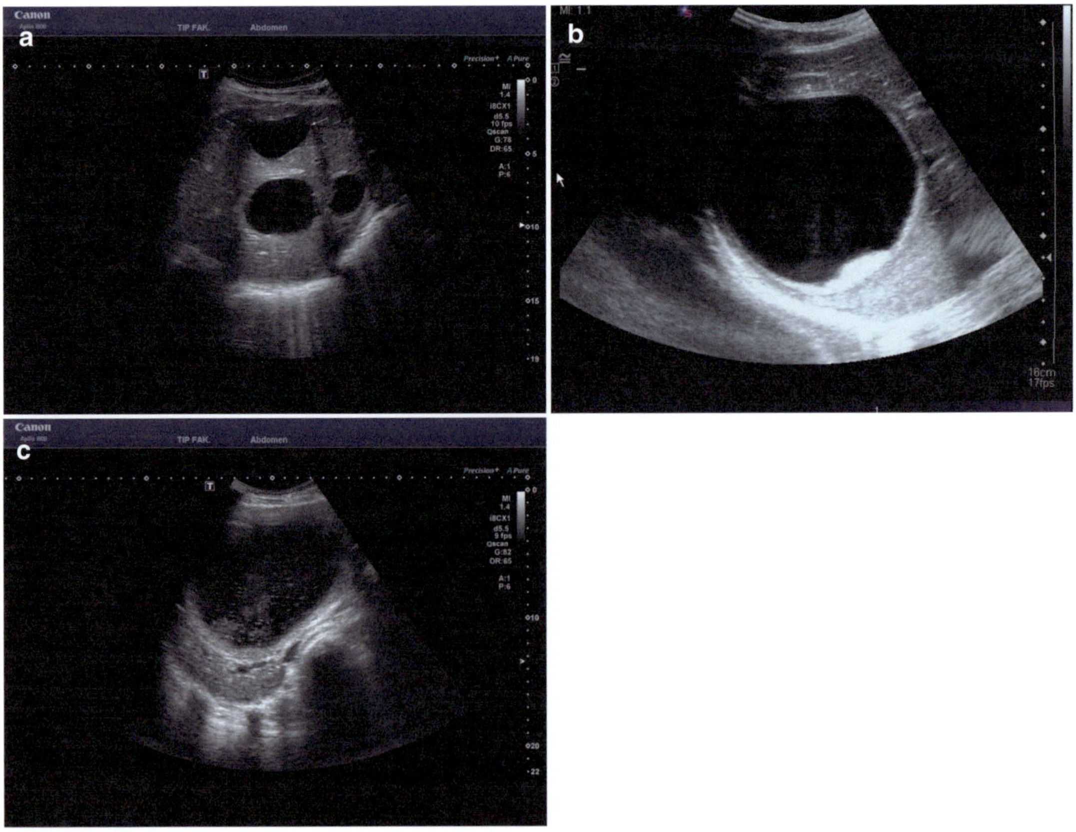

Fig. 4.1 (**a**) Ultrasonography of multiple CE 1 hydatid cysts located inside the right lobe of the liver. The presence of a wall (arrow) is useful for differentiating from a simple liver cyst. (**b**) CE1 hydatid cyst with a thick wall and hydatid sand forming a layer. (**c**) Mobile echoes in a CE1 hydatid cyst (“snowflake sign”). (From the archives of Istanbul University Faculty of Medicine)

content. Degenerated membranes may be observed as spiral-shaped linear structures within the cyst, creating a "ball of wool" appearance (Fig. 4.4). In CE5 category, the prominent feature is calcification, which can vary in extent (Fig. 4.5). The content of the cysts with partially calcified walls are easier to visualize by sonography, whereas cysts with heavily calcified walls may be harder to evaluate due to acoustic shadowing and additional imaging modalities may be warranted.

Gharbi et al.'s [6] classification system is very similar to WHO classification, as it is the system the WHO classification is based on. CE1, CE4, and CE5 correspond to Gharbi Type 1, 4, and 5

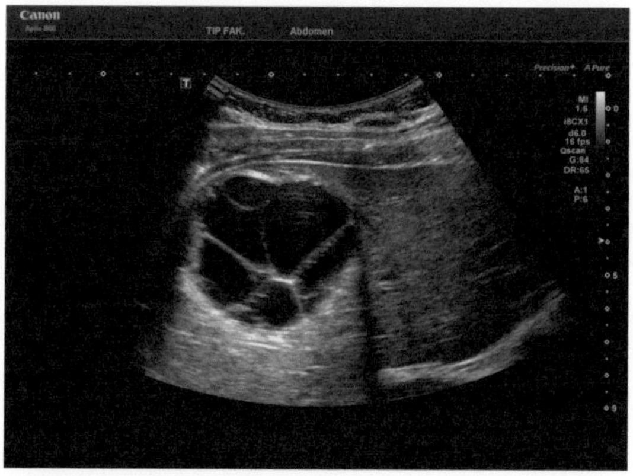

Fig. 4.2 Ultrasonography of CE2 hydatid cyst with multiple daughter cysts creating a multivesicular appearance. (From the archives of Istanbul University Faculty of Medicine)

cysts, CE3a corresponds to Gharbi Type 2 and CE 2 and CE3b correspond to Gharbi Type 3 cysts. The CL category is not included in Gharbi's classification.

Although ultrasonography is the primary screening method of choice, as well as the most commonly used method to monitor the therapy, with the advances in cross-sectional imaging such as computed tomography and magnetic resonance imaging, these modalities have started playing an increasing role in diagnosis, management, planning, and follow up of hydatid disease. Cross-sectional imaging plays a pivotal role in detecting and management of complicated cysts.

As for uncomplicated cysts, computed tomography is the imaging modality which cystic echinococcosis is discovered incidentally [11]. It is also the best modality to detect the number, size, location, and dissemination of the cysts [12]. The sensitivity and specificity of computed tomography in the diagnosis of hydatid cyst has been shown to be 98% [13]. Hydatid cysts appear as low density masses (0–30 Hounsfield Units (HU)), round or oval, of variable size and number with sharply defined borders, with a calcified wall or hyperdense wall representing the pericyst, i.e., the host derived adventitial layer (Fig. 4.6) [14]. A detached laminated membrane can be seen as serpentine hyperdense areas and daughter cysts are appreciated as rounded structures containing lower attenuation when compared with the mother cyst [15, 16].

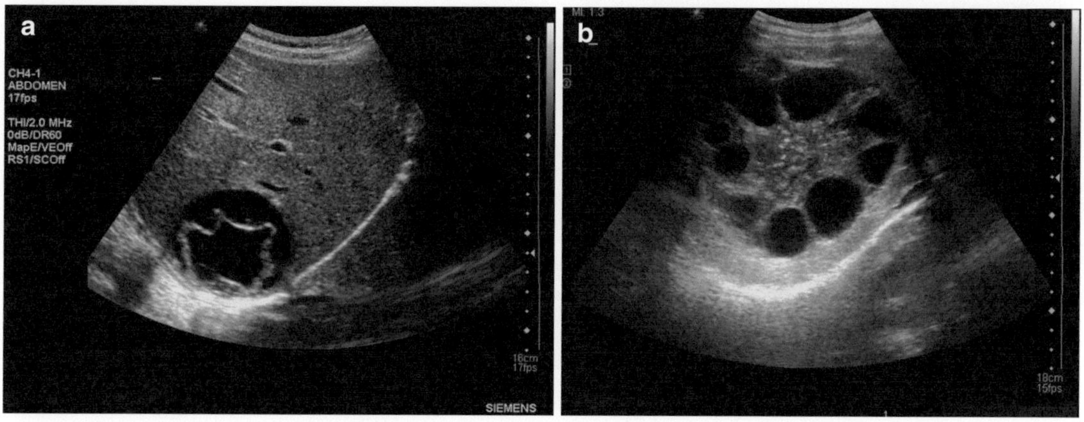

Fig. 4.3 Ultrasonography of CE3 hydatid cyst. (**a**) CE3a hydatid cyst with floating endocyst membrane ("waterlily sign"). (**b**) CE3b hydatid cyst with anechoic daughter cysts interspersed with echogenic material consisting of debris and degenerated daughter cysts. (From the archives of Istanbul University Faculty of Medicine)

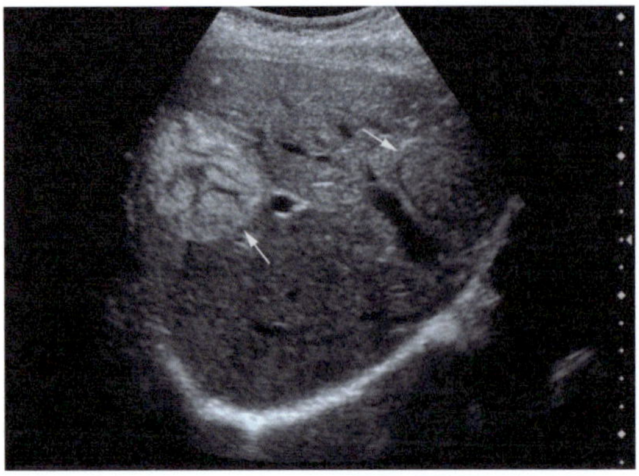

Fig. 4.4 Ultrasonography in CE4 hydatid cyst. A hyperechoic and an isoechoic, solid looking cyst (arrows) with internal linear structures ("ball of wool" appearance). (From the archives of Istanbul University Faculty of Medicine)

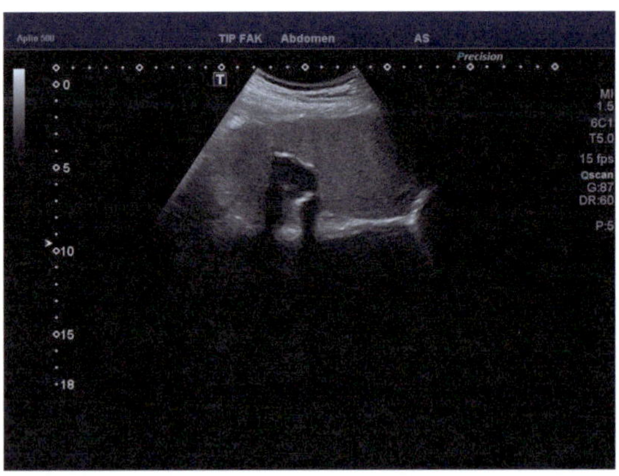

Fig. 4.5 Ultrasonography of a CE5 hydatid cyst with calcified walls. The internal structure is harder to evaluate compared to other stages, due to the presence of calcification. (From the archives of Istanbul University Faculty of Medicine)

Computed tomography is particularly useful in patients for whom ultrasonography is not adequate due to some technical difficulties including obesity, excessive intestinal gas, or incompliance, and in lesions in which ultrasonography is indeterminate [3, 17]. Ultrasonography is usually indeterminate if the matrix fills entire cyst, and the resultant echogenic pattern mimics a solid mass [6, 18]. In these cases, differentiating the lesion from other solid lesions of the liver is crucial.

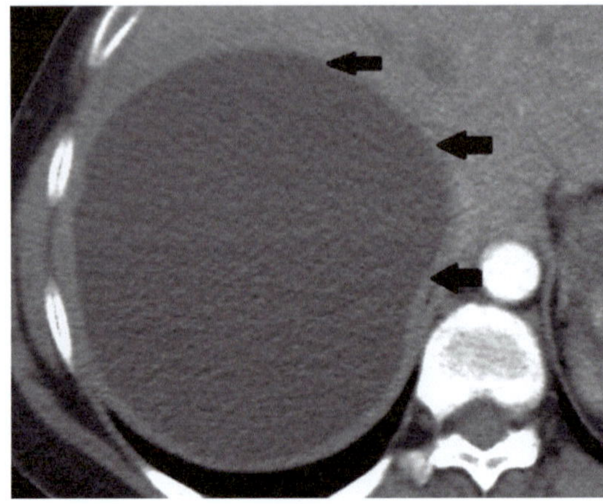

Fig. 4.6 Axial computed tomography image. CE1 cyst appears as a low density (12 HU in this case), round mass sharply defined borders, and hyperdense wall representing pericyst (arrows). (From the archives of Istanbul University Faculty of Medicine)

Besides these, in ultrasonography it is difficult to assess the internal structures of cysts with calcification on their walls whereas computed tomography is an excellent imaging modality to assess calcifications [19]. Calcification on the cyst wall is not necessarily specific for inactive lesions (i.e., CE4 or CE5). Although it is known that calcification increases as the cyst undergoes degeneration, it has been shown that calcification does not always indicate inactivity and can be seen on the walls of active cysts. It is only when the cyst is totally calcified that we can predict inactivity [20]. Despite the accuracy of the modality in detecting calcifications, it should be noted that on computed tomography it is challenging to classify lesions from type I to type 4 since computed tomography is not an optimal method to examine cystic matrix [11].

While there is no widely accepted cyst classification system specific for computed tomography, ultrasonography classification methods may be used for the modality. According to stages determined by WHO-IWGE classification, computed tomography features for different stages of the cyst are as follows [11, 21]:

CE 1 Appears as well-defined, water-attenuation mass with hyperdense wall indicating pericyst. Double line sign is not as frequently depicted as in ultrasonography (Fig. 4.6).

CE2 Type II hydatid cysts are subclassified into three types according to arrangement of daughter cysts. This is determined by the age, number, and arrangement of the daughter cysts.

Type IIA lesions contain round daughter cysts located at the periphery of the cyst. Central matrix is represented as a high-density central area forming the wheel-spoke appearance. The average computed tomography attenuation of the mother cyst is higher than that of daughter cysts.

Type IIB lesions contain larger daughter cysts occupying almost the entire cyst. The high attenuation fluid within the mother cyst that surrounds the daughter cysts looks like septa, creating a "rosette appearance." (Fig. 4.7).

Type IIC lesions appear as relatively high attenuation round or oval masses with scattered calcifications and occasional daughter cysts. These findings represent degeneration of old cysts.

Type II cysts do not enhance with IV contrast. Daughter cysts and endocysts are less clearly visible on computed tomography than they are on ultrasonography and magnetic resonance imaging.

CE3a Detached laminated membrane seen as serpentine hyper attenuating lines in a cyst with liquid content (Fig. 4.8).

CE3b Unilocular cysts with daughter cyst inside a mucinous or solid cyst matrix. Daughter cyst in the solid matrix is not as easily visible as they are on ultrasonography and magnetic resonance imaging (Fig. 4.9).

CE4 Heterogenous solid lesions with degenerative, canalicular structure of the cyst content which is specific to this stage. Canalicular structure is not always clearly seen, which leads to incorrect staging of CE4 lesion as CE 1 (Fig. 4.10).

CE5 Dead cysts with partial or total calcifications. Calcified or hyperdense lesions on computed tomography.

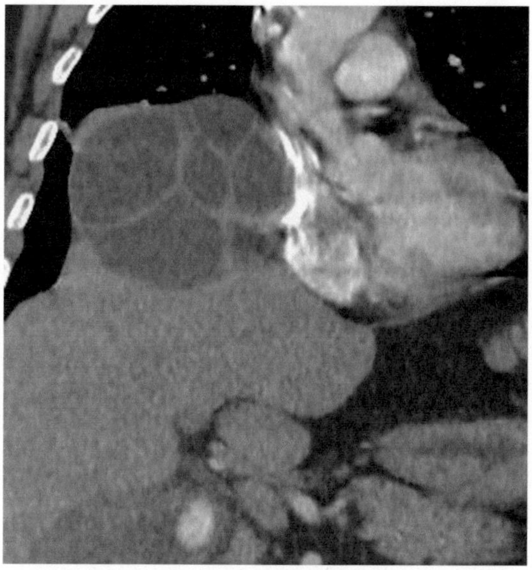

Fig. 4.7 Multiplanar reformatted (MPR) computed tomography image in coronal plane. CE2 cyst appears as multiseptated multivesicular cyst representing daughter vesicles. In this case, the cyst originated from the dome of the liver and compressed lung parenchyma by transdiaphragmatic extension of the cyst. (From the archives of Istanbul University Faculty of Medicine)

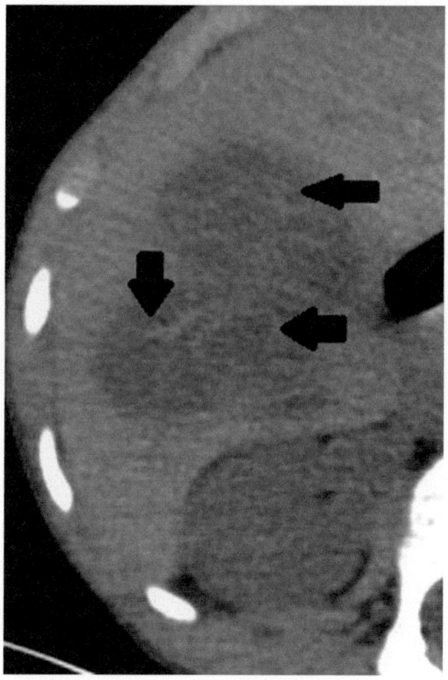

Fig. 4.8 Axial computed tomography image. Detached membranes are appreciated as hyperdense serpentine lines within the cyst (CE3a) (arrows). (From the archives of Istanbul University Faculty of Medicine)

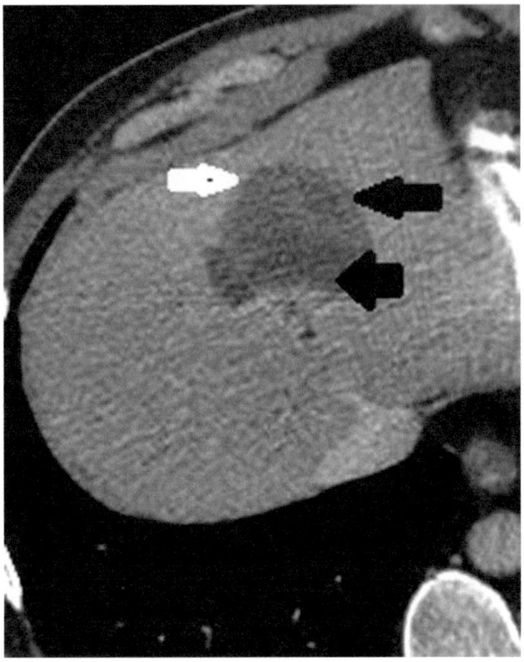

Fig. 4.9 Axial computed tomography image. C3b cyst appears as daughter cysts within a solid matrix. This stage is not always well appreciated on computed tomography. Daughter cysts are seen as hypodense structures (black arrows) and the matrix is higher in density (white arrows). (From the archives of Istanbul University Faculty of Medicine)

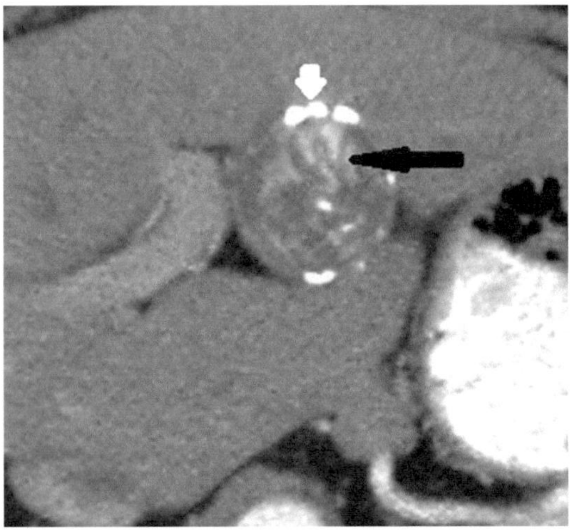

Fig. 4.10 Axial computed tomography image. C4 cyst is a heterogenous solid cyst with degenerative, canalicular structure of the cyst content (black arrows) and some calcification on the wall (white arrow). (From the archives of Istanbul University Faculty of Medicine)

Computed tomography is particularly important in planning management owing to its ability to accurately reveal a vascular map showing vascular relations of the cyst. Especially, computed tomography angiography can accurately demonstrate hepatic artery, portal vein, hepatic vein, and inferior vena cava and lesions neighborhood [22].

In contrast to computed tomography, magnetic resonance imaging, the modality with the highest soft tissue contrast, is successful in demonstrating the cystic component of the lesions. Active and transitional cyst stages are very well appreciated on magnetic resonance imaging compared to ultrasonography, the gold standard modality. On the other hand, solid lesions and pseudolesions are weakly demonstrated on magnetic resonance imaging [11]. This is particularly important for treatment planning for those lesions inaccessible with ultrasonography.

Once again, there are no cyst staging/classification systems specific for magnetic resonance imaging and ultrasonography classification methods are used for the modality. According to stages determined by WHO-IWGE classification, magnetic resonance imaging features for different stages of the cyst are as follows [11, 21]:

CE 1 Lesion appears hypointense on T1W and hyperintense with hypointense rim on T2W with specific double line sign (Fig. 4.11). Double line sign is not as frequently depicted as in ultrasonography.

CE2 Daughter cysts may appear hypointense or isointense relative to the maternal matrix on T1W and T2W images (Figs. 4.12a, b and 4.13).

CE3a Daughter cysts may appear hypointense or isointense relative to the maternal matrix on T1W and T2W images. Detached, floating membranes within the cyst which create "serpent sign" or "snake sign" appear hypointense with all sequences (Fig. 4.14a–c).

CE3b Unilocular cysts with daughter cyst inside a mucinous or solid cyst matrix (Fig. 4.15).

CE4 Heterogenous solid lesions with degenerative, canalicular structure of the cyst content which is specific to this stage.

CE5 Dead cysts with partial or total calcifications. Hypointense on magnetic resonance imaging due to calcified content. Calcifications are not as clearly demonstrated as in computed tomography.

Other than this, WHO-IWGE have proposed another classification based on medical imaging morphology. That was divided into four categories [23, 24].

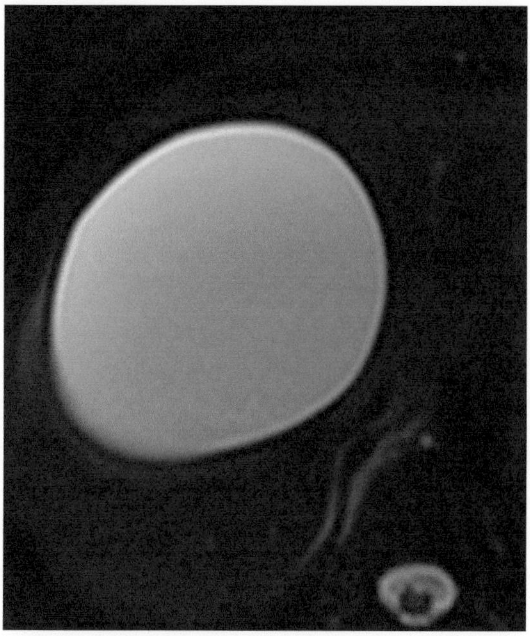

Fig. 4.11 CE1 cyst on axial T2W magnetic resonance imaging. Lesion appears hyperintense with hypointense rim on T2W with specific double line sign. (From the archives of Istanbul University Faculty of Medicine)

Type I: Simple cyst with no architecture: A well-circumscribed, unilocular cystic lesion. It might have internal septations or hydatid sand. Cyst wall and septa frequently enhance on post-contrast computed tomography. On magnetic

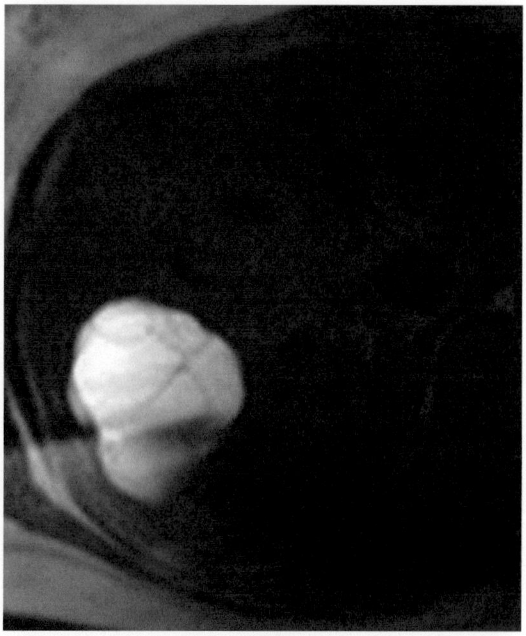

Fig. 4.13 CE2 cyst on T2W magnetic resonance imaging appears as a multivesicular, multiseptated cyst. Daughter cysts may appear hypointense or isointense relative to the maternal matrix on T1W and T2W images. (From the archives of Istanbul University Faculty of Medicine)

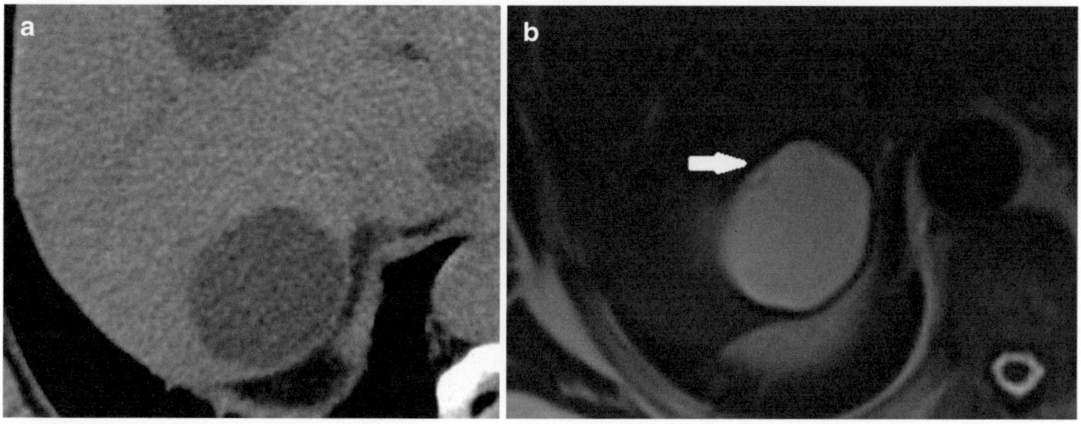

Fig. 4.12 (**a**) Computed tomography image and (**b**) axial T2W magnetic resonance imaging of a patient with multiple liver hydatid cysts at different stages. In segment 7 daughter cyst is clearly depicted in CE2 cyst on magnetic resonance imaging image (white arrow on (**b**)) whereas it is not visible on (**a**). (From the archives of Istanbul University Faculty of Medicine)

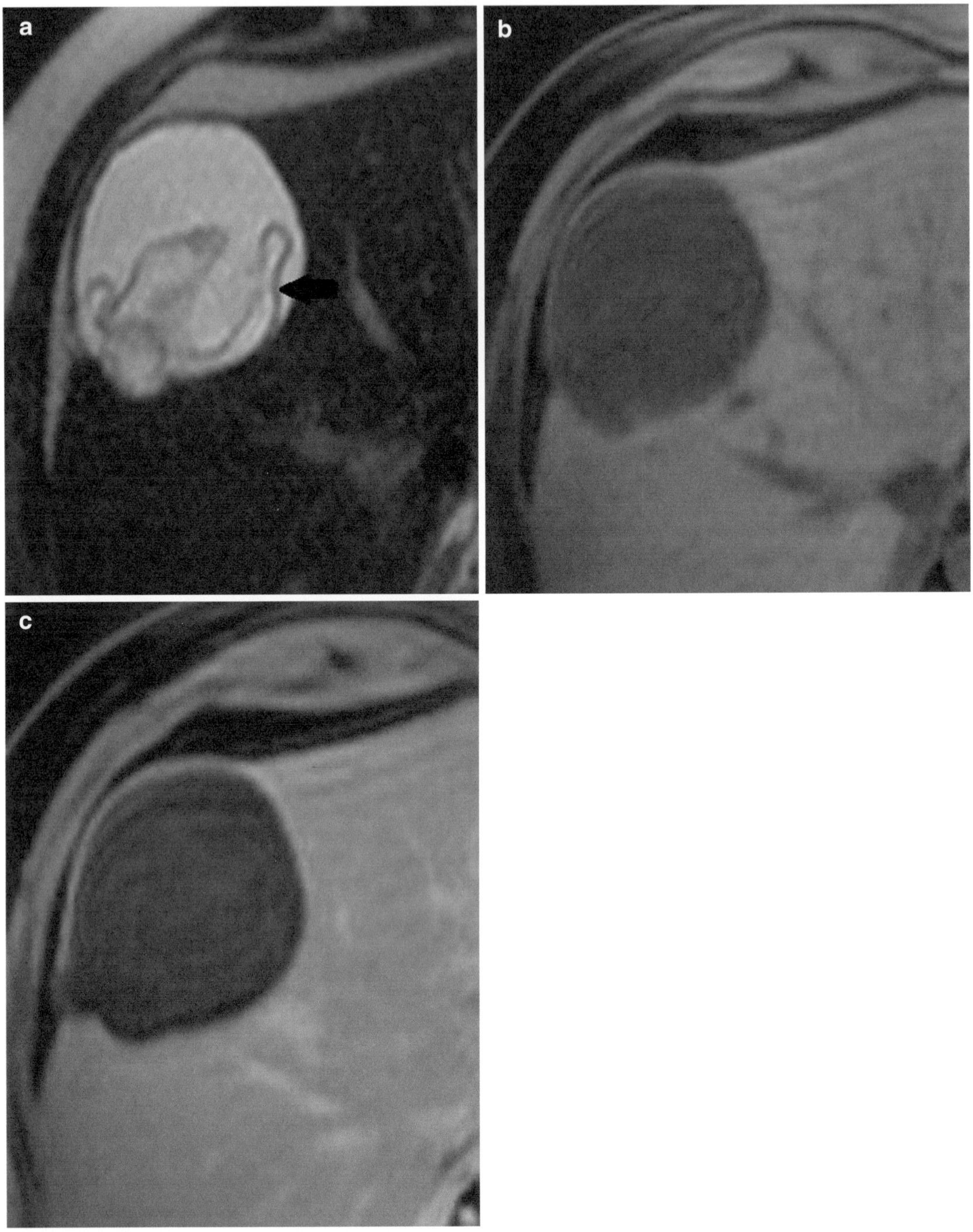

Fig. 4.14 (**a**) Axial T2W, (**b**) axial pre-contrast T1W, and (**c**) axial post-contrast T1W images. CE3a cyst with detached membranes (arrows on (**a**)) create "serpent sign" or "snake sign" and appear hypointense with all sequences and do not enhance after iv contrast administration on post-contrast images (**c**). (From the archives of Istanbul University Faculty of Medicine)

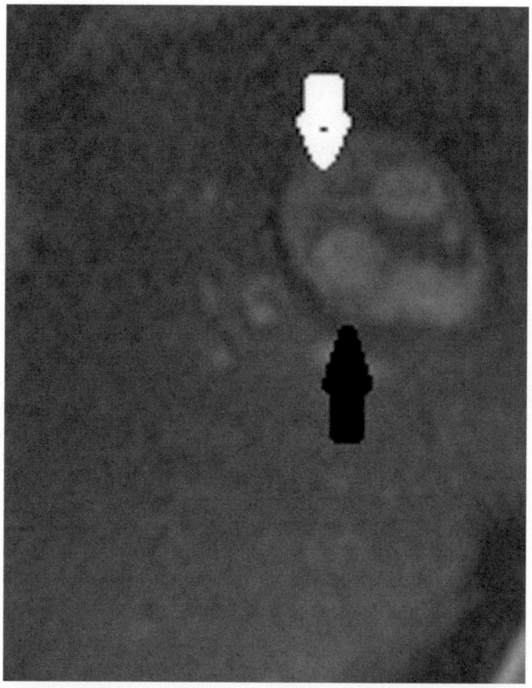

Fig. 4.15 Axial T2W magnetic resonance imaging. CE3b cyst presenting as a unilocular cyst with hyperintense daughter cysts (white arrow) inside a mucinous or solid cyst matrix hypo- or isointense to the parenchyma (black arrow). (From the archives of Istanbul University Faculty of Medicine)

resonance imaging, hydatid cysts appear as a low signal intensity on T1-weighted (W) images, and a high signal intensity on T2W images with a low signal intensity rim. However heterogenous intensity on T2W images are not uncommon. Diffusion weighted imaging (DWI) is an magnetic resonance imaging sequence based on diffusion of water molecules in biological tissues quantitatively. Several studies have shown that with higher b values DWI can differentiate between simple cysts and CE 1 hydatid cyst. On high b-value DWI, CE cysts appear hyperintense, whereas benign cysts do not [23, 25].

Type II: Cyst with daughter cysts and matrix: lesions under this category are active and this is the phase which may lead to dissemination of the disease [24].

Type III: Calcified cysts: Inactive dead phase of hydatid disease. It appears as a calcified lesion with posterior acoustic shadowing on ultrasound imaging, a round hyperdense lesion on computed tomography, and a low signal intensity area on a magnetic resonance imaging. Computed tomography is the gold standard imaging modality used to diagnose calcified hydatid cysts [26].

Type IV: Complicated hydatid cyst: This is a hydatid cyst with a rupture or superinfection, which may be seen in both Types I and II [26].

4.1.3 Complications

Ultrasonography and particularly computed tomography and magnetic resonance imaging are important in detecting and diagnosing complications in hydatid cysts. Complications can be divided in two, being either mechanical or infectious. Mechanical complications include mass effect, rupture which may be contained, communicating, or direct, and exophytic growth [21, 23]. Mass effect is mainly due to increase in size. When a cyst reaches a large size, it may compress neighboring biliary duct and may cause focal dilatation. Compression of portal vein can result in focal atrophy or portal hypertension, compression of hepatic veins or inferior vena cava can result in Budd-Chiari syndrome [27].

Rupture may be related to minor trauma. However, the natural history of hepatic hydatid cysts leads to rupture as a complication in about 35% [28]. The cyst's size is also an important factor, as a study demonstrated that cysts measuring >7.5 cm had a specificity of 73% and sensitivity of 79% for predicting biliary-cystic communications [29]. Communicating rupture is the most common type (15%), followed by contained rupture (12%) and direct rupture (6%) [28]. Contained rupture occurs due to increased pressure in the cyst when the endocyst ruptures but the pericyst is still intact. It also is part of the natural course of the degeneration of the cyst or may be in response to therapy. On computed tomography and magnetic resonance imaging, it is appreciated as serpentine linear hyperdense structures in the cyst.

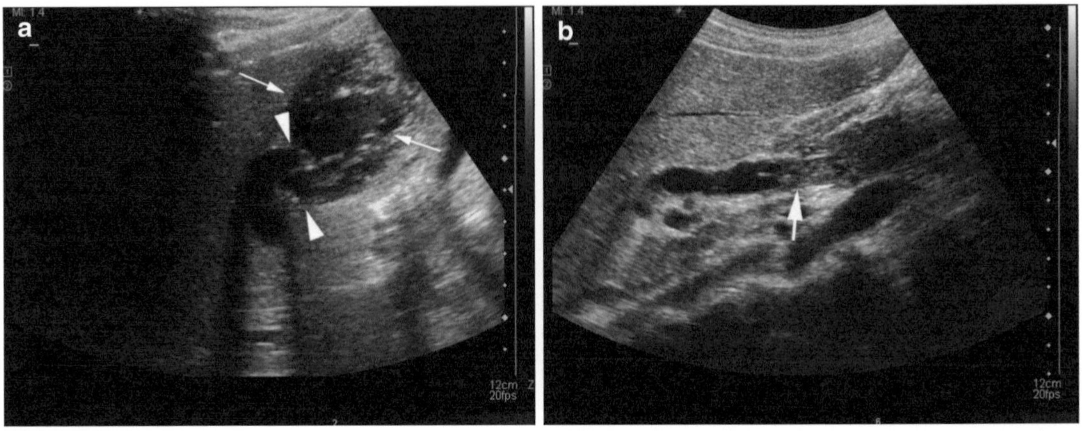

Fig. 4.16 Ultrasonography imaging in hydatid cyst complication: (**a**) Communicating rupture of a hydatid cyst into a bile duct. A beak-like protrusion can be demonstrated, connecting the duct to the cyst (arrowheads). The cyst is deformed (arrows). (**b**) Debris and hydatid sand in the dilated common bile duct (arrow)

Communicating rupture is perforation of the cyst into the bile ducts or to the bronchioles or rarely into vessels such as portal vein, inferior vena cava, or adjacent viscera, and passage of the content into these structures. This is mainly seen in lesions in active phase (i.e., CE1 and CE2) [26]. Ultrasonography can be useful (Fig. 4.16a, b) and computed tomography has diagnostic sensitivity of 85.7% [30]. Magnetic resonance imaging as well as magnetic resonance cholangiopancreatography, a special type of magnetic resonance imaging that produces detailed images of the hepatobiliary and pancreatic systems, including the liver, gallbladder, bile ducts, pancreas, and pancreatic duct, are effective and noninvasive diagnostic modalities proved to be useful [30]. Indirect signs of communicating rupture are deformity of cyst, defect in the wall, presence of beak-like projection extending from cyst wall (Fig. 4.17), dilated bile ducts (Fig. 4.18a–c), and daughter vesicles or germinative membranes of cysts in bile ducts on magnetic resonance cholangiopancreatography, presence of fat inside the cysts on computed tomography, and magnetic resonance imaging in cysts with biliary tree communication. Cyst content in the biliary ducts appears as hyperdense material on computed tomography, and the content is hyperintense on T1W images and appears as filling defect on T2W images on magnetic resonance imaging (Fig. 4.17d). Direct sign is demonstrating the connection between the cyst and biliary ducts itself (Fig. 4.18b) but is difficult to appreciate on imaging due to high intracystic pressure than bile ducts [26, 31]. Hepatocyte-specific contrast agent, which dif-

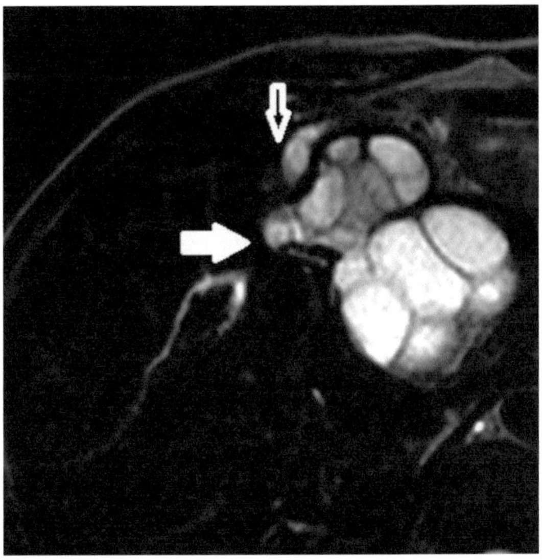

Fig. 4.17 Axial T2W MRI. A CE2 lesion with daughter vesicles is ruptured into biliary tract. Beak sign (solid arrow) is an indirect sign of the rupture and adjacent to the cyst a focal dilated biliary duct is seen (empty arrow). In ERCP, detached membranes were demonstrated in common bile duct as well

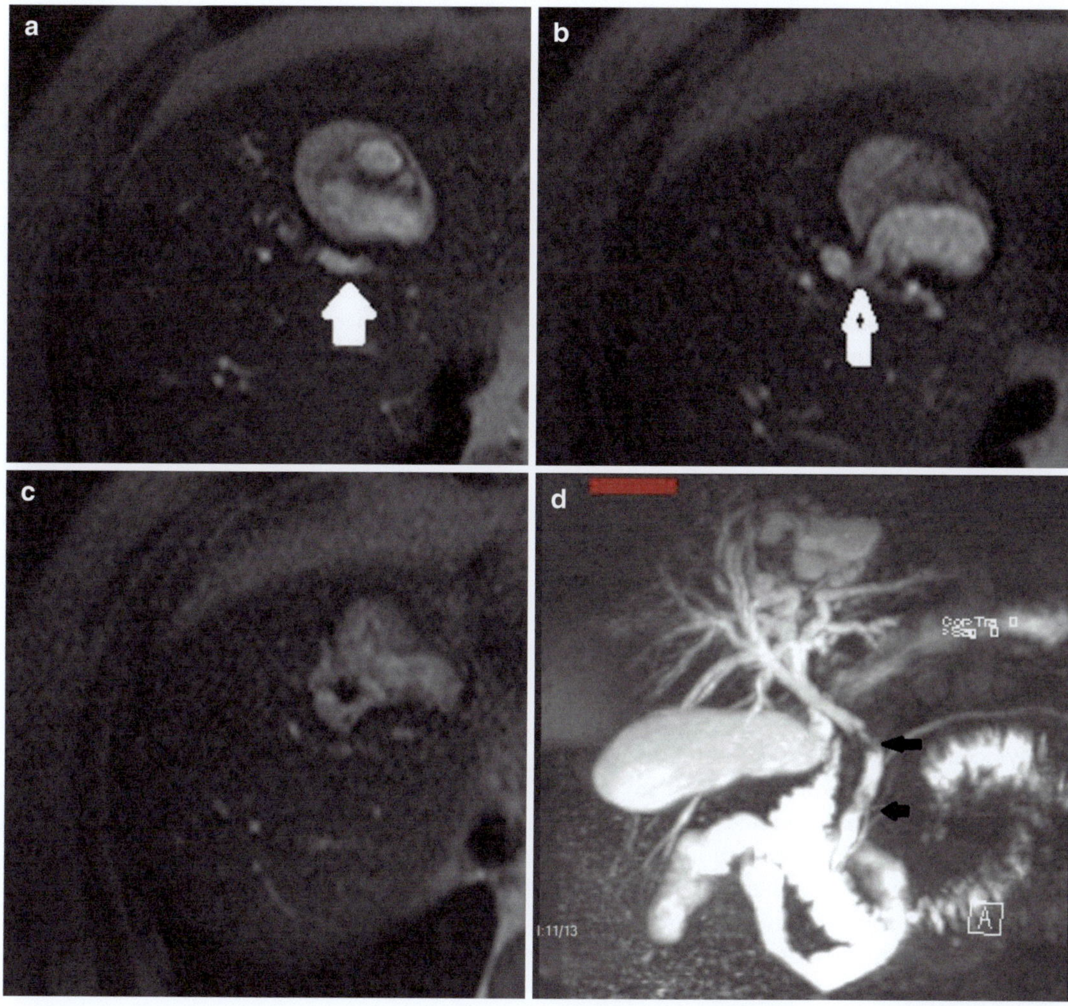

Fig. 4.18 (**a–c**) Axial T2W images from bottom to top demonstrate CE3b lesion. On the most caudal slice. (**a**) Cyst is seen with an adjacent biliary duct which is focally dilated. On the other slices (**b**, **c**) the connection between the cyst and the biliary duct is seen. On (**d**) maximum intensity projection (MIP) slice magnetic resonance cholangiopancreatography filling defects are noted in the common bile duct (arrows) and they were demonstrated to be detached membranes on endoscopic retrograde cholangiopancreatography

fers from the other contrast agents by being excreted from biliary tract along with the kidneys, may be used to demonstrate the connection [32]. Gold standard in diagnosing biliary duct connection is endoscopic retrograde cholangiopancreatography, which has a sensitivity of 86–100% in the diagnosis of intrabiliary rupture owing to its ability to directly demonstrate filling defects of laminated hydatid. However, since endoscopic retrograde cholangiopancreatography carries risk of postoperative biliary fistulae, magnetic resonance imaging should be preferred over endoscopic retrograde cholangiopancreatography for the diagnosis [33].

Direct rupture occurs when both endocyst and pericyst lose their integrity, which causes spillage of the content in peritoneal cavity or sometimes trans-diaphragmatically into pleural cavities. Direct rupture is the most fatal one among all the complications leading to anaphylaxis, hydatid dissemination, and recurrences after surgery [34]. Peripherally located cysts are most prone to rupture. Computed tomography and magnetic resonance imaging are the modalities to detect the extent of hydatosis in these cases. Computed tomography and magnetic resonance imaging demonstrate daughter cysts within the peritoneal cavity. Hydatid cysts may masquerade other cystic masses, but presence of peripheral calcifications, floating membranes, or daughter cysts within cysts are useful clues for differential diagnosis. When rupture occurs into hallow viscera, then this leads to air-fluid levels within the cyst [31]. The identification of parietal calcifications or direct passage of oral contrast from bowel into hepatic hydatid cyst are key imaging features for the diagnosis. When the rupture occurs into bronchial tree, then computed tomography demonstrates pleural effusion, elevation of the diaphragm, lung consolidation, or laminated atelectasis at the lung bases [31].

Exophytic growth occurs via the bare area of the liver with transdiaphragmatic migration to the lung or mediastinum, or via the gastrohepatic ligament into the peritoneal cavity.

Infection affects ruptured hydatid cysts. It mostly occurs in cysts showing intrabiliary rupture [3]. Superinfection may lead to liver abscess, with fever and jaundice being the most common symptoms. Early detection of superinfection is considered mandatory as it may be fatal. Computed tomography findings are similar to those of ultrasonography and are nonspecific since they may be seen in hepatic abscesses due to other causative agents. Computed tomography may demonstrate multiple confluent intrahepatic lesions with poorly defined margins, with hyperdense rim and presence of air and air-fluid level [3].

4.2 Alveolar Echinococcosis

Alveolar echinococcosis is a parasitic disease caused by another cestode in the same genus, called *Echinococcus alveolaris* (*E. alveolaris*) or *Echinococcus multilocularis* (*E. multilocularis*). The final hosts of this parasite are foxes and dogs, but humans can be accidental intermediary hosts by ingesting eggs due to interactions with contaminated pets or by eating contaminated vegetables/fruits [35]. Like cystic echinococcosis, the liver is the most frequently involved organ, followed by the lungs [27].

Despite belonging to the same genus, cystic echinococcosis and alveolar echinococcosis exhibit distinct biological behaviors, leading to variations in disease severity and distinct imaging manifestations. Cystic echinococcosis, unless ruptured, tends to have an expansile growth pattern due to its well-defined laminar structure. Alveolar echinococcosis does not have an encapsulated structure, and as a result exhibits an invasive growth [30], mimicking malignant masses and leading to extensive organ damage.

Ultrasound examination may demonstrate a mass with irregular margins (due to its invasive nature) with heterogenous echogenicity (due to the presence of microcysts, fibrosis, necrosis, and calcifications). Additionally, an encircling hyperechoic fibrotic zone surrounding the lesion may be discerned (Fig. 4.19) [27]. Hemangiomas and invasive tumors may be included in the differential diagnosis, so other cross-sectional modalities may be required for the definitive diagnosis.

Computed tomography is the main modality to characterize the lesions, detect the location and dissemination, and demonstrate the calcifications which are one of the typical features of lesions [36]. It is particularly useful in determining the relations of the lesions with the vascular structures and the biliary tree [37]. Moreover, extrahepatic involvement can be readily shown which is

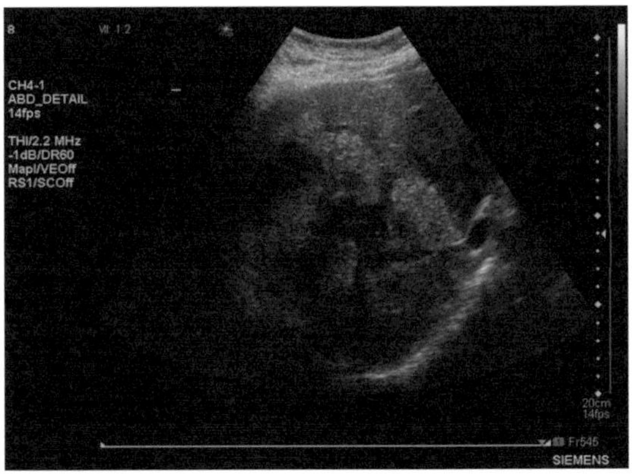

Fig. 4.19 Ultrasonography in alveolar echinococcosis. A heterogenous mass lesion, with mostly hyperechoic peripheral areas and hypoechoic central portion, which mimics a solid liver mass

important to predict stage and therefore resectability of the lesion [38].

On unenhanced computed tomography, alveolar echinococcosis lesions appear as tumor-like masses with irregular borders, heterogeneous internal structures, and multiple distributed calcific foci (Fig. 4.20a). While no significant intralesional enhancement is seen on contrast-enhanced computed tomography, mild enhancement can be seen in the peripheral fibro-inflammatory tissue on delayed phase. Hypodense necrotic areas can sometimes be seen as large cystic cavities within the center of the lesion. Atrophy and capsular retraction secondary to vascular and biliary involvement can be detected in the affected liver lobe [39]. On some occasions the disease may show lobar involvement [40].

In pseudocystic hepatic lesions, computed tomography shows a large necrotic central area surrounded by an irregular ringlike region of fibrous tissue that is often partly calcified. If the parasitic process has infiltrated the hilum, the intrahepatic bile ducts typically appear dilated. Computed tomography appearances of alveolar echinococcosis lesions have been reported to resemble primary hepatic neoplasms such as cholangiocarcinoma, biliary cystadenoma, and biliary cystadenocarcinoma, as well as hepatic metastases [41]. However, the computed tomography findings of hypoattenuation, calcification, and absence of contrast enhancement in a hepatic lesion help identify it as alveolar echinococcosis.

Magnetic resonance imaging can be complimentary to computed tomography in diagnosis of alveolar echinococcosis. Calcifications are not demonstrated well on magnetic resonance imaging (Fig. 4.20b). On the other hand, it is very useful for characterizing different components of the lesion as well as demonstrating its relation with the vascular structures and the biliary tract [39].

Since MRI is very useful in showing extension to neighboring organs, it should be included in the preoperative imaging [42]. The typical MRI finding of alveolar echinococcosis in the liver is a mass lesion with infiltrative characteristics having irregular borders and internal heterogeneity with necrotic areas in the center. These show hypo- or isointensity on T1-weighted images and hypo-, iso-, or hyperintense signal features on T2-weighted images [39] (Fig. 4.21a–e). Hemorrhage and fat tissue within the lesion is not an expected finding. Alveolar echinococcosis lesions are hypovascular masses that develop diffuse areas of necrosis as they grow. The absence of contrast uptake in a large proportion of the mass is an important in differential diagnosis of the lesions. Mild peripheral contrast enhancement is seen on gadolinium-enhanced

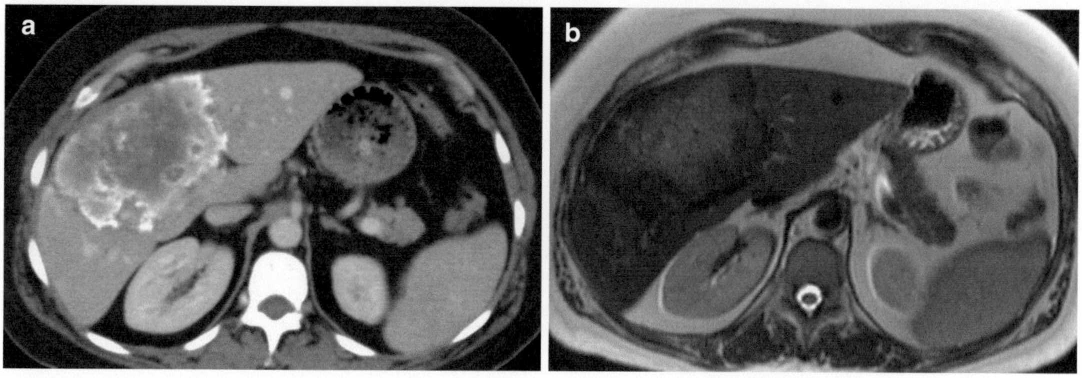

Fig. 4.20 (a) Axial contrast enhanced CT demonstrate an echinococcus alveolaris (EA) lesion appear as tumor-like masses with irregular borders, multiple distributed calcific foci. While no significant intralesional enhancement is seen on contrastenhanced CT, mild enhancement can be seen in the peripheral fibro-inflammatory tissue. (b) Axial T2W MRI image of the same patient demonstrating the lesion, but the calcification at the periphery is not visible although they are clearly shown on CT

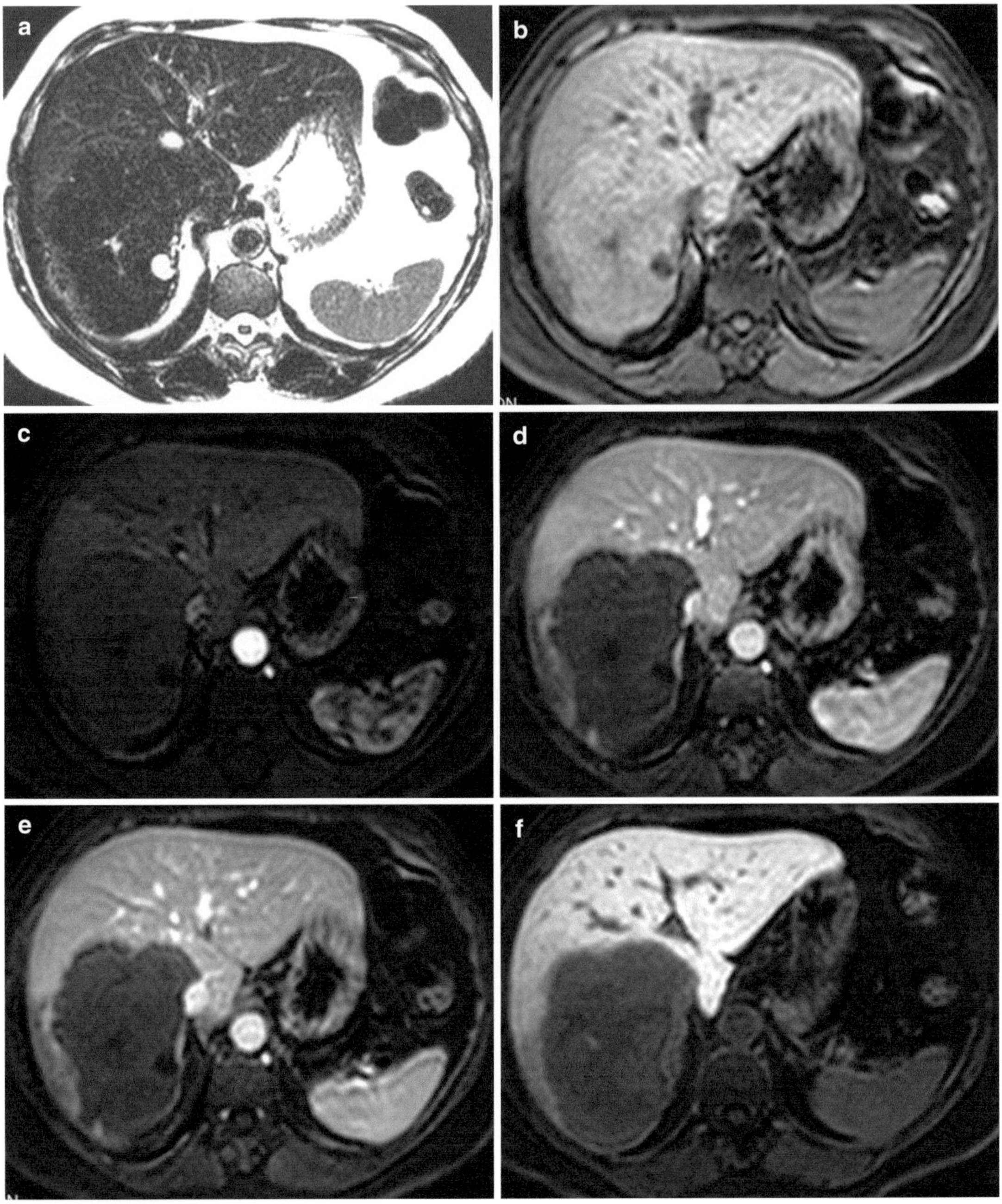

Fig. 4.21 Magnetic resonance imaging of a female patient demonstrating the *Echinococcus alveolaris* lesion. On T2W image (**a**) the lesion is iso-hypointense to normal liver parenchyma. On T1W pre-contrast image (**b**) the lesion is isointense. On dynamic series (**c**–**e**) the lesion does not enhance. The 20-min delay image (**f**) after administration of hepatocyte specific contrast material shows the lesion is hypointense

T1-weighted images [36]. Perilesional fibroinflammatory tissue is considered to be responsible for such contrast enhancement. However, perilesional areas with intense and delayed contrast enhancement can be seen in some patients [42]. Vascular structures can be visualized by dynamic contrast MRI. Invasion and thrombosis of the portal vein can lead to lobar atrophy [43].

Magnetic resonance cholangiopancreatography has replaced percutaneous cholangiography, a more invasive imaging modality that can be complicated with cholangitis and septic shock. Magnetic resonance cholangiopancreatography has an important impact on planning of surgery or interventional palliative procedures by showing biliary tract compression, distortion, and biliary tract invasion by the parasitic mass. In addition, it can show the connection between the biliary tracts and necrotic pseudocystic cavities [39].

References

1. Rinaldi F, Brunetti E, Neumayr A, Maestri M, Goblirsch S, Tamarozzi F. Cystic echinococcosis of the liver: a primer for hepatologists. World J Hepatol. 2014;6(5):293–305. https://doi.org/10.4254/wjh.v6.i5.293. PMID: 24868323 PMCID: PMC4033287.
2. McManus DP, Zhang W, Li J, Bartley PB. Echinococcosis. Lancet. 2003;362(9392):1295–304. https://doi.org/10.1016/S0140-6736(03)14573-4. PMID: 14575976.
3. Pedrosa I, Saíz A, Arrazola J, Ferreirós J, Pedrosa CS. Hydatid disease: radiologic and pathologic features and complications. Radiographics. 2000;20(3):795–817. https://doi.org/10.1148/radiographics.20.3.g00ma06795. PMID: 10835129.
4. Polat P, Kantarci M, Alper F, Suma S, Koruyucu MB, Okur A. Hydatid disease from head to toe. Radiographics. 2003;23(2):475–94; quiz 536–7. https://doi.org/10.1148/rg.232025704. PMID: 12640161.
5. Turgut AT, Akhan O, Bhatt S, Dogra VS. Sonographic spectrum of hydatid disease. Ultrasound Q. 2008;24(1):17–29. https://doi.org/10.1097/RUQ.0b013e318168f0d1. PMID: 18362529.
6. Gharbi HA, Hassine W, Brauner MW, Dupuch K. Ultrasound examination of the hydatic liver. Radiology. 1981;139(2):459–63. https://doi.org/10.1148/radiology.139.2.7220891. PMID: 7220891.
7. Group WHOIW. International classification of ultrasound images in cystic echinococcosis for application in clinical and field epidemiological settings. Acta Trop. 2003;85(2):253–61. https://doi.org/10.1016/s0001-706x(02)00223-1. PMID: 12606104.
8. Ozturk G, Uzun MA, Ozkan OF, Kayaalp C, Tatli F, Eren S, Aksungur N, Coker A, Bostancı EB, Oter V, Kaya E, Tasar P. Turkish HPB Surgery Association consensus report on hepatic cystic echinococcosis (HCE). Turk J Surg. 2022;38(2):101–20. https://doi.org/10.47717/turkjsurg.2022.5757. PMID: 36483170 PMCID: PMC9714645.
9. Rawla P, Sunkara T, Muralidharan P, Raj JP. An updated review of cystic hepatic lesions. Clin Exp Hepatol. 2019;5(1):22–9. https://doi.org/10.5114/ceh.2019.83153. PMID: 30915403 PMCID: PMC6431089.
10. Junghanss T, da Silva AM, Horton J, Chiodini PL, Brunetti E. Clinical management of cystic echinococcosis: state of the art, problems, and perspectives. Am J Trop Med Hyg. 2008;79(3):301–11. PMID: 18784219.
11. Stojkovic M, Rosenberger K, Kauczor HU, Junghanss T, Hosch W. Diagnosing and staging of cystic echinococcosis: how do CT and MRI perform in comparison to ultrasound? PLoS Negl Trop Dis. 2012;6(10):e1880. https://doi.org/10.1371/journal.pntd.0001880. PMID: 23145199 PMCID: PMC3493391.
12. Kalinova K. Imaging (ultrasonography, computed tomography) of patients with hydatid liver disease. Bulg J Vet Med. 2007;10(1):45–51.
13. de Diego CJ, Lecumberri Olaverri FJ, Franquet Casas T, Ostiz ZS. Computed tomography in hepatic echinococcosis. AJR Am J Roentgenol. 1982;139(4):699–702. https://doi.org/10.2214/ajr.139.4.699. PMID: 6981931.
14. Scherer U, Weinzierl M, Sturm R, Schildberg FW, Zrenner M, Lissner J. Computed tomography in hydatid disease of the liver: a report on 13 cases. J Comput Assist Tomogr. 1978;2(5):612–7. https://doi.org/10.1097/00004728-197811000-00018. PMID: 711950.
15. Moro P, Schantz PM. Echinococcosis: a review. Int J Infect Dis. 2009;13(2):125–33. https://doi.org/10.1016/j.ijid.2008.03.037. PMID: 18938096.
16. Pakala T, Molina M, Wu GY. Hepatic echinococcal cysts: a review. J Clin Transl Hepatol. 2016;4(1):39–46. https://doi.org/10.14218/jcth.2015.00036. PMID: 27047771 PMCID: PMC4807142.
17. Rozanes I, Güven K, Acunaş B, Emre A. Cystic echinococcal liver disease: new insights into an old disease and an algorithm for therapy planning. Cardiovasc Intervent Radiol. 2007;30(6):1112–6. https://doi.org/10.1007/s00270-007-9081-y. PMID: 17533534.
18. Beggs I. The radiology of hydatid disease. AJR Am J Roentgenol. 1985;145(3):639–48. https://doi.org/10.2214/ajr.145.3.639. PMID: 3895873.
19. Brunetti E, Tamarozzi F, Macpherson C, Filice C, Piontek MS, Kabaalioglu A, Dong Y, Atkinson N, Richter J, Schreiber-Dietrich D, Dietrich CF. Ultrasound and cystic echinococcosis. Ultrasound Int Open. 2018;4(3):E70–8. https://

doi.org/10.1055/a-0650-3807. PMID: 30364890 PMCID: PMC6199172.
20. Conchedda M, Caddori A, Caredda A, Capra S, Bortoletti G. Degree of calcification and cyst activity in hepatic cystic echinococcosis in humans. Acta Trop. 2018;182:135–43. https://doi.org/10.1016/j.actatropica.2018.02.027. PMID: 29486175.
21. Devi MA, Venumadhav T, Sukanya B, Manmada RT, Gopal P, Rammurti S. Role of imaging in diagnosis, predicting biological activity and in treatment plan of hydatid disease. Open J Intern Med. 2018;8(03):177–95. https://doi.org/10.4236/ojim.2018.83018.
22. Liu W, Jiang Y, Wang J, Zhang T, Xiong X, Li L, Li H. Imaging diagnostic criteria of hepatic echinococcosis in China. Radiol Infect Dis. 2020;7(2):35–42. https://doi.org/10.1016/j.jrid.2020.04.005.
23. Marrone G, Crino F, Caruso S, Mamone G, Carollo V, Milazzo M, Gruttadauria S, Luca A, Gridelli B. Multidisciplinary imaging of liver hydatidosis. World J Gastroenterol. 2012;18(13):1438–47. https://doi.org/10.3748/wjg.v18.i13.1438. PMID: 22509075 PMCID: PMC3319939.
24. Polat P, Atamanalp SS. Hepatic hydatid disease: radiographics findings. Eurasian J Med. 2009;41(1):49–55. PMID: 25610064 PMCID: PMC4261660.
25. Inan N, Arslan A, Akansel G, Anik Y, Sarisoy HT, Ciftci E, Demirci A. Diffusion-weighted imaging in the differential diagnosis of simple and hydatid cysts of the liver. AJR Am J Roentgenol. 2007;189(5):1031–6. https://doi.org/10.2214/ajr.07.2251. PMID: 17954636.
26. Mehta P, Prakash M, Khandelwal N. Radiological manifestations of hydatid disease and its complications. Trop Parasitol. 2016;6(2):103–12. https://doi.org/10.4103/2229-5070.190812. PMID: 27722098 PMCID: PMC5048696.
27. Calame P, Weck M, Busse-Cote A, Brumpt E, Richou C, Turco C, Doussot A, Bresson-Hadni S, Delabrousse E. Role of the radiologist in the diagnosis and management of the two forms of hepatic echinococcosis. Insights Imaging. 2022;13(1):68. https://doi.org/10.1186/s13244-022-01190-y. PMID: 35394226 PMCID: PMC8994011.
28. Marti-Bonmati L, Menor SF. Complications of hepatic hydatid cyst: ultrasound, computed tomography, magnetic resonance diagnosis. Gastrointest Radiol. 1990;15:119–25.
29. Kilic M, Yoldas O, Koc M, Keskek M, Karakose N, Ertan T, Gocmen E, Tez M. Can biliary-cyst communication be predicted before surgery for hepatic hydatid disease: does size matter? Am J Surg. 2008;196(5):732–5. https://doi.org/10.1016/j.amjsurg.2007.07.034. PMID: 18513700.
30. Kern P, Menezes da Silva A, Akhan O, Müllhaupt B, Vizcaychipi KA, Budke C, Vuitton DA. The Echinococcoses: diagnosis, clinical management and burden of disease. Adv Parasitol. 2017;96:259–369. https://doi.org/10.1016/bs.apar.2016.09.006. PMID: 28212790.
31. Greco S, Cannella R, Giambelluca D, Pecoraro G, Battaglia E, Midiri M, Brancatelli G, Vernuccio F. Complications of hepatic echinococcosis: multimodality imaging approach. Insights Imaging. 2019;10(1):113. https://doi.org/10.1186/s13244-019-0805-8. PMID: 31792750 PMCID: PMC6889260.
32. Kulali F, Acar A, Semiz-Oysu A, Canbak T, Tolan K, Bukte Y. Misleading findings of liver-specific MR contrast agent for radiological diagnosis of cystobiliary communication in hydatid cysts. Radiol Med. 2019;124(6):460–6. https://doi.org/10.1007/s11547-019-01000-4. PMID: 30725396.
33. Dolay K, Akbulut S. Role of endoscopic retrograde cholangiopancreatography in the management of hepatic hydatid disease. World J Gastroenterol. 2014;20(41):15253–61. https://doi.org/10.3748/wjg.v20.i41.15253. PMID: 25386073 PMCID: PMC4223258.
34. Alghofaily KA, Saeedan MB, Aljohani IM, Alrasheed M, McWilliams S, Aldosary A, Neimatallah M. Hepatic hydatid disease complications: review of imaging findings and clinical implications. Abdom Radiol (NY). 2017;42(1):199–210. https://doi.org/10.1007/s00261-016-0860-2. PMID: 27487777.
35. Eckert J, Deplazes P. Biological, epidemiological, and clinical aspects of echinococcosis, a zoonosis of increasing concern. Clin Microbiol Rev. 2004;17(1):107–35. https://doi.org/10.1128/CMR.17.1.107-135.2004. PMID: 14726458 PMCID: PMC321468.
36. Kantarci M, Bayraktutan U, Karabulut N, Aydinli B, Ogul H, Yuce I, Calik M, Eren S, Atamanalp SS, Oto A. Alveolar echinococcosis: spectrum of findings at cross-sectional imaging. Radiographics. 2012;32(7):2053–70. https://doi.org/10.1148/rg.327125708. PMID: 23150858.
37. Liu W, Delabrousse E, Blagosklonov O, Wang J, Zeng H, Jiang Y, Wang J, Qin Y, Vuitton DA, Wen H. Innovation in hepatic alveolar echinococcosis imaging: best use of old tools, and necessary evaluation of new ones. Parasite. 2014;21:74. https://doi.org/10.1051/parasite/2014072. PMID: 25531446 PMCID: PMC4273719.
38. Bresson-Hadni S, Vuitton DA, Bartholomot B, Heyd B, Godart D, Meyer JP, Hrusovsky S, Becker MC, Mantion G, Lenys D, Miguet JP. A twenty-year history of alveolar echinococcosis: analysis of a series of 117 patients from eastern France. Eur J Gastroenterol Hepatol. 2000;12(3):327–36. https://doi.org/10.1097/00042737-200012030-00011. PMID: 10750654.
39. Bulakci M, Kartal MG, Yilmaz S, Yilmaz E, Yilmaz R, Sahin D, Asik M, Erol OB. Multimodality imaging in diagnosis and management of alveolar echinococcosis: an update. Diagn Interv Radiol. 2016;22(3):247–56. https://doi.org/10.5152/dir.2015.15456. PMID: 27082120 PMCID: PMC4859741.

40. Karcaaltincaba M, Sirlin CB. CT and MRI of diffuse lobar involvement pattern in liver pathology. Diagn Interv Radiol. 2011;17(4):334–42. https://doi.org/10.4261/1305-3825.DIR.4033-10.0. PMID: 21053176.
41. Didier D, Weiler S, Rohmer P, Lassegue A, Deschamps JP, Vuitton D, Miguet JP, Weill F. Hepatic alveolar echinococcosis: correlative US and CT study. Radiology. 1985;154(1):179–86. https://doi.org/10.1148/radiology.154.1.3880602. PMID: 3880602.
42. Bresson-Hadni S, Delabrousse E, Blagosklonov O, Bartholomot B, Koch S, Miguet JP, Mantion GA, Vuitton DA. Imaging aspects and non-surgical interventional treatment in human alveolar echinococcosis. Parasitol Int. 2006;55(Suppl):S267–72. https://doi.org/10.1016/j.parint.2005.11.053. PMID: 16403670.
43. Balci NC, Tunaci A, Semelka RC, Tunaci M, Ozden I, Rozanes I, Acunas B. Hepatic alveolar echinococcosis: MRI findings. Magn Reson Imaging. 2000;18(5):537–41. https://doi.org/10.1016/s0730-725x(00)00145-4. PMID: 10913715.

Laboratory and Serological Findings of Liver Hydatid Disease

5

Akande Oyebanji Azeez, Aliyu Aminu, Mustapha Ahmed Yusuf, and Muhammad Ibrahim Getso

5.1 Introduction

Hydatid disease, also referred to as hydatidosis or cystic echinococcosis (CE) is a parasitic zoonosis (a disease transmitted from animal source to humans) caused by the cestode parasite (*Tinea echinococcus* or tapeworm), *Echinococcus granulosus* [1–3]. It is a serious and potentially fatal disease that is still endemic in many parts of the world especially in rural communities where rearing of domestic animals is still common [4–6]. Various studies have shown that there are four main species of *Tinea echinococcus* that are associated with infections in humans; these include *E. granulosus* (commonly affecting the liver) and *E. multilocularis* with E. *alveolaris,* known to cause infection in the lungs [7, 8] as well as *E. vogeli* that has been linked with a rare polycyst echinococcosis [9]. In its life cycle, *Tinea echinococcus* is known to require both definitive and intermediate hosts for its survival and propagation. Carnivores such as dogs, wolves, cats, and foxes are believed to be the primary carriers and definitive hosts of the tapeworm [10, 11]. Whereas, the recognized intermediate hosts are mainly herbivorous animals including sheep, goats and cattle as well as horses, deer, and camels depending on geographical locations or endemic areas of the disease [12–14]. Humans, who are regarded as accidental hosts, presumably become infected by ingesting ova (metacestode stage) of *Echinococcus* tapeworms from soil, food, or water contaminated by feces of infected carnivores (particularly dogs) or close contact with them [5, 11, 15].

Clinical evidences indicate that infections caused by *E. granulosus* (cystic echinococcosis) and *E. multilocularis* (alveolar echinococcosis), the two main species of medical and public health importance, often begin and proceed without signs and symptoms [7, 16]. However, the disease becomes manifest later in life (usually

A. O. Azeez (✉)
Department of Medical Microbiology and Parasitology, Bayero University, Kano, Nigeria
e-mail: aoazeez.mcp@buk.edu.ng

A. Aminu
Department of Medical Microbiology and Parasitology, Bayero University, Kano, Nigeria

Aminu Kano Teaching Hospital, Kano, Nigeria

Department of Medical Microbiology and Parasitology, Faculty of Basic Clinical Sciences, Bayero University, Kano, Nigeria
e-mail: aaminu.mcb@buk.edu.ng

M. A. Yusuf
Department of Medical Microbiology and Parasitology, Faculty of Clinical Sciences, Bayero University, Kano, Nigeria

Aminu Kano Teaching Hospital, Kano, Nigeria
e-mail: mayusuf.mcp@buk.edu.ng

M. I. Getso
Department of Medical Microbiology and Parasitology, Faculty of Basic Clinical Sciences, Bayero University, Kano, Nigeria
e-mail: migetso.med@buk.edu.ng

S. Akbulut (ed.), *Hydatid Disease*, https://doi.org/10.1007/978-3-031-97277-5_5

between 10 and 15 years following infection) with attendant complications including anaphylactic shock from ruptured cyst [17–19].

Over the years, it has been observed that the need for laboratory and other diagnosis (especially radiological (imaging) procedures) of hydatid disease was often based on high index of suspicion for the disease and the exigency for surgical operations proposed for infected individuals. In most cases, diagnosis of cystic echinococcosis apart from clinical suspicion has always involved imaging procedures such as plain X-ray, ultrasonography, computed tomography scan, and magnetic resonance imaging, which often reveal the size, shape, and location of the cyst [20–22]. The detailed overview of the radiological diagnosis of hydatid disease has been described in Chap. 4 of this book.

5.1.1 Laboratory Diagnosis of Hydatid Disease

The laboratory diagnosis of cystic hydatid disease is a useful tool employed to compliment the clinical evaluation or data of patient under review. Generally, differential diagnosis of echinococcosis infection mainly depends on the combined evaluation of the following: (i) physical examination of the patient; (ii) history of the disease including previous exposure to risk factors of the disease, that is, history of contact with animals such as sheep or dogs; (iii) life history in endemic areas and travel history; (iv) diagnostic radiologic findings; (v) histopathological examination of biopsy materials or tissues/cysts [23, 24]; and (vi) Casoni skin test. However, this is no longer in use due to associated complications [25, 26]. Serological tests are classified into two main categories as follows:

5.1.1.1 General Serological Tests for Diagnosing Hydatid Disease

These tests are mainly used for the diagnosis of cystic hydatid disease of the liver and lungs. They include:

- Indirect immunofluorescence assay (IFA).
- Direct and indirect hemagglutination assay (DHA and IHA).
- Latex agglutination test (LAT).
- Immunoelectrophoresis (IF) with antigen 5 identification capacity, and which usually confirms about 80%–95% of hepatic (cystic) hydatidosis. The method is also known to confirm about 65% of pulmonary (alveolar) hydatidosis cases [27, 28].

5.1.1.2 Special Serological Techniques

However, in cases of other suspected anatomical localizations of hydatidosis outside the liver and lungs, or when cysts are calcified, the following serological techniques, in most cases, are recommended for use. They include:

- Enzyme-linked immunosorbent assay (ELISA)
- Western blot
- Polymerase chain reaction (PCR) [29–33]

5.1.2 Serological Diagnosis of Hydatid Disease

5.1.2.1 Enzyme-Linked Immunosorbent Assay (ELISA)

The use of serological (blood) tests such as ELISA for the detection of parasite-specific IgG antibody in serum or plasma of infected individuals has been recognized as a useful supplemental test (and in many cases as confirmatory test) to the clinical and radiologic diagnosis of hydatid disease of the liver and lungs. The serological differentiation of *E. granulosus* (CE) and *E. multilocularis* (AE) has been made possible due to the introduction of species-specific antigens that can detect corresponding specie-specific antibodies produced against them by the hosts [5]. More so, the anti-echinococcus ELISA (IgG) can similarly detect both parasite species antigens and thus can serve as a screening test for hydatid disease [28, 34].

Aside the detection of IgG antibodies (which suggests current or past infection with *Echinococcus* parasites) produced by infected hosts against the tapeworms, the ELISA kits or assays are also designed to detect markers as expressed by these parasitic helminthes. More so, it has been suggested that IgE antibodies can be employed as a useful marker or measurement of therapeutic outcome (i.e., success or failure) in hydatic patients with pre-treatment specific IgE antibodies [15].

Interpretation of Echinococcus IgG ELISA Test

Positive IgG ELISA Test

This suggests current or past infection with *Echinococcus* parasites. However, false-positive result may occur due to cross-reaction with other helminthes that share similar antigenic components with *Echinococcus* species. It may also occur in patients with chronic immune disorders or suffering from carcinomas [15, 28]. Therefore, it is often recommended that such results should be considered alongside clinical findings, and results obtained from imaging examinations as well as history of exposure by the patient and histopathological evaluation [35, 36].

In addition, since sero-conversion (indicating rise in antibody titer) between acute and convalescent or following surgical removal of hydatid cyst samples appears to provide strong evidence of infection or re-infection, hence it is advised that testing of a patient at both stages should be performed at the same laboratory to aid reliable conclusive decision by the care givers [15, 28].

Negative IgG ELISA Test

This indicates no significant level of *Echinococcus* IgG antibody detected in the serum of the patient. In this situation, and in order to rule out false negative result, it is usually advised that testing be repeated between 10 and 14 days so that reliable conclusion can also be made on the diagnosis [22].

5.1.2.2 Serological Techniques for Diagnosing Hydatid Disease in Conventional Laboratory Setting

1. **Enzyme-Linked Immunosorbent Assay (ELISA)**: Which includes Emc- and Em2-ELISA for IgG against antigen (Ag) 5 or Ag B. This is known to produce the best results for cystic echinococcosis diagnosis.
2. **Immunoelectrophoresis (IE):** Especially if arc 5 is present (because it involves precipitation for antigen 5). In certain cases, the disease was confirmed by a high antibody titer to hydatid antigen using counter immunoelectrophoresis with over 90% [26] and was considered as a reliable and sensitive diagnostic tool for diagnosing hydatid disease [22].
3. **Indirect Hemagglutination (IHA) Test:** Although IHA has a good sensitivity, false positives have been observed to regularly occur during testing. Other useful serological methods for diagnosing hydatid disease include:
   (a) Bentonite-flocculation test
   (b) Radio-immunoassay
   (c) Compliment-fixation test (CFT)

Note: It should be noted that hemagglutination and flocculation tests have been shown to be more sensitive than the complement-fixation test. However, a combination of hemagglutination and bentonite-flocculation tests appears to give high-level sensitivity result. On the other hand, and depending on the test method adopted, as well as other parameters that are needed to be considered, about 10% of all patients with liver hydatid cysts do not usually produce detectable serum IgG antibodies and thus may present false-negative results [28].

5.1.2.3 Serological Techniques for Field Diagnosis of Hydatid Disease

The following serological techniques are found to be useful for field diagnosis of hydatid disease. These include:

1. Coagglutination test to detect the circulatory antigen in the host
2. Latex agglutination technique
3. Hydatid antigen dot immunobinding assay (Immunoblots) [15, 28, 37]

However, it should be noted that hitherto, there is no single standardized sensitive and specific serologic test for detection of antibodies against *Tinea echinococcus* infection [1, 34]. Therefore, healthcare givers handling cases of hydatidosis need to adopt the laboratory tests that are considered sensitive enough for detection of the disease and are readily available in their environment to compliment clinical and radiological analysis so as to enhance management of the disease and good prognosis. Histopathological procedures are also strongly recommended for confirmatory purposes [36].

5.1.3 Serological Cross-Reactions

It has been observed that one important shortcoming of serological testing for cystic or alveolar echinococcosis (like many viral diseases) is the issue of serological cross-reactivity that may occur during testing. Several reports have shown that there are common antigens in hydatid fluid that may be concurrently circulating in the blood of infected hosts which are found in several helminth species, any of which may be infecting the host at the same time (including *Tinea saginata, Schitosoma mansoni, Fasciola hepatica, Ascaris hunbricoides*, and many others) or cysticerosis. This is because hydatid antigens share common components with these helminthes antibodies and thus can produce cross-reactions with them when an individual is infected by any of these helminths [14, 22, 33].

5.1.4 Remedy for *Serological* Cross-Reactions

Although serological testing has a broad range of sensitivity but specificity, and in most cases is dependent on the purity of the antigen utilized [14]. Therefore, in order to improve the specificity of the serological tests employed for diagnosing hydatid disease of the liver or lung, the use of dialyzed hydatid fluid antigen or antigenic fractions obtained from other sources such as carbohydrate and protein antigen have been suggested. This is predicated on the belief that dialyzed antigens lose many nonspecific components which may interfere in the test or cause cross-reactions. Hence, its use is more preferred to crude whole antigen as it is the case with hydatid antigen. More so, a combination of two serological reactions (e.g., hemagglutination and flocculation tests) is thought to increase the specificity (>97%) and reliability of the diagnostic tests for cystic echinococcosis [15].

5.1.5 Sensitivity of Serological Tests Used for the Diagnosis of Hydatid Disease

Generally, imaging techniques—such as X-ray, computed tomography scan, as well as magnetic resonance imaging—can be used to visually confirm the parasitic cyst formation in an infected host, while serological tests are employed to confirm such parasitic infection [20, 28, 36, 38]. Notably, however, diagnostic reports have shown that there are no major differences in the serological and laboratory diagnosis of alveolar echinococcosis and cystic hydatidosis as both diagnoses are based on the same principles and procedures guiding imaging and serological tests [27, 28, 39].

However, various reports from different centers around the world indicate that sensitivity of serological tests use for hydatidosis is high (80%–100%) especially for liver cysts, but low (50%–56%) for lung or alveolar cysts, and about 25%–56% for other organs of the human body [33]. On the other hand, many investigators have emphasized that the importance of these tests lies mainly in the follow-up of treated patients especially following surgery. This is because, an increase in serological titer results when observed during follow-up of patients after surgical operation, indicating recurrence of disease. Whereas, a decrease in titer is an indication of disease resolution [40–42]. It is believed that these results will guide the care givers on appropriate decision to be taken on such patients that will enhance good prognosis of the condition.

Studies on serological tests and their usage for diagnosing hydatid disease revealed that latex agglutination test (LAT), indirect hemagglutination assay (IHA), indirect immunosorbent assay (ELISA), Western blotting, and polymerase chain reaction (PCR) are among the most commonly used serological tests; whereas ELISA, IHA, and PCR are known to provide the best results for recurrence of the disease during the postoperative period. Although high sensitivity, specificity, and efficacy can be achieved with ELISA, it has been reported that in practice the most frequently used test is IHA, because it is known to be cheaper and easier to use [8, 43].

However, studies have shown that in some cases of Hydatid disease, low level of Echinococcus IgG antibody positivity could be observed particularly when the cystst remains intact. On the other hand, ruptured cyst in a host was found to be the only factor that significantly increased the Echinococcus IgG seropositivityy [27]. Furthermore, detectable immune responses have been severally associated with the physical location, integrity, and vitality of larval cysts [27]. For instance, a cyst in the liver presumably is more likely to cause antibody response than cysts in the lungs, brain, or spleen. On the other hand, cysts that are located in the bone reportedly tend to invoke detectable antibodies, more regularly. Moreover, the integrity of the cysts appears to, in addition, affect the stimulation of immune (antibody) response in infected host. Whereas calcified, dead, or intact hyaline cysts are known to generally cause less antibody response in individuals suffering from hydatidosis [18, 22].

Above all, aside histopathological procedures, imaging techniques such as ultrasound, computed tomography scan, and magnetic resonance imaging are generally believed to still constitute more sensitive diagnostic methods for hydatidosis than serological modes of diagnosis [33, 44].

5.1.6 Other Laboratory Tests Used for Diagnosing Hydatidosis

The other laboratory tests used for diagnosing hydatid disease are as follows:

5.1.6.1 Casoin Skin Test

This is a hypersensitivity-based skin reactive test occasionally employed for diagnosing echinococcosis. It was once used as a major diagnostic tool for the detection of hydatid disease in infected individuals. However, the test is no longer in use because of its low sensitivity and well-known allergic complications in several cases where it was adopted for use [25, 26]. The test has been replaced with more modern and more specific and safer methods such as serological and molecular methods [28].

5.1.6.2 Fine-Needle Aspiration

Fine-needle aspiration cytology has been proposed as a useful diagnostic method for hydatidosis because it is believed that it does not cause any complications as in Casoin skin test [20, 45].

5.1.6.3 Hematological and Biochemical Tests

Some other laboratory tests such as hematological/biochemical liver tests are considered unsuitable for diagnosing hydatidosis due to their low sensitivity. In certain studies for instance, hematological tests presented a low sensitivity with eosinophilia found to be present only in about 35–40% of the patients. Similarly, liver function test results were found to be abnormal in only about 40–45% of the infected patients, whereas elevation of serum alkaline phosphatase was common among the patients [22, 26].

5.1.7 Histopathological Evaluation of Hydatid Cyst

The hydatid cyst formation is believed to always start as a fluid-filled, cyst-like structure (referred to as Type 1 lesion), which may proceed to become a Type II lesion, that is, if daughter cysts and/or matrix develop [46, 47]. These investigators also reported that in some instances, the Type II lesion becomes hypermature and perhaps due to starvation dies to form a mummified, inert calcified Type III lesion. Furthermore, hydatid cysts have been described as slow-growing fluctuant, painless masses; known to resemble any slow-growing benign tumor in the body [33, 48]. Although the diagnosis of hydatid cyst is often based on radiologic imaging and serological tests, it is, however, generally recommended that a definitive diagnosis using histopathological procedures should also be adopted during evaluation of the disease [23, 36, 49]. Accordingly, various studies have shown that histopathological evaluation often shows three layers of hydatid cysts as follows:

Layer One: This is recognized as the innermost germinal layer, which appears thin and translucent on gross. It is believed that the embryonic tapeworm's scolices develop from an out pouching of the germinal layer and form hydatid sand that settles into the dependent parts of the cysts. The cyst thus appears crystal clear, as it is thought to be a transudate of serum that contains proteins known to be antigenic.

Layer Two: The section is designated as the middle laminated membrane that usually appears white. It is about 2 mm thick and found to be selectively permeable to nutrients but not to bacterial cells.

Layer Three: This outer layer is also referred to as pericyst. It is described as a rigid protective layer with a few millimeters in thickness and represents the host's response to the presence of parasite.

It should be noted that the presence of scolices (an accellular, thick lamellar cyst wall), and the surrounding host reaction, which is composed of the inflammatory fibrous tissue, that forms a dense pseudo-capsule around the cyst, often confirms the diagnosis of hydatid cyst [33, 46, 47].

5.1.8 Polymerase Chain Reaction (PCR)

The use of polymerase chain reaction (PCR) for the detection of nucleic acids of *E. granulosus* (CE) and *E. multilocularis* (AE) is being recognized as a reliable confirmation to hydatid disease. Although the adoption of this modern technology is not yet widespread, it is hoped that many laboratories and healthcare centers (especially in hydatid disease endemic areas) will, in no distant future, have access to this highly sensitive, specific, and reliable testing technique that will ultimately improve laboratory diagnosis, management, and prognosis of hydatid disease.

5.1.9 Importance of Diagnostic Tests for Hydatid Disease in Different Organs of the Body

Hydatid disease is a well-recognized parasitic infection associated with *Echinococcus granulosus* (cystic echinococcosis) and *E. multilocularis* (alveolar echinococcosis). It is found to frequently occur in the liver and lungs respectively. The disease is believed to be chronic and the associated cyst formation can localize in different organs including some rare locations of the body [8, 33, 50–54]. Hence, Sultana et al. [33] were of the opinion that high index suspicion was required to diagnose the disease especially in such rare locations.

Various reports have shown that although hydatid cyst develops most frequently in the liver (65%) and the lungs (≥25%), about 10% of the remaining cases may be found in other locations such as the spleen, kidneys, muscle, brain, bones including cranium, heart, eye, and pancreas [8, 33, 40, 55–57]. However, soft tissue hydatid disease is believed to represent less than 3% of all hydatid disease cases [58]. Other extremely rare cases of the body where hydatid cyst has also been detected include head and neck (occipitocervical hydatid cyst) [33, 40, 47, 56, 58] including pre-auricular area [53] as well as submandibular and intracranial sites [42]. Consequently, formation of hydatid cyst in any of these locations can result in serious complications including cyst rupture if not detected and completely treated early [7].

Therefore, Sultana et al. [33] had recommended that hydatid cyst diagnostic tests should also be considered in other organs where hydatid cyst occurrence is presumed to be rare when there is high level of suspicion. The suggestion was necessitated due to increasing detection of hydatidosis as earlier stated, in areas considered to be extremely rare for its development and also due to the possibility of benign swellings of echinococcosis in those areas [16].

In view of the aforementioned, the knowledge of hydatid disease epidemiology and its anatomical distribution can help in the general control and prevention of the disease. In addition, early detection of hydatid disease in any location (including unusual site) can lead to its adequate management via appropriate medication and surgery, which may prevent fatal anaphylaxis when carefully handled [33, 59].

References

1. Botezatu C, Mastalier B, Patrascu T. Hepatic hydatid cyst—diagnose and treatment algorithm. J Med Life. 2018;11(3):203–9. https://doi.org/10.25122/jml-2018-0045. PMID: 30364592 PMCID: PMC6197524.
2. Cemil B, Tun K, Gurcay AG, Uygur A, Kaptanoglu E. Cranial epidural hydatid cysts: clinical report and review of the literature. Acta Neurochir. 2009;151(6):659–62. https://doi.org/10.1007/s00701-009-0276-7. PMID: 19319472.
3. Yagmur Y, Akbulut S. Epidemiology of hydatid disease. Turkiye Klinikleri J Gen Surg-Special Topics. 2010;3(2):6–8.
4. Darabi M, Varedi P, Mohebi AR, Mahmoodi S, Varedi P, Nabavizadeh SA, Erfan A, Ostadali Makhmalbaf A, Saedi D, Saadat Mostafavi SR, Mousavi SM. Hydatid cyst of the parotid gland. Oral Maxillofac Surg. 2009;13(1):33–5. https://doi.org/10.1007/s10006-008-0138-0. PMID: 18979123.
5. Eckert J, Deplazes P. Biological, epidemiological, and clinical aspects of echinococcosis, a zoonosis of increasing concern. Clin Microbiol Rev. 2004;17(1):107–35. https://doi.org/10.1128/cmr.17.1.107-135.2004. PMID: 14726458 PMCID: PMC321468.
6. Himsworth CG, Jenkins E, Hill JE, Nsungu M, Ndao M, Andrew Thompson RC, Covacin C, Ash A, Wagner BA, McConnell A, Leighton FA, Skinner S. Emergence of sylvatic Echinococcus granulosus as a parasitic zoonosis of public health concern in an indigenous community in Canada. Am J Trop Med Hyg. 2010;82(4):643–5. https://doi.org/10.4269/ajtmh.2010.09-0686. PMID: 20348513 PMCID: PMC2844548.
7. Gomez IGC, López-Andújar R, Belda Ibáñez T, Ramia Ángel JM, Moya Herraiz Á, Orbis Castellanos F, Pareja Ibars E, San Juan Rodríguez F. Review of the treatment of liver hydatid cysts. World J Gastroenterol. 2015;21(1):124–31. https://doi.org/10.3748/wjg.v21.i1.124. PMID: 25574085 PMCID: PMC4284328.
8. Yagmur Y, Akbulut S. Unusual location of hydatid cysts: a case report and literature review. Int Surg. 2012;97(1):23–6. https://doi.org/10.9738/cc85.1. PMID: 23101997 PMCID: PMC3723195.
9. Knapp J, Chirica M, Simonnet C, Grenouillet F, Bart JM, Sako Y, Itoh S, Nakao M, Ito A, Millon L. Echinococcus vogeli infection in a hunter, French Guiana. Emerg Infect Dis. 2009;15(12):2029–31. https://doi.org/10.3201/eid1512.090940. PMID: 19961693 PMCID: PMC3044547.
10. Bonelli P, Masu G, Dei Giudici S, Pintus D, Peruzzu A, Piseddu T, Santucciu C, Cossu A, Demurtas N, Masala G. Cystic echinococcosis in a domestic cat (Felis catus) in Italy. Parasite. 2018;25:25. https://doi.org/10.1051/parasite/2018027. PMID: 29727269 PMCID: PMC5935470.
11. Sréter T, Széll Z, Egyed Z, Varga I. Echinococcus multilocularis: an emerging pathogen in Hungary and Central Eastern Europe? Emerg Infect Dis. 2003;9(3):384–6. https://doi.org/10.3201/eid0903.020320. PMID: 12643838 PMCID: PMC2958538.
12. Celik C, Sasmaz M, Ucan H. Spinal hydatid cyst. Türkiye Klinikleri tıp Bilimleri Dergisi. 2010;30(3):1073–7.
13. Dziri C. Hydatid disease—continuing serious public health problem: introduction. World J Surg. 2001;25(1):1–3. https://doi.org/10.1007/s002680020000. PMID: 11213146.
14. Shaw J, Bornman P, Krige J. Hydatid disease of the liver. S Afr J Surg. 2006;44(2):70–7.
15. Gottstein B, Wang J, Blagosklonov O, Grenouillet F, Millon L, Vuitton DA, Müller N. Echinococcus metacestode: in search of viability markers. Parasite. 2014;21:63. https://doi.org/10.1051/parasite/2014063. PMID: 25429386 PMCID: PMC4245873.
16. Tsaroucha AK, Polychronidis AC, Lyrantzopoulos N, Pitiakoudis MS, Karayiannakis AJ, Manolas KJ, Simopoulos CE. Hydatid disease of the abdomen and other locations. World J Surg. 2005;29(9):1161–5.

https://doi.org/10.1007/s00268-005-7775-3. PMID: 16088421.
17. Kayaalp C, Bostanci B, Yol S, Akoglu M. Distribution of hydatid cysts into the liver with reference to cystobiliary communications and cavity-related complications. Am J Surg. 2003;185(2):175–9. https://doi.org/10.1016/s0002-9610(02)01202-3. PMID: 12559452.
18. Sözüer EM, Ok E, Arslan M. The perforation problem in hydatid disease. Am J Trop Med Hyg. 2002;66(5):575–7. https://doi.org/10.4269/ajtmh.2002.66.575. PMID: 12201593.
19. Yahya AI, Shwereif HE, Ekheil MA, Thoboot AS, Algader KA, Gyaed FO, Aldarat AS. The role of emergency surgery in hydatid liver disease. World J Emerg Surg. 2014;9:12. https://doi.org/10.1186/1749-7922-9-12. PMID: 24479989 PMCID: PMC4012255.
20. Hadad GM, Abdel Salam RA, Emara S. Validated and optimized high-performance liquid chromatographic determination of tizoxanide, the main active metabolite of nitazoxanide in human urine, plasma and breast milk. J Chromatogr Sci. 2012;50(6):509–15. https://doi.org/10.1093/chromsci/bms041. PMID: 22525879.
21. Kumar MJ, Toe K, Banerjee RD. Hydatid cyst of liver. Postgrad Med J. 2003;79(928):113–4. https://doi.org/10.1136/pmj.79.928.113. PMID: 12612332 PMCID: PMC1742618.
22. Niścigorska J, Sluzar T, Marczewska M, Karpińska E, Boroń-Kaczmarska A, Morańska I, Daniel B. Parasitic cysts of the liver—practical approach to diagnosis and differentiation. Med Sci Monit. 2001;7(4):737–41. PMID: 11433203.
23. Bal N, Kocer NE, Arpaci R, Ezer A, Kayaselcuk F. Uncommon locations of hydatid cyst. Saudi Med J. 2008;29(7):1004–8. PMID: 18626530.
24. Losanoff JE, Richman BW, Jones JW. Primary hydatid cyst of the axilla. ANZ J Surg. 2004;74(5):393–4. https://doi.org/10.1111/j.1445-1433.2004.03002.x. PMID: 15144273.
25. Doty JE, Tompkins RK. Management of cystic disease of the liver. Surg Clin North Am. 1989;69(2):285–95. https://doi.org/10.1016/s0039-6109(16)44786-9. PMID: 2648617.
26. Safioleas MC, Misiakos EP, Kouvaraki M, Stamatakos MK, Manti CP, Felekouras ES. Hydatid disease of the liver: a continuing surgical problem. Arch Surg. 2006;141(11):1101–8. https://doi.org/10.1001/archsurg.141.11.1101. PMID: 17116803.
27. Aydin M, Adiyaman G, Dogruman-Al F, Kustimur S, Ozkan S. [Determination of anti-echinococcus IgG antibodies by ELISA in patients with suspected hydatid cyst]. Turkiye Parazitol Derg. 2012;36(2):61–4. https://doi.org/10.5152/tpd.2012.16. PMID: 22801907.
28. Biava MF, Dao A, Fortier B. Laboratory diagnosis of cystic hydatic disease. World J Surg. 2001;25(1):10–4. https://doi.org/10.1007/s002680020002. PMID: 11213147.
29. Ito A, Nakao M, Kutsumi H, Lightowlers MW, Itoh M, Sato S. Serodiagnosis of alveolar hydatid disease by western blotting. Trans R Soc Trop Med Hyg. 1993;87(2):170–2. https://doi.org/10.1016/0035-9203(93)90476-7. PMID: 8337719.
30. Ito A, Schantz PM, Wilson JF. Em18, a new serodiagnostic marker for differentiation of active and inactive cases of alveolar hydatid disease. Am J Trop Med Hyg. 1995;52(1):41–4. https://doi.org/10.4269/ajtmh.1995.52.41. PMID: 7531957.
31. Kern P, Frosch P, Helbig M, Wechsler JG, Usadel S, Beckh K, Kunz R, Lucius R, Frosch M. Diagnosis of Echinococcus multilocularis infection by reverse-transcription polymerase chain reaction. Gastroenterology. 1995;109(2):596–600. https://doi.org/10.1016/0016-5085(95)90350-x. PMID: 7542219.
32. Sako Y, Nakao M, Nakaya K, Yamasaki H, Gottstein B, Lightowers MW, Schantz PM, Ito A. Alveolar echinococcosis: characterization of diagnostic antigen Em18 and serological evaluation of recombinant Em18. J Clin Microbiol. 2002;40(8):2760–5. https://doi.org/10.1128/jcm.40.8.2760-2765.2002. PMID: 12149326 PMCID: PMC120647.
33. Sultana N, Hashim TK, Jan SY, Khan Z, Malik T, Shah W. Primary cervical hydatid cyst: a rare occurrence. Diagn Pathol. 2012;7:157. https://doi.org/10.1186/1746-1596-7-157. PMID: 23157817 PMCID: PMC3523054.
34. Moro P, Schantz PM. Echinococcosis: a review. Int J Infect Dis. 2009;13(2):125–33. https://doi.org/10.1016/j.ijid.2008.03.037. PMID: 18938096.
35. Langer JC, Rose DB, Keystone JS, Taylor BR, Langer B. Diagnosis and management of hydatid disease of the liver. A 15-year North American experience. Ann Surg. 1984;199(4):412–7. https://doi.org/10.1097/00000658-198404000-00007. PMID: 6712316 PMCID: PMC1353359.
36. Pedrosa I, Saíz A, Arrazola J, Ferreirós J, Pedrosa CS. Hydatid disease: radiologic and pathologic features and complications. Radiographics. 2000;20(3):795–817. https://doi.org/10.1148/radiographics.20.3.g00ma06795. PMID: 10835129.
37. Wen H, Vuitton L, Tuxun T, Li J, Vuitton DA, Zhang W, McManus DP. Echinococcosis: advances in the 21st century. Clin Microbiol Rev. 2019;32(2) https://doi.org/10.1128/cmr.00075-18. PMID: 30760475 PMCID: PMC6431127.
38. Bakhtiar NM, Akbarzadeh A, Casulli A, Mahami-Oskouei M, Ahmadpour E, Nami S, Rostami A, Spotin A. Therapeutic efficacy of nanocompounds in the treatment of cystic and alveolar echinococcoses: challenges and future prospects. Parasitol Res. 2019;118(9):2455–66. https://doi.org/10.1007/s00436-019-06416-5. PMID: 31402401.
39. Eris FN, Akisu C, Aksoy U. Evaluation of two ELISA and two indirect hemagglutination tests for serodiagnosis of pulmonary hydatid disease. Korean J

Parasitol. 2009;47(4):427–9. https://doi.org/10.3347/kjp.2009.47.4.427. PMID: 19967097 PMCID: PMC2788728.

40. Eroğlu A, Atabekoğlu S, Kocaoğlu H. Primary hydatid cyst of the neck. Eur Arch Otorrinolaringol. 1999;256(4):202–4. https://doi.org/10.1007/s004050050140. PMID: 10337512.
41. Fujimoto Y, Ito A, Ishikawa Y, Inoue M, Suzuki Y, Ohhira M, Ohtake T, Kohgo Y. Usefulness of recombinant Em18-ELISA to evaluate efficacy of treatment in patients with alveolar echinococcosis. J Gastroenterol. 2005;40(4):426–31. https://doi.org/10.1007/s00535-004-1559-7. PMID: 15870979.
42. Guney O, Ozturk K, Kocaogullar Y, Eser O, Acar O. Submandibular and intracranial hydatid cyst in an adolescent. Laryngoscope. 2002;112(10):1857–60. https://doi.org/10.1097/00005537-200210000-00029. PMID: 12368630.
43. Beyhan YE, Babür C, Mungan M, Özkan AT. [Evaluation of cystic echinococcosis suspected patients applied to National Parasitology Reference Laboratory of Public Health Institution of Turkey between 2009-2013]. Turkiye Parazitol Derg. 2015;39(1):17–21. https://doi.org/10.5152/tpd.2015.3646. PMID: 25917579.
44. Wuestenberg J, Gruener B, Oeztuerk S, Mason RA, Haenle MM, Graeter T, Akinli AS, Kern P, Kratzer W. Diagnostics in cystic echinococcosis: serology versus ultrasonography. Turk J Gastroenterol. 2014;25(4):398–404. https://doi.org/10.5152/tjg.2014.7112. PMID: 25254522.
45. Bulut N, Dagıstanlı S. A case of ruptured pulmonary hydatid cyst of the liver and review of the literature. Case Rep Radiol. 2017;2017:7639056. https://doi.org/10.1155/2017/7639056. PMID: 29098107 PMCID: PMC5642873.
46. Izci Y, Tüzün Y, Seçer HI, Gönül E. Cerebral hydatid cysts: technique and pitfalls of surgical management. Neurosurg Focus. 2008;24(6):E15. https://doi.org/10.3171/foc/2008/24/6/e15. PMID: 18518745.
47. Katilmiş H, Oztürkcan S, Ozdemir I, Adadan Güvenç I, Ozturan S. Primary hydatid cyst of the neck. Am J Otolaryngol. 2007;28(3):205–7. https://doi.org/10.1016/j.amjoto.2006.08.007. PMID: 17499142.
48. Giebultowicz J, Polanska-Plachta M, Wroczynski P, Zaborowski P, Polanski JA. How echinococcosis affects potential cancer markers in plasma: galectin-3, sN-cadherin and sE-cadherin? A preliminary report. Diagn Pathol. 2012;7:17. https://doi.org/10.1186/1746-1596-7-17. PMID: 22340429 PMCID: PMC3355024.
49. Kemp C, Roberts A. Infectious diseases: echinococcosis (hydatid disease). J Am Acad Nurse Pract. 2001;13(8):346–7. https://doi.org/10.1111/j.1745-7599.2001.tb00047.x. PMID: 11930567.
50. Akbulut S, Senol A, Ekin A, Bakir S, Bayan K, Dursun M. Primary retroperitoneal hydatid cyst: report of 2 cases and review of 41 published cases. Int Surg. 2010;95(3):189–96. PMID: 21066995.
51. Polat P, Kantarci M, Alper F, Suma S, Koruyucu MB, Okur A. Hydatid disease from head to toe. Radiographics. 2003;23(2):475–94; quiz 536–7. https://doi.org/10.1148/rg.232025704. PMID: 12640161.
52. Rochidi Y, Raji A, Elhattab Y. A rare localization of hydatidosis: a cervical hydatid cyst. Fr ORL. 2007;92:315–7.
53. Tekin M, Osma U, Yaldiz M, Topcu I. Preauricular hydatid cyst: an unusual location for echinoccosis. Eur Arch Otorrinolaringol. 2004;261(2):87–9. https://doi.org/10.1007/s00405-003-0650-7. PMID: 12856191.
54. Versaci A, Scuderi G, Rosato A, Angiò LG, Oliva G, Sfuncia G, Saladino E, Macrì A. Rare localizations of echinococcosis: personal experience. ANZ J Surg. 2005;75(11):986–91. https://doi.org/10.1111/j.1445--2197.2005.03588.x. PMID: 16336394.
55. Aleksic-Shihabi A, Vidolin EP. Cystic echinococcosis of the heart and brain: a case report. Acta Med Okayama. 2008;62(5):341–4. https://doi.org/10.18926/amo/30968. PMID: 18985095.
56. Celik A, Turanli M, Kutun S, Delibasi T, Mengi N, Comert E, Aslan S, Altundag K, Cetin A. Unusual location of hydatid cyst: soft tissue mass in the neck. Eur Arch Otorrinolaringol. 2006;263(12):1147–50. https://doi.org/10.1007/s00405-006-0122-y. PMID: 16858578.
57. Kireşi DA, Karabacakoğlu A, Odev K, Karaköse S. Uncommon locations of hydatid cysts. Acta Radiol. 2003;44(6):622–36. https://doi.org/10.1080/02841850312331287749. PMID: 14616207.
58. Cangiotti L, Muiesan P, Begni A, de Cesare V, Pouchè A, Giulini SM, Tiberio G. Unusual localizations of hydatid disease: a 18 year experience. G Chir. 1994;15(3):83–6. PMID: 8060784.
59. Akbulut S. Parietal complication of the hydatid disease: comprehensive literature review. Medicine (Baltimore). 2018;97(21):e10671. https://doi.org/10.1097/md.0000000000010671. PMID: 29794743 PMCID: PMC6392988.

Medical Treatment of Hydatid Disease

6

Veysel Umman and Sami Akbulut

6.1 Introduction

The *Echinococcus* genus of parasites is the source of hydatid disease, a zoonotic infection. It is found all across the world and is indigenous to areas where cattle and sheep are raised. The main method of transmission is drinking water or eating food tainted with dog excrement. Alveolar echinococcosis, caused by *Echinococcus multilocularis*, and cystic echinococcosis, caused by *Echinococcus granulosus*, are the two main kinds of this parasite infection. In humans, the infection causes cysts to grow in the liver and lungs, though other organs may also be affected. Untreated cysts can grow and cause a variety of health issues, including organ failure and, in extreme circumstances, possibly fatal consequences.

Surgery has historically been used to treat hydatid illness, with consideration for the size, location, and stage of the cyst. A variety of pharmacological treatments as well as percutaneous therapies, such as PAIR (puncture, aspiration, injection, and re-aspiration), have nevertheless been made possible by ongoing research and developments. This chapter examines the many medical therapies that are currently available, their single application or integration with PAIR, and their application before surgery.

6.2 Epidemiology of the Disease

Hydatid disease is a global public health issue that affects millions of individuals in several regions, including the Middle East, Africa, Turkey, India, South America, Southern and Central Europe, China, Australia, and New Zealand [1, 2]. Since people and animals often interact closely in rural areas, there is a high risk of transmission there. According to a recent study, the disease is spreading into metropolitan areas along with a rise in urbanization [3].

Disease prevalence is determined by objective factors such as appropriate animal hosts, human-animal contact, and environmental circumstances conducive to egg survival. The World Health Organization (WHO) recognizes hydatid disease as a serious tropical disease that affects people worldwide and deems cystic echinococcosis and alveolar echinococcosis neglected diseases because of their prevalence in underserved areas [4].

Liver involvement is the most frequent issue in cystic echinococcosis, occurring in around 75% of instances. Hydatid cysts may also poten-

V. Umman (✉)
Department of Surgery, Division of Organ Transplantation and HPB Surgery, Yeditepe University Faculty of Medicine, Istanbul, Türkiye

S. Akbulut
Department of Surgery and Liver Transplantation Institute, Inonu University Faculty of Medicine, Malatya, Türkiye

Department Biostatistics and Medical Informatics, Inonu University Faculty of Medicine, Malatya, Türkiye

S. Akbulut (ed.), *Hydatid Disease*, https://doi.org/10.1007/978-3-031-97277-5_6

tially develop in the lungs and many other organs, such as the pancreas, spleen, and heart, brain, bone, and soft tissue, although this occurrence is rare and mainly found in endemic regions [2, 5–9]. In sub-clinic cases with diagnosis at advanced stage, severe complications such as cystocutaneous fistula, subcutaneous rupture, cystosubcutaneous abscess, cysto-cutaneo-bronchial fistula, and cysto-cutaneo-bronchio-biliary fistula can be observed [10]. Due to the disease's high incidence and ability to seriously harm the liver and other organs, it is imperative to conduct cutting-edge research, closely monitor patients with tight follow-up routines, administer prolonged treatments, and move promptly to modify treatment plans. Healthcare professionals, particularly those specializing in hepatopancreaticobiliary care, need to be kept abreast of the epidemiology, pathophysiology, and appropriate interventions for echinococcosis.

6.3 Symptoms and Clinical Diagnosis

The two primary forms of echinococcosis are cystic and alveolar. While the alveolar type is more invasive and mostly affects the liver, the cystic form mostly affects the lungs and liver. Regarding viability and host involvement, *E. granulosus* and *E. multilocularis* can last extended periods in intermediate hosts by employing an evasive tactic to the host's immune response. The location of the involved organ, its size, the site of the cyst within the organ, the rupture of the cyst, and impact of the cyst on the affected organ all influence the symptoms that appear. While the majority of people exhibit no symptoms at all, some may experience discomfort, notice a lump, or experience issues with the affected organ's functionality due to the growth of the cyst, which may result in a local inflammatory response with fibrosis around the cyst. Alternatively, they may be systemic inflammatory reactions such as asthma or anaphylaxis.

Hydatid disease is diagnosed using clinical signs, imaging methods, serologic testing, and analysis of aspirated hydatid fluid. Ultrasonography is the preferred initial diagnostic method due to its excellent sensitivity and specificity in detecting hepatic hydatid cysts, as well as its cost-effectiveness and safety. Therefore, in highly endemic areas, ultrasonography may be preferred as the first diagnostic method before serologic testing. Identifying distinctive ultrasonographic features, such as hydatid sand, daughter cysts, and a double-layered wall, can significantly improve diagnostic accuracy. Magnetic resonance imaging and computed tomography are secondary methods commonly reserved for detecting hydatid disease with inconclusive ultrasound findings. As up to 60% of patients may be asymptomatic, incidental and early detection may also be possible with the wider use of computed tomography and magnetic resonance imaging for evaluation of other medical conditions [11]. In addition, clinical signs of complications may include cholangitis, pancreatitis, bile duct rupture, damage to surrounding organs, peritoneal or pleural space injury, cyst rupture, and systemic or local inflammatory response [12]. However, a definitive diagnosis typically requires histopathologic examination. A precise diagnosis depends on the viability and stage of the hydatid cyst, which can affect its contents. By employing morphological, histological, and molecular techniques to directly assess the cyst content retrieved through percutaneous puncture or surgery, the diagnosis can also be verified [13].

6.4 Significance of Medical Treatment

Although surgery has the potential to eliminate all cysts and result in a complete cure, the medical approach is used to treat hydatid disease, particularly in patients for whom surgery is inappropriate or as adjuvant therapy before and after surgery. In patients with multiple cysts situated in different organs, medical treatment should be the initial treatment. The optimal treatment plan is typically personalized and based on factors such as the cyst's size and location, complications, previous medication response, and the advanced disease stage. It has been shown that small cysts with a size of less than 5 cm in the liver and lung respond favorably to benzimidazole treatment alone [14].

Pharmacological therapy employs the use of benzimidazoles—mainly albendazole and mebendazole, which are commonly used as preoperative medications. It has been shown that they can reduce intracystic pressure and soften the cyst, making surgical removal easier. It is also essential to have medical treatment before and after PAIR treatment.

For cystic echinococcosis, there is no "optimal" course of treatment, and no clinical study has examined all available approaches, especially the efficacy of drug therapy versus surgery, which has not been compared in a controlled trial. Therefore, the use of medical therapy alone, in conjunction with surgery, or PAIR requires clinical decision-making depending on the specifics of each patient. While the Gharbi classification was developed earlier and may still be preferred by radiologists, the WHO has outlined a more relevant grouping of cysts into three categories: active (CE1 and CE2), transitional (CE3), and inactive (CE4 and CE5). This new classification better delineates the characteristics of cystic lesions. To standardize treatment, the WHO-Informal Working Group on Echinococcosis (WHO-IWGE) recommended a stage-specific approach for treating liver cystic echinococcosis in their 1996 consensus report. According to this consensus, especially for active CE1 and CE2 cysts, which have a size less than 5 cm, albendazole can be used alone. Whereas for cysts greater than 5 cm, PAIR and albendazole can be used together. After surgery or PAIR, benzimidazoles should be administered to prevent recurrence [13, 15, 16].

The recommendation for the use of albendazole alone can be in line with the studies reporting the effectivity of drug therapy is inversely proportional to the diameter of the cyst. Although the majority of these cysts can become inactive after a couple of years of drug treatment, medical therapy alone has been reported to have a high recurrence rate, with cysts becoming activated in 25–50% of cases during 2-year follow-ups [12, 17–20].

The efficacy of the drug therapy decreases further as the stage progresses from early CE1 stage to CE2 and transitional CE3 cysts. Regarding how long medical treatment for cystic echinococcosis should last, opinions differ. After a few months of treatment, ultrasonography exams are used to assess the effectiveness of the therapy. The therapy is deemed successful if the ultrasound scan reveals indications of membrane detachment (C3a) or conversion to an inactive cyst (C4). It is important to remember that cysts in the dormant CE4 stage can revert to the transitional stage and therefore close follow-up and controls are necessary.

6.5 Agents Used in Medical Treatment and Their Effects

6.5.1 Mebendazole

Mebendazole was the first drug to be used in the benzimidazole group and has a comparable effect to albendazole; however, it is often used as a secondary option due to its lower absorption.

Mechanism of Action Mebendazole binds and prevents the development of microtubules and induces their depolymerization, which specifically and permanently prevents the uptake of glucose and other nutrients, consequently interrupting mobilization and reproduction, and resulting in the death of the parasite [21]. Furthermore, a reduction in cyst mass was observed in animal trials after administering 25 mg/kg of mebendazole.

Dosage and Administration For a period of three to 6 months, the recommended oral dosage of mebendazole is 40 to 50 mg per kg per day. Mebendazole has a half-life of around 3–6 h and is rapidly processed in the liver and eliminated through bile and urine. When the medication is taken with a fat-rich meal as opposed to an empty stomach, the rate of absorption is considerably higher. It is advised to begin mebendazole preoperatively 4 days before surgery and to continue medication for 3 months [15].

Adverse Effects and Management Mebendazole side effects most frequently include appetite loss, headache, tinnitus, diarrhea, flatulence, nausea, and vomiting, but elevated liver enzymes are the most important to monitor [22, 23].

For patients who have received higher doses or longer durations of treatment, neutropenia and thrombocytopenia should also be monitored [24].

6.5.1.1 Efficacy and Clinical Study Data

Individual differences in serum levels are very common, and they are unrelated to oral dosage. In an animal study, it was shown that all parasites died when protoscoleces were incubated with mebendazole at higher concentrations (5 and 10 μg/ml), and mebendazole showed dose-dependent efficacy against germinal cells [25].

To modify drug dosages, maintaining patient compliance, and preventing harmful side effects is crucial. The WHO recommends monitoring serum drug levels after 1 month of chemotherapy and every 3 months thereafter [26].

In an effort to improve the solubility and bioavailability of mebendazole, it was suspended in oil and its potential to enhance the cysticidal activity of the drug against secondary cysts of *Echinococcus granulosus* in mice was investigated. It was observed that oral preparations given to mice not only increased mebendazole's bioavailability but also greatly strengthened their anti-hydatid cystic properties [27].

6.5.2 Albendazole

The mainstay of treatment for both cystic and alveolar echinococcosis is albendazole; long-term intervention is especially important for cystic echinococcosis.

Mechanism of Action Unlike mebendazole, which is broken down into metabolites and does not have high activity, albendazole is rapidly metabolized to albendazole sulfoxide. This is a potent prodrug that selectively inhibits cytoplasmic microtubules and is effective on the germinative membrane. The course of action is similar, as depolymerization of microtubules inhibits glycogen and nutrient uptake, causing energy depletion and disruption of metabolic pathways leading to parasite death.

Dosage and Administration To maximize the bioavailability of albendazole, it is recommended to take it with food, preferably a fatty meal. The dosage is 10–15 mg/kg/day, split into two doses, with a maximum of 400 mg taken orally twice daily with food due to its poor absorption. Treatment is administered in multiple 1-month courses spaced 14 days apart. Typically, treatment lasts for 3–6 months. Preoperative chemotherapy with albendazole should commence at least 4 days before surgery and continue for 1 month [15, 28].

Adverse Effects and Management Even though side effects are brief and minimal, it is imperative to regularly monitor liver function tests because albendazole has the potential to cause hepatotoxicity. Long-term therapies should be closely monitored for symptoms such as nausea, headaches, temporary hair loss, or bone marrow depression [29, 30]. During the follow-up, it is particularly important to monitor hemogram parameters and liver function tests. Due to its excretion via bile, it should be used with caution in patients with biliary obstruction.

Efficacy and Clinical Study Data It was shown that intracytoplasmic concentrations of albendazole were more significant than systemic levels. Furthermore, it was discovered that there was no discernible variation between the albendazole intracystic and blood levels; hence, a greater dosage did not necessarily translate into higher intracystic levels [31, 32].

Long-term exposure to drugs in the germinative layer improves the efficacy of the treatment, even if the length of medication administration does not increase intracystic drug concentration [33]. Following spontaneous or traumatic cyst rupture, possibly during surgery, albendazole inhibits the development of subsequent echinococcosis.

Although the combination of albendazole with PAIR is suggested by WHO and well-documented in many studies [15, 16], a new study suggests an alternative to the instillation of toxic topic scolicides for the treatment of medium-sized unilocular CE1 cysts. For cysts with unilocular echinococcal cysts, percutaneous drainage coupled with oral praziquantel under albendazole treatment is suggested as a safe and less intrusive alternative to PAIR [34].

It is unclear how long albendazole should be taken by people not receiving surgical therapy. However, there are instances where albendazole has been taken consistently for up to 20 years without any reported side effects [13].

A meta-analysis of 33 studies on the therapeutic management of human cystic echinococcus found that the use of benzimidazole medications before or after surgery, or in combination with PAIR or surgery, resulted in superior treatment outcomes. Albendazole chemotherapy was identified as the main pharmacological intervention for medical care. However, the combined use of albendazole and praziquantel showed greater effectiveness in reducing cysts and killing scolices, and was more likely to result in a cure or improvement compared to using albendazole alone [35].

Although albendazole has curative potential, its poor solubility and absorption create a quest for combination therapies to increase its efficacy. As a result, a lot of research focuses on how to enhance the dissolution and bioavailability of albendazole by repositioning drugs in conjunction with nanotechnology. In the realm of nanotechnology-based formulations, albendazole has been subject to more research than other medications. To enhance treatment effectiveness and reduce negative impacts, researchers are investigating the use of nanomaterials, including metal nanoparticles, nanopolymers, solid lipid nanoparticles, and nanocrystals, for the development of antiparasitic medications [36]. A study investigated creating chewable tablets containing self-microemulsfying drug delivery system. It was shown that the systemic bioavailability was 1.3 times higher compared to commercial tablets [37]. Benzamidazole pharmacokinetic effects can be enhanced by the use of nanoparticle enantiomers of albendazole and nanostructured drug delivery devices. Higher death rates against protoscoleces and microcysts are demonstrated by these nanocompounds, along with enhanced scolicidal and cysticidal actions. Future studies may employ electromagnetically generated nanodrugs and recently formed entities to create more precise formulations [38, 39].

6.5.3 Praziquantel

It is commonly employed to treat alveolar echinococcosis. Typically, it is administered alongside albendazole or mebendazole. Although praziquantel has demonstrated substantial effectiveness in certain studies, its use should be exercised with caution due to potential side effects.

Mechanism of Action Praziquantel is a synthetic isoquinoline-pyrazine derivative that works by making cell membranes more susceptible to calcium, hence impairing ion transport in both cestodes and nematodes. Additionally, it is believed to bind to the actin and myosin light chains, changing the fluidity of their membranes and lowering the glutathione concentration [40, 41].

Dosage and Administration Due to its protoscolicidal effect, praziquentel can be used to treat cystic echinococcosis either on its own or in conjunction with albendazole. It has been shown that combination treatment with albendazole and praziquantel is more effective than albendazole alone [42]. When administered in conjunction with benzimidazoles, a once-weekly dose of 40 mg/kg of the isoquinoline derivative praziquantel is recommended.

Adverse Effects and Management Praziquantel rarely causes negative side effects with mild and temporary adverse affects such as abdominal pain, diarrhea, vomiting, headache, and dizziness [41]. Toxicities are dose-dependent and can be minimized by taking the medication with food.

Efficacy and Clinical Study Data Praziquantel is the most often paired medication with albendazole, and it has been discovered that praziquantel considerably raises the blood level of albendazole. It has not been demonstrated that this combination raises intracystic concentrations, though. As such, there is limited data to support the therapeutic effectiveness of combining these two medications [33, 42].

A retrospective evaluation of 70 patients compared two groups: group A received a combination chemotherapy consisting of albendazole and praziquantel following the surgical excision of cysts, while group B received chemotherapy alone; no surgery was performed. There was no significant difference in treatment outcomes or shrinkage or extended calcification of cysts between groups [43].

A more recent study looked at the side effects of 57 hydatidosis patients who received combination therapy with praziquantel and albendazole. Albendazole monotherapy and combination therapy have not been compared in a previous randomized study. According to the study, the side effects of combination praziquantel and albendazole therapy are minor, and reversible when the medication is stopped. Therefore, it was concluded that using this combination treatment is both safe and practicable [44].

6.5.4 Mefloquine

Mefloquine, a synthetic analog of quinine, is utilized for the treatment and prevention of malaria. It was reported to have in vitro activity against *E. multilocularis* as well [25]. Mefloquine side effects might range from mild nausea, vomiting, and upset stomach to more serious ones including neurological and psychiatric problems, and therefore it is not advised for those with a history of psychological disorders [45].

Mefloquine administered in vitro to *E. multilocularis* metacestodes caused the metacestode tissue to collapse and the cellular germinal layer to quickly separate from the acellular laminated layer. The medication demonstrated parasiticidal action when injected into animals after being treated in vitro with parasites [46].

The effects of mebendazole, mefloquine, and nitazoxanide on *E. granulosus* protoscoleces, germinal cells, and mice infected with secondary hydatid cysts were examined in both in vivo and in vitro settings. It was found that mebendazole and mefloquine were active against the protoscoleces and germinal cell [25].

In another animal study, mice infected via oral uptake of eggs followed by 12 weeks of treatment were later examined for the presence of parasite lesions in a liver squeeze chamber and by PCR. It was seen that parasite load was significantly reduced in mefloquine-treated animals [47].

In addition, researchers are using drug repurposing strategies and investigating new medications that target energy metabolism, and mefloquine has been demonstrated to be effective against *E. multilocularis* metacestodes in both in vitro screening and in vivo testing. The drug target *E. multilocularis* ferritin, which binds mefloquine, was found, and oral therapy inhibited the development of the parasite in mice. For people with benzamidazole toxicity or resistance problems, mefloquine could be an option, albeit using it might have negative neurological effects. Additional research on its mechanism of action and cyst penetration may yield more information [48].

6.5.5 Amphotericin B

Benzimidazoles require a longer course of treatment and carry a risk of reactivation since they are parasitostatic rather than parasiticidal for *E. multilocularis* [49]. Combination treatments may help reduce treatment time and improve efficacy. Amphotericin B is commonly employed as a part of a combination treatment plan. Nonetheless, an in vitro investigation shows that it is possible to apply albendazole and amphotericin in that order, and that albendazole can set the stage for amphotericin's accelerated vesicle destruction [50]. When given simultaneously, there was a delay in the vesicle destruction, and therefore sequential therapy instead of combination therapy was recommended.

Although the exact mechanism of action is yet unknown, the most significant one seems to be the selective and irreversible binding to sterols in cell membranes, which creates transmembrane channels [50]. Additionally, it has been demonstrated that amphotericin forms stable complexes with membrane phospholipids, altering membrane fluidity, and affecting anion transport as well as membrane-bound enzymes [51–53].

Individuals with alveolar echinococcosis often require lifelong benzimidazole therapy. However, amphotericin B has shown effectiveness in reducing the disease's progression. In three cases with extensive disease, it was shown that the medication successfully halted the parasite's development, suggesting it could be a viable salvage treatment option for those resistant to or sensitive to benzimidazole [54]. However, its long-term use is limited due to nephrotoxicity, hypokalemia, and the need for intravenous administration.

6.5.6 Nitazoxanide

Nitazoxanide was not as successful in early tests with sheep and mice as it was in in vitro trials [55–57]. The objective of a later study by Mai et al. [58] was to examine the effects of six medications, alone or in combination, on *Echinococcus multilocularis* metacestodes in vitro [58]. Among the medications utilized were rifampin, miltefosine, nitazoxanide, albendazole, itraconazole, and ivermectin. Following a 5-week cultivation period, the *Echinococcus multilocularis* metacestodes were split into 17 groups, each containing 120–140 cysts. Six weeks, 3 months, and 6 months following the end of the medication were used to monitor the effects. Injecting cysts into female BALB/c mice and monitoring their vitality 8 weeks later allowed researchers to assess the survivability of metacestodes. The findings demonstrated that in the in vitro culture system, albendazole, itraconazole, nitazoxanide, and albendazole plus nitazoxanide all suppressed or killed the parasites ($P < 0.05$). When it came to eliminating parasites, a high dose of nitazoxanide (10 microg/ml) worked better than others. In the four groups receiving combined therapy with nitazoxanide and albendazole, cysts did not develop again after the drugs were stopped for 3 weeks, 3 months, or 6 months. In the groups receiving combination therapy, there was no sign of cyst regrowth.

A report of seven cases assessed the efficacy of nitazoxanide as a treatment for individuals with disseminated cystic echinococcosis who did not respond to antiparasitic and surgical interventions. In this study, five patients had bone involvement. Three patients who received albendazole and nitazoxanide for 3–24 months exhibited good improvement. It was concluded that the long-term safety profile was favorable [59].

Later a study investigated the effects of nitazoxanide and tizoxanide on *E. granulosus* in vivo and in vitro. As nitazoxanide is rapidly and completely converted to tizoxanide after oral administration [60], both substances were investigated in this study and the significant protoscolicidal effects of tizoxanide and nitazoxanide were validated [25]. However, neither medication significantly slowed the development of cysts or germinal cells in mice. Since nitazoxanide did not affect the in vivo formation of hydatid cysts, it was concluded that it was ineffective on germinal cells, but only affected the protoscoleces. Pharmacological properties and side effect profiles of the antihelminthic drugs mentioned above are given in Table 6.1.

6.6 Other Experimental Drugs

Pharmaceutical companies are often hesitant to participate in the development of new medications due to the low return on investment and the uncommon nature of this disease. The possibility of drug repurposing or exploring existing pharmaceuticals for potential therapeutic efficacy in new applications is an interesting treatment option for several illnesses which is also tried for echinococcus infections [61].

Similar to mefloquine, another antimalarial drug used for this purpose is atovaquone. A study evaluated the effectiveness of combination treatment with atovaquone and albendazole on *E.*

Table 6.1 Pharmacological features and side effect profiles of the drugs used for Echinococcus infection

Drug name	Drug class	Effecting organism	Mechanism of action	Dosage	Effectivity	Adverse effects	Combination therapy	Research and remarks
Albendazole	Benzimidazoles	*Echinococcus multilocularis* *E. granulosus*	Via albendazole sulfoxide causes depolymerization of microtubules, inhibiting glycogen and nutrient uptake	10–15 mg/kg/day (400 mg) taken orally twice daily × 2–3 years with 14 days break every 1 month cycle	The most effective and primarily used drug	Hepatotoxicity, nausea, headaches, temporary hair loss, bone marrow depression	Mostly combined drug and used with PAIR, Praziquantel, Ivermectin, Nitazoxanide, Mefloquine	Nano formulations to increase bioavailability
Mebandazole	Antihelmintic; Benzimidazoles	*E. multilocularis* *E. granulosus*	Causes depolymerization of microtubules, inhibiting glycogen and nutrient uptake	50–70 mg/kg/day × 6–24 months	Fairly good	Appetite loss, headache, tinnitus, diarrhea, flatulence, nausea, vomiting, and, but elevated liver enzymes is the most important. Neutropenia and thrombocytopenia at high doses	–	Less effective than albendazole.
Praziquantel	Antihelmintic; Isoquinolines	*E. multilocularis* *E. granulosus*	Not certain, disrupts calcium ion homeostasis	20–75 mg/kg/day	Slight improvement in combination with albendazole	Abdominal pain, diarrhea, vomiting, headache, and dizziness	Albendazole	Increases efficacy of albendazole
Mefloquine	Antimalarial; quinine	*E. multilocularis*	Not certain, disrupts energy metabolism, binds ferritin inhibiting transport	Studied in mice	Slight improvement	Neurotoxicity, elevation of liver enzyme levels	Albendazole	Can be an option for people with benzamidazole toxicity

Amphotericin B	Antifungal; polyene	*E. multilocularis*	Not certain, selective and irreversible binding to sterols in cell membranes.	0.5 mg/kg of body	Fairly good	Nausea, vomiting, fever, irregular heart beats, hypertension or hypotension, hypoxia, nephrotoxicity	–	Salvage treatment for alveolar echinococcosis patients with intolerance or resistance to benzimidazoles
Nitazoxanide	Antiprotozoal; thiazolides	*E. multilocularis*, *E. granulosus*	Not certain, via tizoxanid inhibits enzyme-dependent anaerobic energy metabolism	500 mg/12 h	Fairly good	Abdominal pain, diarrhea, headache, nausea	Albendazole	Well tolerated, long term safe

multilocularis in both in vivo studies and culture. The length of time it took for the parasites to be eliminated under both aerobic and anaerobic circumstances was measured after treating protoscoleces with 50 μM albendazole and/or atovaquone. In the in vivo study, the combined therapy lasted significantly less time in the livers of mice infected orally with parasite eggs compared to atovaquone alone under aerobic conditions. When albendazole was used to treat alveolar echinococcus in mice, atovaquone enhanced its effects [62].

The high prices, long-term medication use, and high recurrence rates suggest the need for more effective treatment alternatives. In an experimental setting, fenbendazole therapy for *Echinococcus multilocularis* infected mice showed comparable effectiveness to albendazole treatment. Transmission electron microscopy revealed drug-induced changes in the parasites' germinal layer. These findings suggest the potential use of fenbendazole in alveolar echinococcus chemotherapy [63].

Another study assessed the combination of ivermectin and flubendazole's in vitro effectiveness against *Echinococcus granulosus* protoscoleces and metacestodes. The maximal protoscolicidal action was demonstrated by the combination. Hydatid cysts were eliminated faster with the combination of flubendazole and ivermectin than with either medication alone. The treated cysts' germinal layer experienced degenerative alterations and lost its multicellular structure, as demonstrated by ultrastructural examinations. The combination had a positive effect on *E. granulosus*, as demonstrated by the results [64].

Like malignancies, hepatic alveolar echinococcosis spreads by budding or penetrating into tissues and persistently stimulating angiogenesis. There have been suggestions that some signal transduction protein kinase molecules may be connected to *Echinococcus*'s capacity for unrestricted proliferation [61]. Bortezomib, a proteasome inhibitor, has shown high anti-metacestodal activity in *E. multilocularis* metacestodes. Treatment led to an accumulation of ubiquinated proteins and parasite death. Bortezomib treatment reduced parasite weight in experimentally infected mice but induced adverse effects like diarrhea and neurological symptoms. The proteasome was identified as a drug target in *E. multilocularis* [65].

Anticancer medication sorafenib inhibits the development of tumor cells and angiogenesis by targeting tyrosine kinases, including platelet-derived growth factor receptors and intravascular growth factor receptors (VEGFR-2 and VEGFR-3) [61]. In multilocular echinococcosis, Xiao et al. [66] investigated the effectiveness of various doses of the anti-cancer medication sorafenib using animal models. As a possible medication for the treatment of echinococcosis, sorafenib has shown a substantial effect on cyst inhibition, according to the findings [66].

Imatinib, a tyrosine kinase inhibitor, was discovered to be effective in eliminating the *Echinococcus protoscolex* in vitro. At doses as low as 10 m, imatinib was also shown to effectively block the production of proto tapeworm vesicles by parasitic stem cells. Consequently, in anti-echinococcosis chemotherapy, imatinib may also prove to be a viable substitute for albendazole [67].

A study assessed the impact of resveratrol on hydatid cysts and protoscolices of *Echinococcus granolosus*. Viability, gamma-glutamyl transpeptidase, and caspase-3 activity all significantly varied when exposed to different doses of resveratrol in vitro. It was discovered that resveratrol t reated protoscolices showed loss of viability, increased gamma-glutamyl transpeptidase, and caspase-3 activity with significant differences compared to all control groups. Still, concentration had no apparent effect on tissue, and it was safe without significantly inducing apoptosis [68]. More clinical studies are required to establish its dose and effectivity.

6.7 Conclusion

Medications are often used in conjunction with surgery or alternative treatments to effectively treat hydatid disease. Close monitoring of the patient is essential for successful therapy. New and, most importantly, better options are being intensively investigated as alternatives to the benzimidazoles currently employed in the chemotherapeutic treatment of hydatid disease. Drug repurposing has been utilized to investigate antimalarials as potential treatments for generally neglected disorders. Additionally, protease inhibitors and nanoparticles have shown promise in increasing albendazole concentrations. Further research is necessary to enhance our understanding of the efficacy of current medications, establish a standardized and uniform treatment approach, and identify new medical therapy options.

References

1. Budke CM, Deplazes P, Torgerson PR. Global socioeconomic impact of cystic echinococcosis. Emerg Infect Dis. 2006;12(2):296–303. https://doi.org/10.3201/eid1202.050499. PMID: 16494758 PMCID: PMC3373106.
2. Kankilic N, Aydin MS, Günendi T, Göz M. Unusual hydatid cysts: cardiac and pelvic-Ilio femoral hydatid cyst case reports and literature review. Braz J Cardiovasc Surg. 2020;35(4):565–72. https://doi.org/10.21470/1678-9741-2019-0153. PMID: 32865381 PMCID: PMC7454634.
3. Ghatee MA, Nikaein K, Taylor WR, Karamian M, Alidadi H, Kanannejad Z, Sehatpour F, Zarei F, Pouladfar G. Environmental, climatic and host population risk factors of human cystic echinococcosis in southwest of Iran. BMC Public Health. 2020;20(1):1611. https://doi.org/10.1186/s12889-020-09638-w. PMID: 33109152 PMCID: PMC7590804.
4. Torgerson PR, Devleesschauwer B, Praet N, Speybroeck N, Willingham AL, Kasuga F, Rokni MB, Zhou XN, Fèvre EM, Sripa B, Gargouri N, Fürst T, Budke CM, Carabin H, Kirk MD, Angulo FJ, Havelaar A, de Silva N. World Health Organization estimates of the global and regional disease burden of 11 foodborne parasitic diseases, 2010: a data synthesis. PLoS Med. 2015;12(12):e1001920. https://doi.org/10.1371/journal.pmed.1001920. PMID: 26633705 PMCID: PMC4668834.
5. Akbulut S, Yavuz R, Sogutcu N, Kaya B, Hatipoglu S, Senol A, Demircan F. Hydatid cyst of the pancreas: report of an undiagnosed case of pancreatic hydatid cyst and brief literature review. World J Gastrointest Surg. 2014;6(10):190–200. https://doi.org/10.4240/wjgs.v6.i10.190. PMID: 25346801 PMCID: PMC4208043
6. García-Díez AI, Ros Mendoza LH, Villacampa VM, Cózar M, Fuertes MI. MRI evaluation of soft tissue hydatid disease. Eur Radiol. 2000;10(3):462–6. https://doi.org/10.1007/s003300050077. PMID: 10756997.
7. Ok E, Sözüer EM. Solitary subcutaneous hydatid cyst: a case report. Am J Trop Med Hyg. 2000;62(5):583–4. https://doi.org/10.4269/ajtmh.2000.62.583. PMID: 11289668.
8. Yaghan RJ. Hydatid disease of the breast: a case report and literature review. Am J Trop Med Hyg. 1999;61(5):714–5. https://doi.org/10.4269/ajtmh.1999.61.714. PMID: 10586899.
9. Yörükoğlu Y, Zengin M, Dolgun A, Nazliel K, Salman E, Paşaoğlu E, Yucel E. Primary muscular hydatid cyst causing arterial insufficiency: case report and literature review. Angiology. 1993;44(5):399–401. https://doi.org/10.1177/000331979304400509. PMID: 8480918.
10. Akbulut S. Parietal complication of the hydatid disease: comprehensive literature review. Medicine (Baltimore). 2018;97(21):e10671. https://doi.org/10.1097/md.0000000000010671. PMID: 29794743 PMCID: PMC6392988.
11. Pawłowski Z, Eckert J, Vuitton D, Ammann R, Kern P, Craig P, Dar K, Rosa F, Filice C, Gottstein B, Grimm F, Macpherson C, Sato N, Todorov T, Uchino J, Wv S, Wen H. Echinococcosis in humans: clinical aspects, diagnosis and treatment. In: Eckert J, Gemmell M, Meslin F, Pawłowski Z, editors. WHO/OIE manual on echinococcosis in humans and animals: a public health problem of global concern. Paris: Office International des Epizooties; 2001. p. 20–66.
12. Öztürk G, Uzun MA, Özkan ÖF, Kayaalp C, Tatlı F, Eren S, Aksungur N, Çoker A, Bostancı EB, Öter V, Kaya E, Taşar P. Turkish HPB Surgery Association consensus report on hepatic cystic echinococcosis (HCE). Turk J Surg. 2022;38(2):101–20. https://doi.org/10.47717/turkjsurg.2022.5757. PMID: 36483170 PMCID: PMC9714645.
13. Brunetti E, Kern P, Vuitton DA. Expert consensus for the diagnosis and treatment of cystic and alveolar echinococcosis in humans. Acta Trop. 2010;114(1):1–16. https://doi.org/10.1016/j.actatropica.2009.11.001. PMID: 19931502.
14. Doğru D, Kiper N, Ozçelik U, Yalçin E, Göçmen A. Medical treatment of pulmonary hydatid disease: for which child? Parasitol Int. 2005;54(2):135–8. https://doi.org/10.1016/j.parint.2005.02.003. PMID: 15866475.
15. Guidelines for treatment of cystic and alveolar echinococcosis in humans. WHO Informal Working Group on Echinococcosis. Bull World Health Organ. 1996;74(3):231–42. PMID: 8789923 PMCID: PMC2486920.
16. Arif SH, Shams Ul B, Wani NA, Zargar SA, Wani MA, Tabassum R, Hussain Z, Baba AA, Lone RA. Albendazole as an adjuvant to the standard surgical management of hydatid cyst liver. Int J

Surg. 2008;6(6):448–51. https://doi.org/10.1016/j.ijsu.2008.08.003. PMID: 18819855.
17. Group WIW. International classification of ultrasound images in cystic echinococcosis for application in clinical and field epidemiological settings. Acta Trop. 2003;85(2):253–61. https://doi.org/10.1016/S0001-706X(02)00223-1.
18. Stojkovic M, Rosenberger KD, Steudle F, Junghanss T. Watch and wait management of inactive cystic echinococcosis—does the path to inactivity matter—analysis of a prospective patient cohort. PLoS Negl Trop Dis. 2016;10(12):e0005243. https://doi.org/10.1371/journal.pntd.0005243. PMID: 27992434 PMCID: PMC5207761.
19. Stojkovic M, Zwahlen M, Teggi A, Vutova K, Cretu CM, Virdone R, Nicolaidou P, Cobanoglu N, Junghanss T. Treatment response of cystic echinococcosis to benzimidazoles: a systematic review. PLoS Negl Trop Dis. 2009;3(9):e524. https://doi.org/10.1371/journal.pntd.0000524. PMID: 19787039 PMCID: PMC2745697.
20. Vuitton DA. Benzimidazoles for the treatment of cystic and alveolar echinococcosis: what is the consensus? Expert Rev Anti-Infect Ther. 2009;7(2):145–9. https://doi.org/10.1586/14787210.7.2.145. PMID: 19254162.
21. Hemphill A, Müller J. Alveolar and cystic echinococcosis: towards novel chemotherapeutical treatment options. J Helminthol. 2009;83(2):99–111. https://doi.org/10.1017/s0022149x0928936x. PMID: 19296876.
22. Mebendazole. LiverTox: clinical and research information on drug-induced liver injury. Bethesda: National Institute of Diabetes and Digestive and Kidney Diseases; 2012.
23. Thakur RK, Patel SP. Mebendazole. StatPearls. Treasure Island: StatPearls Publishing Copyright © 2024, StatPearls Publishing LLC; 2024.
24. Wilson JF, Rausch RL. Mebendazole and alveolar hydatid disease. Ann Trop Med Parasitol. 1982;76(2):165–73. https://doi.org/10.1080/00034983.1982.11687523. PMID: 7092370.
25. Liu C, Zhang H, Yin J, Hu W. In vivo and in vitro efficacies of mebendazole, mefloquine and nitazoxanide against cyst echinococcosis. Parasitol Res. 2015;114(6):2213–22. https://doi.org/10.1007/s00436-015-4412-4. PMID: 25773183.
26. Un M, Yaman SS, Erbas O. Hydatid cyst and treatment. Demiroglu Sci Univ Florence Nightingale J Transplant. 2020;5(2):070–8. https://doi.org/10.5606/dsufnjt.2020.018.
27. Liu CS, Zhang HB, Jiang B, Yao JM, Tao Y, Xue J, Wen AD. Enhanced bioavailability and cysticidal effect of three mebendazole-oil preparations in mice infected with secondary cysts of Echinococcus granulosus. Parasitol Res. 2012;111(3):1205–11. https://doi.org/10.1007/s00436-012-2954-2. PMID: 22661241.
28. Bildik N, Cevik A, Altintaş M, Ekinci H, Canberk M, Gülmen M. Efficacy of preoperative albendazole use according to months in hydatid cyst of the liver. J Clin Gastroenterol. 2007;41(3):312–6. https://doi.org/10.1097/01.mcg.0000225572.50514.e6. PMID: 17426473.
29. Veraldi S, Francia C, La Vela V, Nazzaro G, Barbareschi M. Telogen effluvium after oral albendazole. Clin Exp Dermatol. 2012;37(5):565–7. https://doi.org/10.1111/j.1365-2230.2011.04267.x. PMID: 22420487.
30. Xing F, Ye H, Yang J, Chan JF, Seto WK, Pai PM, Yuen KY, Hung DL. Fatal pancytopenia due to albendazole treatment for strongyloidiasis. IDCases. 2018;12:112–6. https://doi.org/10.1016/j.idcr.2018.04.002. PMID: 29942764 PMCID: PMC6010955.
31. Aktan AO, Yalin R. Preoperative albendazole treatment for liver hydatid disease decreases the viability of the cyst. Eur J Gastroenterol Hepatol. 1996;8(9):877–9. PMID: 8889454.
32. Dervenis C, Delis S, Avgerinos C, Madariaga J, Milicevic M. Changing concepts in the management of liver hydatid disease. J Gastrointest Surg. 2005;9(6):869–77. https://doi.org/10.1016/j.gassur.2004.10.016. PMID: 15985246.
33. Falagas ME, Bliziotis IA. Albendazole for the treatment of human echinococcosis: a review of comparative clinical trials. Am J Med Sci. 2007;334(3):171–9. https://doi.org/10.1097/MAJ.0b013e31814252f8. PMID: 17873530.
34. Richter J, Lindner AK, Geisel D, Fleckenstein FN, Torsello GF, Millet Pascual-Leone B, Ivanov O, Zöllner C, Wilde AB, Equihua MG. Percutaneous drainage and combined praziquantel-albendazole therapy: a novel approach for the treatment of simple echinococcal liver cysts. Z Gastroenterol. 2023; https://doi.org/10.1055/a-2084-3735. PMID: 37729942.
35. Velasco-Tirado V, Alonso-Sardón M, Lopez-Bernus A, Romero-Alegría Á, Burguillo FJ, Muro A, Carpio-Pérez A, Muñoz Bellido JL, Pardo-Lledias J, Cordero M, Belhassen-García M. Medical treatment of cystic echinococcosis: systematic review and meta-analysis. BMC Infect Dis. 2018;18(1):306. https://doi.org/10.1186/s12879-018-3201-y. PMID: 29976137 PMCID: PMC6034244.
36. Cheraghipour K, Rouzbahani AK, Fallahi S, Taherpour F, Moradifard F, Shakib P, Lashgarian HE, Marzban A. Recent advances in therapeutic strategies against hydatid cysts using nanomaterials: a systematic review. Lett Drug Design Discov. 2023;20(9):1185–93. https://doi.org/10.2174/1570180819666220922104034.
37. Sawatdee S, Atipairin A, Sae Yoon A, Srichana T, Changsan N, Suwandecha T. Formulation development of albendazole-loaded self-microemulsifying chewable tablets to enhance dissolution and bioavailability. Pharmaceutics. 2019;11(3) https://doi.org/10.3390/pharmaceutics11030134. PMID: 30897738 PMCID: PMC6471907.
38. Bakhtiar NM, Akbarzadeh A, Casulli A, Mahami-Oskouei M, Ahmadpour E, Nami S, Rostami A,

Spotin A. Therapeutic efficacy of nanocompounds in the treatment of cystic and alveolar echinococcoses: challenges and future prospects. Parasitol Res. 2019;118(9):2455–66. https://doi.org/10.1007/s00436-019-06416-5. PMID: 31402401.

39. Pensel P, Paredes A, Albani CM, Allemandi D, Sanchez Bruni S, Palma SD, Elissondo MC. Albendazole nanocrystals in experimental alveolar echinococcosis: enhanced chemoprophylactic and clinical efficacy in infected mice. Vet Parasitol. 2018;251:78–84. https://doi.org/10.1016/j.vetpar.2017.12.022. PMID: 29426481.
40. Cupit PM, Cunningham C. What is the mechanism of action of praziquantel and how might resistance strike? Future Med Chem. 2015;7(6):701–5. https://doi.org/10.4155/fmc.15.11. PMID: 25996063.
41. Jaoko WG, Muchemi G, Oguya FO. Praziquantel side effects during treatment of Schistosoma mansoni infected pupils in Kibwezi, Kenya. East Afr Med J. 1996;73(8):499–501. PMID: 8898462.
42. Mohamed AE, Yasawy MI, Al Karawi MA. Combined albendazole and praziquantel versus albendazole alone in the treatment of hydatid disease. Hepato-Gastroenterology. 1998;45(23):1690–4. PMID: 9840130.
43. Haralabidis S, Diakou A, Frydas S, Papadopoulos E, Mylonas A, Patsias A, Roilides E, Giannoulis E. Long-term evaluation of patients with hydatidosis treated with albendazole and praziquantel. Int J Immunopathol Pharmacol. 2008;21(2):429–35. https://doi.org/10.1177/039463200802100223. PMID: 18547489.
44. Alvela-Suárez L, Velasco-Tirado V, Belhassen-Garcia M, Novo-Veleiro I, Pardo-Lledías J, Romero-Alegría A, Pérez del Villar L, Valverde-Merino MP, Cordero-Sánchez M. Safety of the combined use of praziquantel and albendazole in the treatment of human hydatid disease. Am J Trop Med Hyg. 2014;90(5):819–22. https://doi.org/10.4269/ajtmh.13-0059. PMID: 24615131 PMCID: PMC4015571.
45. Tickell-Painter M, Maayan N, Saunders R, Pace C, Sinclair D. Mefloquine for preventing malaria during travel to endemic areas. Cochrane Database Syst Rev. 2017;10(10):Cd006491. https://doi.org/10.1002/14651858.CD006491.pub4. PMID: 29083100 PMCID: PMC5686653.
46. Küster T, Stadelmann B, Hermann C, Scholl S, Keiser J, Hemphill A. In vitro and in vivo efficacies of mefloquine-based treatment against alveolar echinococcosis. Antimicrob Agents Chemother. 2011;55(2):713–21. https://doi.org/10.1128/aac.01392-10. PMID: 21135182 PMCID: PMC3028781.
47. Rufener R, Ritler D, Zielinski J, Dick L, da Silva ET, da Silva AA, Joekel DE, Czock D, Goepfert C, Moraes AM, de Souza MVN, Müller J, Mevissen M, Hemphill A, Lundström-Stadelmann B. Activity of mefloquine and mefloquine derivatives against Echinococcus multilocularis. Int J Parasitol Drugs Drug Resist. 2018;8(2):331–40. https://doi.org/10.1016/j.ijpddr.2018.06.004. PMID: 29933218 PMCID: PMC6020078.
48. Lundström-Stadelmann B, Rufener R, Hemphill A. Drug repurposing applied: activity of the antimalarial mefloquine against Echinococcus multilocularis. Int J Parasitol Drugs Drug Resist. 2020;13:121–9. https://doi.org/10.1016/j.ijpddr.2020.06.002. PMID: 32636148 PMCID: PMC7389337.
49. Ammann RW, Hirsbrunner R, Cotting J, Steiger U, Jacquier P, Eckert J. Recurrence rate after discontinuation of long-term mebendazole therapy in alveolar echinococcosis (preliminary results). Am J Trop Med Hyg. 1990;43(5):506–15. https://doi.org/10.4269/ajtmh.1990.43.506. PMID: 2240375.
50. Reuter S, Beisler T, Kern P. Combined albendazole and amphotericin B against Echinococcus multilocularis in vitro. Acta Trop. 2010;115(3):270–4. https://doi.org/10.1016/j.actatropica.2010.04.009. PMID: 20412782.
51. Abu-Salah KM. Amphotericin B: an update. Br J Biomed Sci. 1996;53(2):122–33. PMID: 8757689
52. Cohen BE. A sequential mechanism for the formation of aqueous channels by amphotericin B in liposomes. The effect of sterols and phospholipid composition. Biochim Biophys Acta. 1992;1108(1):49–58. https://doi.org/10.1016/0005-2736(92)90113-z. PMID: 1643081.
53. Hartsel S, Bolard J. Amphotericin B: new life for an old drug. Trends Pharmacol Sci. 1996;17(12):445–9. https://doi.org/10.1016/s0165-6147(96)01012-7. PMID: 9014498.
54. Reuter S, Buck A, Grebe O, Nüssle-Kügele K, Kern P, Manfras BJ. Salvage treatment with amphotericin B in progressive human alveolar echinococcosis. Antimicrob Agents Chemother. 2003;47(11):3586–91. https://doi.org/10.1128/aac.47.11.3586-3591.2003. PMID: 14576122 PMCID: PMC253803.
55. Gavidia CM, Gonzalez AE, Lopera L, Jayashi C, Angelats R, Barron EA, Ninaquispe B, Villarreal L, Garcia HH, Verastegui MR, Gilman RH. Evaluation of nitazoxanide and oxfendazole efficacy against cystic echinococcosis in naturally infected sheep. Am J Trop Med Hyg. 2009;80(3):367–72. PMID: 19270283 PMCID: PMC2673094.
56. Stettler M, Rossignol JF, Fink R, Walker M, Gottstein B, Merli M, Theurillat R, Thormann W, Dricot E, Segers R, Hemphill A. Secondary and primary murine alveolar echinococcosis: combined albendazole/nitazoxanide chemotherapy exhibits profound antiparasitic activity. Int J Parasitol. 2004;34(5):615–24. https://doi.org/10.1016/j.ijpara.2004.01.006. PMID: 15064126.
57. Walker M, Rossignol JF, Torgerson P, Hemphill A. In vitro effects of nitazoxanide on Echinococcus granulosus protoscoleces and metacestodes. J Antimicrob Chemother. 2004;54(3):609–16. https://doi.org/10.1093/jac/dkh386. PMID: 15282238.
58. Mai HA, Wang YH, Zhao JM, Bai L. [Effect of different drugs on Echinococcus multilocularis

metacestodes in vitro]. Zhongguo Ji Sheng Chong Xue Yu Ji Sheng Chong Bing Za Zhi. 2011;29(6):419–24. PMID: 24822340.
59. Pérez-Molina JA, Díaz-Menéndez M, Gallego JI, Norman F, Monge-Maillo B, Ayala AP, López-Vélez R. Evaluation of nitazoxanide for the treatment of disseminated cystic echinococcosis: report of five cases and literature review. Am J Trop Med Hyg. 2011;84(2):351–6. https://doi.org/10.4269/ajtmh.2011.10-0513. PMID: 21292913 PMCID: PMC3029196.
60. Hadad GM, Abdel Salam RA, Emara S. Validated and optimized high-performance liquid chromatographic determination of tizoxanide, the main active metabolite of nitazoxanide in human urine, plasma and breast milk. J Chromatogr Sci. 2012;50(6):509–15. https://doi.org/10.1093/chromsci/bms041. PMID: 22525879.
61. Xu X, Qian X, Gao C, Pang Y, Zhou H, Zhu L, Wang Z, Pang M, Wu D, Yu W, Kong F, Shi D, Guo Y, Su X, Hu W, Yan J, Feng X, Fan H. Advances in the pharmacological treatment of hepatic alveolar echinococcosis: from laboratory to clinic. Front Microbiol. 2022;13:953846. https://doi.org/10.3389/fmicb.2022.953846. PMID: 36003932 PMCID: PMC9393627.
62. Enkai S, Kouguchi H, Inaoka DK, Irie T, Yagi K, Kita K. In vivo efficacy of combination therapy with albendazole and atovaquone against primary hydatid cysts in mice. Eur J Clin Microbiol Infect Dis. 2021;40(9):1815–20. https://doi.org/10.1007/s10096-021-04230-5. PMID: 33770336 PMCID: PMC8346398.
63. Küster T, Stadelmann B, Aeschbacher D, Hemphill A. Activities of fenbendazole in comparison with albendazole against Echinococcus multilocularis metacestodes in vitro and in a murine infection model. Int J Antimicrob Agents. 2014;43(4):335–42. https://doi.org/10.1016/j.ijantimicag.2014.01.013. PMID: 24646943.
64. Elissondo MC, Ceballos L, Alvarez L, Sánchez Bruni S, Lanusse C, Denegri G. Flubendazole and ivermectin in vitro combination therapy produces a marked effect on Echinococcus granulosus protoscoleces and metacestodes. Parasitol Res. 2009;105(3):835–42. https://doi.org/10.1007/s00436-009-1469-y. PMID: 19444468.
65. Stadelmann B, Aeschbacher D, Huber C, Spiliotis M, Müller J, Hemphill A. Profound activity of the anti-cancer drug bortezomib against Echinococcus multilocularis metacestodes identifies the proteasome as a novel drug target for cestodes. PLoS Negl Trop Dis. 2014;8(12):e3352. https://doi.org/10.1371/journal.pntd.0003352. PMID: 25474446 PMCID: PMC4256282.
66. Xiao N. [Concept and mechanism innovations provide a sustainable driver for echinococcosis control in China]. Zhongguo Xue Xi Chong Bing Fang Zhi Za Zhi. 2021;33(4):329–33. https://doi.org/10.16250/j.32.1374.2021188. PMID: 34505437.
67. Hemer S, Brehm K. In vitro efficacy of the anticancer drug imatinib on Echinococcus multilocularis larvae. Int J Antimicrob Agents. 2012;40(5):458–62. https://doi.org/10.1016/j.ijantimicag.2012.07.007. PMID: 22947125.
68. Ganji A, Jalali-Mashayekhi F, Hajihossein R, Eslamirad Z, Bayat PD, Sakhaie M. Anti-parasitic effects of resveratrol on protoscolices and hydatid cyst layers. Exp Parasitol. 2022;241:108360. https://doi.org/10.1016/j.exppara.2022.108360. PMID: 35995249.

7 Percutaneous Treatment of Liver Hydatid Disease

Adem Ucar

7.1 Introduction

Hydatid disease, caused by the metacestode *Echinococcus granulosus* has been recognized since Hippocrates [1]. Despite being over 24 centuries old, the disease still poses a major health problem in sheep-raising areas of developing countries, which constitute a significant part of the world. Although hydatid disease can affect any part of the body, it is most commonly found in the liver (50–70%), followed by the lungs (20–30%), and less commonly, the spleen, kidneys, heart, bone, central nervous system, and other organs. In the vast majority of cases, one organ is affected, and there is a single cyst [2–6]. Patients with hepatic hydatid disease (HHD) can remain asymptomatic for years. They are usually detected incidentally when there is an increase in liver size or when the liver is imaged for other reasons. While clinical and laboratory findings can be helpful in diagnosis, the diagnosis is primarily made based on imaging findings such as ultrasonography, computed tomography, or magnetic resonance imaging.

A cyst's viability and pathogenicity depend on its germinal membrane's integrity. All treatments aim to kill the scolices located in the germinal membrane. Treatment of HHD can be divided into four main groups: medical, open or laparoscopic surgery, percutaneous treatment, and follow-up without intervention. While surgery has traditionally been the mainstay of therapy for HHD, long-term studies have shown that percutaneous treatment is also effective in treating the disease. Percutaneous treatment offers several advantages over traditional open surgery, including shorter hospital stays, lower complication rates, and shorter recovery times [7, 8].

It is well-known that the natural course of hydatid disease is relatively slow. Frider et al. [9], in a small patient cohort, found that the lesions in the majority of asymptomatic patients did not progress when they were followed for a long period without any treatment. Active surveillance may be an option for asymptomatic patients with small cysts [9].

A comprehensive classification system is necessary to choose the appropriate treatment method for HHD. In 1981, Gharbi et al. [10] developed a classification system that uses ultrasonography to categorize HHD based on its morphological structure (Table 7.1). This system has been widely used since then. In 2003, the World Health Organization–Informal Working Group on Echinococcosis (WHO-IWGE) developed a standardized international classification system using ultrasonography to replace the previously used classification systems (Table 7.1) [11]. The lesions are classified based on size: small (<5 cm), medium (5–10 cm), or large (>10 cm). There are five types of echinococcosis, which are categorized into three stages. Types CE 1 and CE

A. Ucar (✉)
Department of Radiology, Istinye University, Faculty of Medicine, Istanbul, Türkiye

S. Akbulut (ed.), *Hydatid Disease*, https://doi.org/10.1007/978-3-031-97277-5_7

Table 7.1 World Health Organization and Gharbi Classifications of hydatid cysts

Gharbi Classification[a]	WHO Classification[b]	Stage	Description
Type I	CE Type 1	Active	Unilocular cysts, visible double-layer cyst wall
Type III	CE Type 2	Active	Multilocular cysts
Type II	CE Type 3a	Transitional	Unilocular cysts, detached germinal membrane, the water-lily sign
Type III	CE Type 3b	Transitional	Predominantly solid with daughter cysts
Type IV	CE Type 4	Inactive	Heterogenous cyst material
Type V	CE Type 5	Inactive	Calcified cyst wall, solid appearance

[a] Ref [7]; [b] Ref [8]

2 are active cystic echinococcosis lesions, while CE 3 are transitional lesions containing viable protoscoleces. Types CE 4 and CE 5 are inactive lesions. These two classification systems are essential in determining which patient will benefit from percutaneous treatment [12]. On the other hand, the most significant criticism of these two classifications is that they do not predict the occurrence of symptoms and complications associated with the disease. The location of the cysts may be important in terms of free rupture into the bile duct or peritoneum. This is the most crucial problem posed by hydatid disease. The location of the cyst is also essential in addition to its size. Cysts located near the bile duct or peritoneum are at risk of rupture. The rupture of a hydatid cyst can cause serious complications, including anaphylaxis, peritonitis, cholangitis, and sepsis [8].

Significant progress has been made in developing percutaneous techniques for treating HHD since the late 1980s. In 1985, Mueller et al. reported the first percutaneous aspiration and lavage with silver nitrate and hypertonic saline for recurrent HHD [13]. Following this, Ben Amor et al. [14] from Tunisia developed an interventional procedure called PAIR, which they first tested in sheep and then successfully applied in humans [15]. The percutaneous drainage of HHD involves four sequential steps known as PAIR: puncture (P), aspiration (A), instillation (I), and re-aspiration (R).

Over time, various procedure variants have been developed to improve treatment outcomes. One of the most frequently discussed variants involves using a scolicidal agent. Hypertonic saline destroys scolices by creating a strong osmotic gradient across their outer membrane, causing them to shrink and eventually rupture. The scolicidal effect increases with the solution's strength and exposure duration.

In the percutaneous treatment of hydatid cysts, the scolicidal and sclerosing effect of alcohol, used as the only endocavitary agent is well known [16]. Absolute alcohol, known to have a more substantial scolicidal and sclerosing effect than 20% hypertonic saline, can be used as a scolicidal agent [16, 17]. Other potent scolicidal agents include cetrimide 0.1%, povidone-iodine 10%, silver nitrate, mebendazole, albendazole, and polidocanol. All these scolicidal agents were potentially toxic and, upon absorption, can rarely cause serious adverse effects [18–21]. Additionally, sclerosing cholangitis can result should there be cyst-biliary communication [22].

7.2 PAIR

The PAIR technique is an interventional method that has been primarily described and used for HHD. However, there are reports of its successful application in cases encountered in the spleen, pancreas, kidney, and bone [23–26]. Its application in cysts located in the brain and lungs is contraindicated as no solid healing tissue formed by the foreign body reaction around the cyst. Puncture of the cyst can result in leakage of the cyst contents or spread of the disease. In addition, the tissues surrounding the cyst can be exposed to the direct damaging effect of the scolocidal agent used [7, 27, 28]. The main indications for percutaneous treatment of hydatid disease are summarized in Table 7.2.

Table 7.2 The indications for percutaneous treatment of hydatid disease

	Favorable for PAIR	Not favorable for PAIR
Organ	Liver, pancreas, kidney, bones	Lungs, brain
Cysts size	More than 4 cm	Less than 4 cm
Cysts appearance	Univesicular cysts with regular double laminated layer	Cyst with honeycombing, solid component, ruptured membranes
Cysts location	Inaccessible site to surgery and/or laparoscopy, cysts in different liver segments	Exophytic cysts, cysts located superficially and leaning against the liver capsule along a long segment
Others	Relapse after surgery, high-risk patient with comorbid disease, patients who refuse surgery, cysts that have failed to drug therapy	

The procedure involves several essential steps to treat a cyst. First, the cyst is punctured using the Seldinger technique under ultrasonography or computed tomography (CT) guidance. A thin needle is inserted to remove about 20% of the liquid inside the cyst. Next, a contrast medium (about 10% of the estimated volume of the cyst) is injected into the cyst to perform cystography. This step aims to detect whether the contrast medium leaks outside the cyst, which may indicate a risk of cyst rupture. If no contrast extravasation is observed, scolicidal agents (such as 20% hypertonic saline) are injected into the cyst cavity to kill the parasite and minimize the risk of recurrence. After the scolicidal agent injection, the patient is monitored for 10–15 minutes to allow the scolicidal agent to contact the cyst wall and germinal layer. Finally, all the fluid inside the cyst is aspirated again after the waiting period to remove the remaining cyst contents, dead parasites, and scolicidal agents. PAIR is a good option in the active disease stage, where the cyst consists of a single compartment filled with liquid content, with a bile duct connection excluded. The bile duct connection should be carefully investigated before preoperative imaging and injection of the scolocidal agent. In stage CE2 disease with several large cysts, the PAIR technique can be modified and applied. PAIR should not be applied in cases with a honeycomb appearance and ruptured membranes. It is sufficient to monitor cysts with a completely solid appearance, part or all of which are calcified (CE4 and CE5), without any treatment. PAIR can be successfully applied as an alternative in patients who cannot be reached surgically, patients with high risk due to comorbidities, recurrences after surgery, failure to respond to drug therapy, or patients who refuse surgery [7, 29].

In cases where it is still uncertain whether a hydatid cyst is present, a cyst puncture can be helpful. When inserting a needle into the cyst cavity, one should expect to encounter a clear fluid under high pressure [30]. This fluid can be examined microscopically to determine if live scolices are present. After injecting a scolocidal agent, a second microscopic examination can show whether the scolices have taken up the methylene blue stain, indicating they have been inactivated completely [7]. If the cyst content is yellow, it may suggest a cyst-biliary system relationship, inactive disease, or non-hydatid cyst conditions. To confirm the diagnosis, bilirubin and scolices from the content should be investigated [7, 30].

The PAIR procedure can lead to various complications, such as anaphylaxis, pneumothorax, infection of the cyst cavity, and formation of biliary fistula [31, 32]. Minor allergic reactions and fever may occur in 5% of cases. Allergic reactions can range from mild itching to hives, skin rash, bronchospasm, or even anaphylactic shock, and may occur before, during, or after the procedure. The mixing of cyst fluid into the bloodstream is often the leading cause of anaphylaxis [33].

During percutaneous interventions, there are several important risks that should be considered. These include the spread of cyst fluid into the abdomen, which can result in anaphylactic shock (0.38%), and the development of secondary intra-abdominal hydatidosis (1.27%). It is also important to consider the risk of the scolicidal agent causing cholangitis in cysts related to the biliary tract. There are some people who are hesitant about puncturing hydatid cysts because of the risk of anaphylaxis, despite its effectiveness and widespread use. A study conducted by Neumayr et al. [34] systematically searched MEDLINE to record complications associated with percutaneous treatment. They included 124 studies from 1980 to 2010. The study found that the risk of fatal anaphylaxis is 0.03% per percutaneous treatment and 0.04% per treated cyst [34], while the rate of reversible allergic reactions is 1.7% per percutaneous treatment and 1.8% per treated cyst. The literature review showed that the possibility of anaphylaxis is very rare and not higher than drug-related anaphylaxis. The results suggest that the risk of anaphylaxis associated with percutaneous treatment of hydatid cysts is very low and comparable to the risk of anaphylaxis associated with other medical procedures, such as administering certain medications.

Anaphylaxis is a severe and potentially life-threatening allergic reaction characterized by sudden onset of respiratory distress, hives, and low blood pressure. To reduce the risk of anaphylaxis, certain precautions can be taken. These include administering corticosteroids and antihistamines before the procedure, and having emergency equipment and personnel available in case of anaphylaxis.

Following the PAIR technique, different percutaneous approaches have been defined, such as catheterization techniques such as percutaneous aspiration, injection, and drainage system (PAIDS), modified catheterization techniques (MoCaT), and the percutaneous evacuation of cyst content (PEVAC) method in which the cyst contents are completely drained.

Three key indicators can summarize early-term procedural success. First, the complete separation of the germinal membrane from the pericyst, which are the inner and outer layers of the cyst wall, respectively. The successful separation of these two layers indicates that the cyst has been effectively treated. Second, there is a loss of movement of the scolices in the cyst washing fluid. This loss of movement is a sign that the scolices have been killed. Third, the inability of the scolices to clear methylene blue, a chemical that kills parasites, also indicates that they have been killed.

Several indicators can determine the success of percutaneous treatment in the long term. These include a decrease in the size of the cyst as observed through repeated ultrasounds, which shows that the cyst is shrinking. Additionally, a change in the cyst's content to a solid state indicates it is being replaced by solid tissue. Developing calcifications on the cyst wall is also an indicator of success. Finally, the most desirable outcome is the complete disappearance of the cyst.

The recurrence of the HHD is often due to inadequate administration of scolicidal agents during percutaneous treatment, applying only PAIR to a large cyst that requires catheter drainage, and patients not taking or irregularly taking oral scolicidal medication.

7.2.1 Medical Treatment

Benzimidazole compounds, especially albendazole, are effective in treating hydatid disease. However, when the goal of treatment is defined as a complete cure, which manifests itself with the disappearance of the lesion, the medical treatment appears to be effective in less than 50% of treated patients [1, 8].

Medical treatment is recommended for both surgical and percutaneous treatment of hydatid disease. The primary objective of medical treatment is to prevent disease recurrence. It is crucial to administer medical treatment before and after the procedure to avoid recurrence [35]. The most commonly used scolicidal drugs are benzimidazole derivatives, such as mebendazole and albendazole. Benzimidazole derivatives work by inhibiting the formation of ATP and the use of glucose by the parasite. This results in the death of the parasite [36]. Medical treatment is an important part of the management of hydatid disease. It can help prevent disease recurrence and improve the overall outcome for patients.

Benzimidazole treatment is used for cysts located in the lung and liver, involving two or more organs, in multiple cysts, or peritoneal cysts. Benzimidazole can be used alone, especially for cysts located in the liver or lung that are smaller than 5 cm, CE 1 and CE 3a cysts [37]. WHO recommends starting albendazole 4 h before the procedure and continuing it for 1 month afterward. In other studies, albendazole is also recommended to be used from 4 h to 10 days before percutaneous intervention and from 2 weeks to 6 months after the procedure [11, 38].

Albendazole is not recommended if there is a risk of cyst rupture or early pregnancy. Although it is considered to have a teratogenic potential in pregnancy, clinically, there have been no reports of pregnancy resulting in abnormalities, and there are publications that it can be used after the second trimester if necessary [39].

7.2.2 Modified PAIR (Percutaneous Drainage)

The PAIR procedure involves the use of a catheter with a diameter of 6–8F for aspiration, injection of a sclerosing agent, and re-aspiration stages in difficult puncture situations, particularly near vascular structures. Catheter use ensures a more stable execution of the procedure. It is crucial to ensure that the catheter used is tolerant to alcohol and other sclerosing agents.

During the first stage of the procedure, a 21G Chiba needle is inserted into the cyst, and a thin guidewire is advanced through it. A dilator is then inserted over the guidewire, and finally, a 6–8F catheter alcohol-resistant is inserted to drain the cyst contents. The amount of alcohol to be used is determined by calculating the cyst volume. Cavitography is performed by injecting contrast equal to the volume of alcohol. The relationship between the cyst and the bile duct is shown by taking images in the anterior, lateral, and oblique positions.

Afterward, pure alcohol is given to the cavity in an amount not exceeding 150 cc, which is one-third of the cyst volume. The pure alcohol is left in the cavity for 20 min. During this time, the patient is turned from the supine position to the prone position, the right and left lateral decubitus positions at 5-min intervals for maximum contact between alcohol and the germinal layer. Studies have shown that 20 min of contact during the treatment of simple cysts lined with biliary epithelium with alcohol does not reach critical levels in blood alcohol levels [40].

The alcohol is then aspirated completely, and the cyst cavity is washed with isotonic saline. A sample is then taken for light microscopy. Detection of fluid with imaging in the cyst cavity after early-term surgery should not suggest recurrence.

7.2.3 The Modified Catheter Technique (MoCaT)

Percutaneous treatment of CE2 and CE3b cysts has been a challenging process since the 1990s. Some believe that percutaneous intervention is the best option, as surgery includes longer hospital stays and a higher complication rate for these types of cysts.

A newer technique, called modified catheterization (MoCaT) has been shown to have a lower recurrence rate (3.8%) than PAIR or standard catheterization. However, MoCaT has also been associated with a higher complication rate (3.8%), including cystobiliary fistula and abscess [41].

MoCaT is a treatment for CE types 2 and 3b. Its objective is to completely remove the entire germinal layer and other cyst contents, as well as to damage the germinal membrane. Thicker catheters with a diameter of 12–16F may be needed to eliminate the dense cyst contents and germinal membranes. The catheter should be held in place until the cyst contents are completely drained.

Various studies have demonstrated low recurrence rates for this treatment, which has been used for 30 years [42–44]. Saremi developed a cutting device that obliterated daughter vesicles and membranes, followed by the use of a high-power suction apparatus to evacuate all of the pieces [45]. Saremi and McNamara [46] observed no recurrence in 32 patients over an average follow-up period of 25.5 months [46]. A statistically significant difference was discovered in the post-treatment complication rates between the MoCaT and PAIR. Cavity infection was the most significant and distinguishing variable of the major complications in this study.

Akhan et al. [42] treated six intramuscular hydatid cysts using catheterization or the "modified catheterization technique" with hypertonic saline and alcohol, depending on the type of cyst [42]. After puncturing the cyst cavity, the authors aspirated 35% to 50% of the estimated volume and filled the cavity with hypertonic saline and 5% to 10% less than the aspirated volume. The process was repeated until the endocyst was entirely separated from the pericyst. After this, a 14 Fr pigtail catheter was placed to completely empty the daughter cysts. When cystography showed complete emptying of the cyst cavity, 95% absolute alcohol was instilled into the cavity for sclerosis. The short-term follow-up results showed no relapse, but cavity infection and cellulitis were reported as complications, which were treated with medical treatment.

A study by Khuroo et al. [47] randomized 50 patients and divided them into two arms to receive either percutaneous drainage or surgery (each arm having 25 patients). There were 9 patients with multivesicular cysts in each group. The percutaneous treatment group had a shorter hospital stay and fewer complications. Although there were more patients with the disappearance of cysts in the surgery group, this was not statistically significant. The authors concluded that percutaneous drainage of uncomplicated hepatic hydatid cysts was safe and effective within two years.

The treatment of multivesicular cysts with a percutaneous approach is controversial due to the high recurrence and complication rates reported in different series [48–50]. In the study by Kabaalioğlu et al. [50], the authors reported a success rate of less than 50% for type III cysts treated with the modified PAIR technique (short-term follow-up). In the studies by Giorgio et al. [49, 51], the authors reported recurrence rates of 30% and 50%, respectively, in short-term and long-term follow-ups for CE2 and CE3b hydatid cysts treated with double percutaneous aspiration and alcohol injection techniques.

A study by Rinaldi et al. [52] looked at the best treatment for HHD in patients with CE3b cysts. They retrospectively analyzed the clinical behavior of single hepatic CE3b cysts in 60 patients who received albendazole or were followed with the watch-and-wait approach. They found that albendazole treatment can temporarily inactivate CE3b cysts but is not a permanent solution. The watch-and-wait approach may suit patients with asymptomatic CE3b cysts who do not need surgery and can follow up regularly. However, long-term prospective observation of a larger cohort, after being assigned to clearly different management options is needed to confirm their findings.

There are not enough prospective studies comparing percutaneous and surgical techniques in treating CE2 and CE3b cyst disease of the liver. A correct meta-analysis does not seem possible with the current studies in the medical literature. Despite the ongoing debate, some authors have reported the successful use of different percutaneous drainage techniques in treating CE2 and CE3b cysts. To end the controversy, we need to conduct more high-quality, randomized controlled trials comparing the two therapies and establish evidence-based criteria to allocate the patient to the most appropriate treatment option according to the specific situation.

7.2.4 PAIR Long-Term Results

Golemanov et al. [53] conducted a study on the PAIR technique in 230 patients from Bulgaria, with 348 hepatic hydatid cysts. At the 12-month follow-up, cyst obliteration was observed in cysts classified as CE1 and CE3a but not in those classified as CE2 or CE3b. During the procedure, two patients experienced anaphylaxis, and 25.2% experienced minor complications.

Chen et al. [54] conducted a meta-analysis to compare PAIR and laparoscopic therapies for treating hepatic hydatid disease. The study analyzed data from 38 studies involving 1230 laparoscopic-treated subjects and 41 studies involving 2686 PAIR-treated subjects. The results showed that both forms of treatment were safe and effective alternatives to open surgery. The type and location of the cyst were considered in determining the appropriate form of treatment. Laparoscopic-treated patients had a higher incidence of cyst-biliary leak, whereas patients treated with PAIR plus drug therapy had a higher incidence of fever.

A study by Nayman et al. [55] followed 374 patients with hepatic hydatid cysts for six years. Of these, 480 patients received the PAIR procedure, while 13 with cyst-biliary fistulization had surgery. The study found that the procedure was successful in 469 patients, accounting for a success rate of 97.7%. Two patients developed non-fatal anaphylaxis, while the other 44 (12%) patients developed minor complications.

Bakdik et al. [56] published findings on the PAIR technique using alcohol as a scolicidal agent in 347 patients with 554 liver hydatid cysts. The procedure was successful in 99.5% of patients, with a mean hospital stay of 1.55–2.3 days. However, cyst recurrences were reported in 19 (5.4%) patients. During the procedure, two patients experienced major complications and another 36 developed minor complications.

Sokouti et al. [57] conducted a meta-analysis to compare the outcomes of PAIR therapy in 1650 patients with laparoscopy in 1182 patients. The analysis revealed that the complications of PAIR procedures (18.5%) were slightly lower than those of laparoscopic procedures (18.7%). Additionally, the mortality rate of PAIR was lower than that of the laparoscopy procedure (1.1% vs. 1.8%). However, the recurrence rate of PAIR was higher than that of laparoscopic surgery (5% vs. 3.9%). The researchers concluded that the PAIR procedure had statistically significant benefits in terms of cure rate, postoperative complications, and mortality. Recurrence rates for laparoscopic surgery are lower than for PAIR approaches. Overall, in this study, they found that the PAIR procedure is preferable to laparoscopic surgery, as it has better cure rates, lower postoperative complications, and mortality.

7.2.5 Thermal Ablation

Percutaneous treatments primarily aim to damage the germinal layer completely. In CE2 or CE3b cysts, it may not be possible to damage the germinal layer of the daughter vesicles. Thermal ablation methods may be an alternative interventional method in this group. There are studies that report that ablation, by inserting a radiofrequency (RFA) or microwave ablation antenna into a complex cyst through the transhepatic route-guided by ultrasonography or computed tomography and including the pericyst in the ablation area is successful [58, 59]. This method can be performed through a percutaneous approach, using the same type of needle that is used in liver tumor ablation. Because heat, rather than a chemical agent, destroys the contents of the cyst, the procedure is simpler than the PAIR treatment, since it avoids the need to inject a scolicidal agent [58, 59].

Lamonaca et al. [60] conducted an ex-vivo experimental study on animals to evaluate the effectiveness of RFA in treating cystic echinococcosis. They treated nine liver and eight lung cysts with RFA and achieved a 100% success rate in both cases. The study showed that RFA was effective in treating explanted hydatid cysts. Cyst volume reduction after PAIR or pericysectomy is estimated to be 65% or more at 6–19 months after treatment [61]. Lamonaca observed the same level of volume reduction immediately after RFA treatment. The effectiveness of RFA treatment was confirmed in our patient, which is consistent with the results of the animal model study conducted by Lamonaca et al. [60].

After ablation, the contents of the complex cyst are aspirated. The viability of the scolex is checked by light microscopy. Follow-up after treatment can be done with biochemical and ultrasound imaging. Treatment success can be evaluated according to cyst size, increase in the solid portion, and whether daughter vesicles develop or not.

Long-term results and larger series are needed for the RFA technique involving CE2 and CE3b cysts, despite promising limited case reports in the literature.

7.2.6 PAIR in Pregnancy

PAIR is considered a safe and effective treatment option for hepatic hydatid cysts during pregnancy. However, it is not recommended to use this method in the first trimester. Studies have shown positive results for PAIR application during pregnancy, although the number of cases reviewed is limited. These studies found that the cysts disappeared or significantly decreased in size after PAIR application. The risk of recurrence after PAIR application during pregnancy is not different compared to periods outside of pregnancy.

Noori et al. [62] reported treating four out of five pregnant patients using a combination of surgery, PAIR, and albendazole. However, WHO guidelines advise against using albendazole during the first trimester of pregnancy. In our experience, PAIR alone can be effective, and additional medical treatment can be provided if recurrence occurs after the completion of pregnancy.

Ustunsoz et al. [63] treated six pregnant patients using PAIR with hypertonic saline as a cytotoxic agent. They found this method to be both effective and feasible.

7.2.7 Follow-Up

Hydatid cysts can recur even years after treatment. Assessing the success of treatment can be challenging because the infection's natural course varies greatly. Cysts can grow anywhere from 1 to 50 mm per year, remain stable, rupture spontaneously, or disappear completely [64]. The ideal approach for follow-up is uncertain and should be tailored to each patient's unique circumstances and available resources [53]. Typically, follow-up should include regular imaging with ultrasonography, computed tomography, or magnetic resonance imaging every three to six months, followed by annual follow-up once the results stabilize. To detect any signs of recurrence, follow-up should continue for at least five years [65].

Ultrasonography is a useful method for evaluating the effectiveness of treatment, but it has its limitations. It can detect the complete disappearance of the cyst; decrease in cyst size and volume, an increase in solid component, thickening and irregularity of cyst walls, and decrease in number and size of daughter cysts in multivesicular cysts [33, 66]. However, it is important to consider new cyst formation, increase in cyst size and section, increase in the liquid compartment, and relapse development during ultrasound follow-up. It is important to note that imaging methods cannot determine the viability of cysts [32, 66].

7.2.8 Hepatic Alveolar Echinococcosis

Alveolar echinococcosis is a serious zoonotic infection that can be life-threatening. This disease presents similarities to a tumor-like lesion. Its invasive growth pattern characterizes it, and its symptoms can be similar to malignant tumors due to its ability to invade local tissues and possibly metastasize [67, 68].

Effective treatment for *alveolar echinococcosis* incorporates WHO staging, pharmacological treatment, and a minimum of 10 years of follow-up [69]. The disease causes destructive and infiltrative effects on the liver, leading to liver dysfunction, biliary obstruction, or abscess formation. In these patients, the main treatment is either complete resection of the cyst or, most commonly, liver transplantation. The role of percutaneous treatment is to ensure that the patient is operated on under optimal conditions, and it is usually the palliation of biliary drainage and percutaneous abscess drainage of cystic-infectious lesions until surgery is performed.

Interventional radiology is an effective, minimally invasive procedure that can drain necrotic fluid and abscesses within or outside the liver and treat diseased or clogged bile ducts in hepatic alveolar echinococcosis patients. These procedures can be done as palliative operations or as a bridge to more radical surgeries. In cases of acute infection, percutaneous drainage of cyst contents and/or biliary channels using minimally invasive techniques is absolutely necessary and can even be life-saving. Moreover, percutaneous cyst drainage with albendazole therapy can significantly improve the quality of life in patients who are unable to undergo surgery [70–73].

References

1. Smego RA Jr, Sebanego P. Treatment options for hepatic cystic echinococcosis. Int J Infect Dis. 2005;9(2):69–76. https://doi.org/10.1016/j.ijid.2004.08.001. PMID: 15708321.
2. Ammann RW, Eckert J. Cestodes. Echinococcus. Gastroenterol Clin North Am. 1996;25(3):655–89. https://doi.org/10.1016/s0889-8553(05)70268-5. PMID: 8863045.
3. Guidelines for treatment of cystic and alveolar echinococcosis in humans. WHO Informal Working Group on Echinococcosis. Bull World Health Organ. 1996;74(3):231–42. PMID: 8789923 PMCID: PMC2486920.
4. Akbulut S, Senol A, Ekin A, Bakir S, Bayan K, Dursun M. Primary retroperitoneal hydatid cyst: report of 2 cases and review of 41 published cases. Int Surg. 2010;95(3):189–96. PMID: 21066995.
5. Akbulut S, Senol A, Sezgin A, Cakabay B, Dursun M, Satici O. Radical vs conservative surgery for hydatid liver cysts: experience from single center. World J Gastroenterol. 2010;16(8):953–9. https://doi.org/10.3748/wjg.v16.i8.953. PMID: 20180233 PMCID: PMC2828599.
6. Akbulut S, Sogutcu N, Eris C. Hydatid disease of the spleen: single-center experience and a brief literature review. J Gastrointest Surg. 2013;17(10):1784–95. https://doi.org/10.1007/s11605-013-2303-5. PMID: 23949423.
7. Khuroo MS. Percutaneous drainage in hepatic hydatidosis-the PAIR technique: concept, technique, and results. J Clin Exp Hepatol. 2021;11(5):592–602. https://doi.org/10.1016/j.jceh.2021.05.005. PMID: 34511821 PMCID: PMC8414317.
8. Rozanes I, Güven K, Acunaş B, Emre A. Cystic echinococcal liver disease: new insights into an old disease and an algorithm for therapy planning. Cardiovasc Intervent Radiol. 2007;30(6):1112–6. https://doi.org/10.1007/s00270-007-9081-y. PMID: 17533534.
9. Frider B, Larrieu E, Odriozola M. Long-term outcome of asymptomatic liver hydatidosis. J Hepatol. 1999;30(2):228–31. https://doi.org/10.1016/s0168-8278(99)80066-x. PMID: 10068100.
10. Gharbi HA, Hassine W, Brauner MW, Dupuch K. Ultrasound examination of the hydatic liver. Radiology. 1981;139(2):459–63. https://doi.org/10.1148/radiology.139.2.7220891. PMID: 7220891.
11. Group WIW. International classification of ultrasound images in cystic echinococcosis for application in clinical and field epidemiological settings. Acta Trop. 2003;85(2):253–61. https://doi.org/10.1016/s0001-706x(02)00223-1. PMID: 12606104.
12. Eren S, Kantarcı M. [Percutaneous liver hydatid cyst treatment]. Trd Sem 2015;3:227–36. https://doi.org/10.5152/trs.2015.221.
13. Mueller PR, Dawson SL, Ferrucci JT Jr, Nardi GL. Hepatic echinococcal cyst: successful percutaneous drainage. Radiology. 1985;155(3):627–8. https://doi.org/10.1148/radiology.155.3.3890001. PMID: 3890001.
14. Ben Amor N, Gargouri M, Gharbi HA, Ghorbel A, Golvan YJ, Hammou-Jeddi H, Kilani M, Lahmar S. [Treatment of hepatic hydatid cyst in sheep by echographic puncture]. Tunis Med. 1986;64(4):325–31. PMID: 3538579.
15. Gargouri M, Ben Amor N, Ben Chehida F, Hammou A, Gharbi HA, Ben Cheikh M, Kchouk H, Ayachi K, Golvan JY. Percutaneous treatment of hydatid cysts (Echinococcus granulosus). Cardiovasc Intervent Radiol. 1990;13(3):169–73. https://doi.org/10.1007/bf02575469. PMID: 2121344.
16. Filice C, Pirola F, Brunetti E, Dughetti S, Strosselli M, Foglieni CS. A new therapeutic approach for hydatid liver cysts. Aspiration and alcohol injection under sonographic guidance. Gastroenterology. 1990;98(5 Pt 1):1366–8. https://doi.org/10.1016/0016-5085(90)90358-8. PMID: 2182372.
17. Acunas B, Rozanes I, Celik L, Minareci O, Acunas G, Alper A, Ariogul O, Gökmen E. Purely cystic hydatid disease of the liver: treatment with percutaneous aspiration and injection of hypertonic saline. Radiology. 1992;182(2):541–3. https://doi.org/10.1148/radiology.182.2.1732977. PMID: 1732977.
18. Krige JE, Millar AJ, Rode H, Knobel D. Fatal hypernatraemia after hypertonic saline irrigation of hepatic hydatid cysts. Pediatr Surg Int. 2002;18(1):64–5. https://doi.org/10.1007/s003830200015. PMID: 11793068.
19. Sathyanarayana MV, Shenoy MG, Pai VM, Haque PW, Hegde P, Rajan R, Pahuja A. Metabolic acidosis induced by cetrimide-chlorhexidine solution in hydatid cyst surgery. Indian J Gastroenterol. 1996;15(3):104. PMID: 8840640.
20. Sharafi SM, Sefiddashti RR, Sanei B, Yousefi M, Darani HY. Scolicidal agents for protoscolices of Echinococcus granulosus hydatid cyst: review of literature. J Res Med Sci. 2017;22:92. https://doi.org/10.4103/jrms.JRMS_1030_16. PMID: 28900448 PMCID: PMC5583616.
21. Tripathy S, Sasmal P, Rao PB, Mishra TS, Nayak S. Cetrimide-chlorhexidine-induced multiorgan failure in surgery of pulmonary hydatid cyst. Ann Card Anaesth. 2016;19(3):557–60. https://doi.org/10.4103/0971-9784.185565. PMID: 27397471 PMCID: PMC4971995.
22. Tale S, Bannoth S, Meitei PS, Hanumanthappa MK, Kolli M. Scolicidal agents, and secondary cholangitis. Int J Res Rev. 2019;6:186–91.
23. Akhan O, Islim F, Balci S, Erbahceci A, Akpınar B, Ciftci T, Akinci D. Percutaneous treatment of simple hepatic cysts: the long-term results of PAIR and catheterization techniques as single-session procedures.

Cardiovasc Intervent Radiol. 2016;39(6):902–8. https://doi.org/10.1007/s00270-015-1283-0. PMID: 26714694.
24. Arslan M, Gulek B, Ogur HU, Adamhasan F. Primary hydatid cyst in the posterior thigh, and its percutaneous treatment. Skeletal Radiol. 2018;47(10):1437–42. https://doi.org/10.1007/s00256-018-2910-z. PMID: 29516112.
25. Yattoo GN, Khuroo MS, Zargar SA, Bhat FA, Sofi BA. Case report: percutaneous drainage of the pancreatic head hydatid cyst with obstructive jaundice. J Gastroenterol Hepatol. 1999;14(9):931–4. https://doi.org/10.1046/j.1440-1746.1999.01967.x. PMID: 10535478.
26. Arslan S, Bakdik S, Oncu F, Tolu I, Eryilmaz MA. Successful percutaneous treatment of extrahepatic cystic echinococcosis through PAIR and single puncture catheter techniques. Jpn J Radiol. 2017;35(6):296–302. https://doi.org/10.1007/s11604-017-0633-z. PMID: 28281046.
27. Baboli S, Baboli S, Soleiman MS. Brain hydatid cyst with atypical symptoms in an adult: a case report. Iran J Parasitol. 2016;11(3):422–5. PMID: 28127351 PMCID: PMC5256062.
28. Morar R, Feldman C. Pulmonary echinococcosis. Eur Respir J. 2003;21(6):1069–77. https://doi.org/10.1183/09031936.03.00108403. PMID: 12797504.
29. Khuroo MS, Zargar SA, Mahajan R. Echinococcus granulosus cysts in the liver: management with percutaneous drainage. Radiology. 1991;180(1):141–5. https://doi.org/10.1148/radiology.180.1.2052682. PMID: 2052682.
30. Salama H, Farid Abdel-Wahab M, Strickland GT. Diagnosis and treatment of hepatic hydatid cysts with the aid of echo-guided percutaneous cyst puncture. Clin Infect Dis. 1995;21(6):1372–6. https://doi.org/10.1093/clinids/21.6.1372. PMID: 8749617.
31. Akhan O, Ozmen MN, Dinçer A, Sayek I, Göçmen A. Liver hydatid disease: long-term results of percutaneous treatment. Radiology. 1996;198(1):259–64. https://doi.org/10.1148/radiology.198.1.8539390. PMID: 8539390.
32. Brunetti E, Kern P, Vuitton DA. Expert consensus for the diagnosis and treatment of cystic and alveolar echinococcosis in humans. Acta Trop. 2010;114(1):1–16. https://doi.org/10.1016/j.actatropica.2009.11.001. PMID: 19931502.
33. Khuroo MS, Dar MY, Yattoo GN, Zargar SA, Javaid G, Khan BA, Boda MI. Percutaneous drainage versus albendazole therapy in hepatic hydatidosis: a prospective, randomized study. Gastroenterology. 1993;104(5):1452–9. https://doi.org/10.1016/0016-5085(93)90355-g. PMID: 8482455.
34. Neumayr A, Troia G, de Bernardis C, Tamarozzi F, Goblirsch S, Piccoli L, Hatz C, Filice C, Brunetti E. Justified concern or exaggerated fear: the risk of anaphylaxis in percutaneous treatment of cystic echinococcosis-a systematic literature review. PLoS Negl Trop Dis. 2011;5(6):e1154. https://doi.org/10.1371/journal.pntd.0001154. PMID: 21695106 PMCID: PMC3114754.
35. Arif SH, Shams Ul B, Wani NA, Zargar SA, Wani MA, Tabassum R, Hussain Z, Baba AA, Lone RA. Albendazole as an adjuvant to the standard surgical management of hydatid cyst liver. Int J Surg. 2008;6(6):448–51. https://doi.org/10.1016/j.ijsu.2008.08.003. PMID: 18819855.
36. Nazligul Y, Kucukazman M, Akbulut S. Role of chemotherapeutic agents in the management of cystic echinococcosis. Int Surg. 2015;100(1):112–4. https://doi.org/10.9738/intsurg-d-14-00068.1. PMID: 25594649 PMCID: PMC4301274.
37. Bildik N, Cevik A, Altintaş M, Ekinci H, Canberk M, Gülmen M. Efficacy of preoperative albendazole use according to months in hydatid cyst of the liver. J Clin Gastroenterol. 2007;41(3):312–6. https://doi.org/10.1097/01.mcg.0000225572.50514.e6. PMID: 17426473.
38. Echinococcosis W-IWGo. PAIR: Puncture, Aspiration, Injection, Re-Aspiration: an option for the treatment of cystic echinococcosis. Geneva: WHO; 2001.
39. Zeybek B, Ergenoglu A, Yeniel A, Ulukus M. Hepatic hydatid disease during pregnancy. Ege J Med. 2010;49(3):197–9.
40. Larssen TB, Viste A, Horn A, Haldorsen IS, Espeland A. Single-session alcohol sclerotherapy of symptomatic liver cysts using 10–20 min of ethanol exposure: no recurrence at 2–16 years of follow-up. Abdom Radiol (NY). 2016;41(9):1776–81. https://doi.org/10.1007/s00261-016-0769-9. PMID: 27156079 PMCID: PMC5002038.
41. Akhan O, Salik AE, Ciftci T, Akinci D, Islim F, Akpinar B. Comparison of long-term results of percutaneous treatment techniques for hepatic cystic echinococcosis types 2 and 3b. AJR Am J Roentgenol. 2017;208(4):878–84. https://doi.org/10.2214/ajr.16.16131. PMID: 28199132.
42. Akhan O, Gumus B, Akinci D, Karcaaltincaba M, Ozmen M. Diagnosis and percutaneous treatment of soft-tissue hydatid cysts. Cardiovasc Intervent Radiol. 2007;30(3):419–25. https://doi.org/10.1007/s00270-006-0153-1. PMID: 17295079.
43. Mohan S, Garg SK, Kathuria M, Baijal SS. Mechanical suction through wide bore catheters for nonsurgical management of Gharbi type III hepatic hydatid cysts. Trop Gastroenterol. 2011;32(3):189–95. PMID: 22332334.
44. Schipper HG, Laméris JS, van Delden OM, Rauws EA, Kager PA. Percutaneous evacuation (PEVAC) of multivesicular echinococcal cysts with or without cystobiliary fistulas which contain non-drainable material: first results of a modified PAIR method. Gut. 2002;50(5):718–23. https://doi.org/10.1136/gut.50.5.718. PMID: 11950823 PMCID: PMC1773202.
45. Saremi F. Percutaneous drainage of hydatid cysts: use of a new cutting device to avoid leakage. AJR

Am J Roentgenol. 1992;158(1):83–5. https://doi.org/10.2214/ajr.158.1.1727363. PMID: 1727363.
46. Saremi F, McNamara TO. Hydatid cysts of the liver: long-term results of percutaneous treatment using a cutting instrument. AJR Am J Roentgenol. 1995;165(5):1163–7. https://doi.org/10.2214/ajr.165.5.7572496. PMID: 7572496.
47. Khuroo MS, Wani NA, Javid G, Khan BA, Yattoo GN, Shah AH, Jeelani SG. Percutaneous drainage compared with surgery for hepatic hydatid cysts. N Engl J Med. 1997;337(13):881–7. https://doi.org/10.1056/nejm199709253371303. PMID: 9302302.
48. Bosanac ZB, Lisanin L. Percutaneous drainage of hydatid cyst in the liver as a primary treatment: review of 52 consecutive cases with long-term follow-up. Clin Radiol. 2000;55(11):839–48. https://doi.org/10.1053/crad.2000.0543. PMID: 11069739.
49. Giorgio A, Tarantino L, de Stefano G, Francica G, Mariniello N, Farella N, Perrotta A, Aloisio V, Esposito F. Hydatid liver cyst: an 11-year experience of treatment with percutaneous aspiration and ethanol injection. J Ultrasound Med. 2001;20(7):729–38. https://doi.org/10.7863/jum.2001.20.7.729. PMID: 11444731.
50. Kabaalioğlu A, Ceken K, Alimoglu E, Apaydin A. Percutaneous imaging-guided treatment of hydatid liver cysts: do long-term results make it a first choice? Eur J Radiol. 2006;59(1):65–73. https://doi.org/10.1016/j.ejrad.2006.01.014. PMID: 16513311.
51. Giorgio A, Di Sarno A, de Stefano G, Liorre G, Farella N, Scognamiglio U, Giorgio V. Sonography and clinical outcome of viable hydatid liver cysts treated with double percutaneous aspiration and ethanol injection as first-line therapy: efficacy and long-term follow-up. AJR Am J Roentgenol. 2009;193(3):W186–92. https://doi.org/10.2214/ajr.08.1518. PMID: 19696257.
52. Rinaldi F, De Silvestri A, Tamarozzi F, Cattaneo F, Lissandrin R, Brunetti E. Medical treatment versus "Watch and Wait" in the clinical management of CE3b echinococcal cysts of the liver. BMC Infect Dis. 2014;14:492. https://doi.org/10.1186/1471-2334-14-492. PMID: 25204575 PMCID: PMC4164709.
53. Golemanov B, Grigorov N, Mitova R, Genov J, Vuchev D, Tamarozzi F, Brunetti E. Efficacy and safety of PAIR for cystic echinococcosis: experience on a large series of patients from Bulgaria. Am J Trop Med Hyg. 2011;84(1):48–51. https://doi.org/10.4269/ajtmh.2011.10-0312. PMID: 21212200 PMCID: PMC3005516.
54. Chen X, Cen C, Xie H, Zhou L, Wen H, Zheng S. The comparison of 2 new promising weapons for the treatment of hydatid cyst disease: PAIR and laparoscopic therapy. Surg Laparosc Endosc Percutan Tech. 2015;25(4):358–62. https://doi.org/10.1097/sle.0000000000000177. PMID: 26241297.
55. Nayman A, Guler I, Keskin S, Erdem TB, Borazan H, Kucukapan A, Ozbiner H, Batur A, Ertekin E, Feyzioglu B, Koc O, Kaya HE, Temizoz O, Kartal A, Ozbek O. A novel modified PAIR technique using a trocar catheter for percutaneous treatment of liver hydatid cysts: a six-year experience. Diagn Interv Radiol. 2016;22(1):47–51. https://doi.org/10.5152/dir.2015.15011. PMID: 26574902 PMCID: PMC4712897
56. Bakdik S, Arslan S, Oncu F, Tolu I, Eryilmaz MA. Percutaneous treatment of hepatic cystic echinococcosis: the success of alcohol as a single endocavitary agent in PAIR, catheterization, and modified catheterization techniques. Radiol Med. 2018;123(2):153–60. https://doi.org/10.1007/s11547-017-0820-0. PMID: 29030721.
57. Sokouti M, Sadeghi R, Pashazadeh S, Abadi SEH, Sokouti M, Ghojazadeh M, Sokouti B. A systematic review and meta-analysis on the treatment of liver hydatid cyst using meta-MUMS tool: comparing PAIR and laparoscopic procedures. Arch Med Sci. 2019;15(2):284–308. https://doi.org/10.5114/aoms.2018.73344. PMID: 30899281 PMCID: PMC6425195.
58. Brunetti E, Filice C. Radiofrequency thermal ablation of echinococcal liver cysts. Lancet. 2001;358(9291):1464. https://doi.org/10.1016/s0140-6736(01)06518-7. PMID: 11705538.
59. Brunetti E, Gulizia R, Garlaschelli A, Filice C. Radiofrequency thermal ablation of echinococcal cysts of the liver: results at 12-month follow-up. Am J Trop Med Hyg. 2004;71(4):613.
60. Lamonaca V, Virga A, Minervini MI, Di Stefano R, Provenzani A, Tagliareni P, Fleres G, Luca A, Vizzini G, Palazzo U, Gridelli B. Cystic echinococcosis of the liver and lung treated by radiofrequency thermal ablation: an ex-vivo pilot experimental study in animal models. World J Gastroenterol. 2009;15(26):3232–9. https://doi.org/10.3748/wjg.15.3232. PMID: 19598298 PMCID: PMC2710778.
61. Smego RA Jr, Bhatti S, Khaliq AA, Beg MA. Percutaneous aspiration-injection-reaspiration drainage plus albendazole or mebendazole for hepatic cystic echinococcosis: a meta-analysis. Clin Infect Dis. 2003;37(8):1073–83. https://doi.org/10.1086/378275. PMID: 14523772.
62. Noori IF, Jabbar AS. Hepatic hydatid cyst diseases during pregnancy: diagnosis, management and best practice. Syst Rev Pharm. 2020;11(8)
63. Ustunsoz B, Ugurel MS, Uzar AI, Duru NK. Percutaneous treatment of hepatic hydatid cyst in pregnancy: long-term results. Arch Gynecol Obstet. 2008;277(6):547–50. https://doi.org/10.1007/s00404-007-0498-y. PMID: 17972086.
64. Müfit K, Nejat I, Mercan S, Ibrahim K, Mete UY, Yüksel K. Growth of multiple hydatid cysts evaluated by computed tomography. J Clin Neurosci. 1998;5(2):215–7. https://doi.org/10.1016/s0967-5868(98)90042-x. PMID: 18639016.
65. Brunetti E, White AC Jr. Cestode infestations: hydatid disease and cysticercosis. Infect Dis Clin N Am. 2012;26(2):421–35. https://doi.org/10.1016/j.idc.2012.02.001. PMID: 22632647.

66. Moro PL, Gilman RH, Verastegui M, Bern C, Silva B, Bonilla JJ. Human hydatidosis in the Central Andes of Peru: evolution of the disease over 3 years. Clin Infect Dis. 1999;29(4):807–12. https://doi.org/10.1086/520440. PMID: 10589894.
67. Akbulut S, Cicek E, Kolu M, Sahin TT, Yilmaz S. Associating liver partition and portal vein ligation for staged hepatectomy for extensive alveolar echinococcosis: first case report in the literature. World J Gastrointest Surg. 2018;10(1):1–5. https://doi.org/10.4240/wjgs.v10.i1.1. PMID: 29391928 PMCID: PMC5785687.
68. Akbulut S, Sahin TT. Comment on surgical approaches for definitive treatment of hepatic alveolar echinococcosis: results of a survey in 178 patients. Parasitology. 2020;147(13):1408–10. https://doi.org/10.1017/s0031182020001390. PMID: 32741385 PMCID: PMC10317737.
69. Kern P, Menezes da Silva A, Akhan O, Mullhaupt B, Vizcaychipi KA, Budke C, Vuitton DA. The echinococcoses: diagnosis, clinical management and burden of disease. Adv Parasitol. 2017;96:259–369. https://doi.org/10.1016/bs.apar.2016.09.006. PMID: 28212790.
70. Bulakçı M, Kartal MG, Yılmaz S, Yılmaz E, Yılmaz R, Şahin D, Aşık M, Erol OB. Multimodality imaging in diagnosis and management of alveolar echinococcosis: an update. Diagn Interv Radiol. 2016;22(3):247–56. https://doi.org/10.5152/dir.2015.15456. PMID: 27082120 PMCID: PMC4859741.
71. Eren S, Aydın S, Kantarci M, Kızılgöz V, Levent A, Şenbil DC, Akhan O. Percutaneous management in hepatic alveolar echinococcosis: a sum of single center experiences and a brief overview of the literature. World J Gastrointest Surg. 2023;15(3):398–407. https://doi.org/10.4240/wjgs.v15.i3.398. PMID: 37032805 PMCID: PMC10080597.
72. Kantarci M, Bayraktutan U, Karabulut N, Aydinli B, Ogul H, Yuce I, Calik M, Eren S, Atamanalp SS, Oto A. Alveolar echinococcosis: spectrum of findings at cross-sectional imaging. Radiographics. 2012;32(7):2053–70. https://doi.org/10.1148/rg.327125708. PMID: 23150858.
73. Koroglu M, Akhan O, Gelen MT, Koroglu BK, Yildiz H, Kerman G, Oyar O. Complete resolution of an alveolar echinococcosis liver lesion following percutaneous treatment. Cardiovasc Intervent Radiol. 2006;29(3):473–8. https://doi.org/10.1007/s00270-005-0017-0. PMID: 16228851.

8 Conventional and Minimally Invasive Treatment of Liver Hydatid Disease

Ali Aloun

8.1 Introduction

Echinococcosis infestations are caused by the larval tapeworm of genus *Echinococcus*. It is a serious disease that is distributed globally and it still constitutes a major public health problem. There are two main types of these parasitic infestations in clinical practice: cystic echinococcosis (CE) and alveolar echinococcosis (AE) caused by *Echinococcus granulosus* and *Echinococcus multilocularis*, respectively. Hydatid disease is considered an endemic disease in countries where sheep are commonly reared. Such countries include Mediterranean countries including North Africa, Spain, and Portugal. The common site affected by the disease is the liver and it usually manifests as cystic lesion that may not be symptomatic [1–3].

The most vulnerable organ affected by these parasites is the liver and this is due to the particular pattern of life cycle of *Echinococcus*. Hepatic echinococcosis may be ignored initially because it is asymptomatic in early-middle stages, and this is the main reason that leads to delayed presentation and high mortality. Recently, more patients with asymptomatic liver echinococcosis are diagnosed early and this is due to the improvement and popularization of examination techniques. Regular long-term use of oral benzimidazoles is a feasible treatment for hepatic echinococcosis. However, owing to the potential serious side effects and heavy economic burdens caused by lifelong drug administration, most people prefer to choose more radical surgical approaches [1, 3–5]. Treatment of hydatid cyst is multidisciplinary involving multiple specialties of medical practice. Some of the specialties include internal medicine, parasitology, interventional radiologist, hepatobiliary surgeons, and others. The treatment of hepatic hydatid cyst may be medical therapy, percutaneous drainage, or surgical intervention, which can be via a conventional or laparoscopic approach [1, 5].

The detailed description of medical and percutaneous radiological intervention has been described in previous chapters. This chapter only focuses on the surgical treatment of hepatic hydatid disease which can be through conventional approach or through minimal access approach. Surgical treatment is considered the best therapeutic choice for uncomplicated hydatid disease of the liver. One of the major advantages of surgical treatment is that it results in complete parasitic eradication; it allows for treatment or prevention of complications and it minimizes the risk of relapse [4, 6].

There is controversy regarding the most appropriate surgical technique that can offer total extirpation of the parasites with minimal postoperative complications. Surgical treatment of hepatic hydatid cyst can be considered as radical surgery or conservative surgery. Radical surgical

A. Aloun (✉)
Department of General Surgery, Jordanian Royal Medical Services, Amman, Jordan

S. Akbulut (ed.), *Hydatid Disease*, https://doi.org/10.1007/978-3-031-97277-5_8

techniques include pericystectomy and hepatic resection, whereas conservative surgical techniques are based on the drainage of the cyst contents, sterilization of the residual cavity, and partial cyst resection. Most reports in the literature suggested that recurrence and complication rates tend to be higher with conservative surgery as compared to those with radical surgery. The high recurrences attributed to conservative surgery are usually caused by failure of complete removal of the endocysts and/or their dissemination during operative intervention [7, 8].

The main aims of surgical treatment of hepatic hydatid diseases include the following:

1. Inactivate the scoleces
2. Prevent spillage of cyst contents
3. Eliminate all viable elements of the cyst
4. Manage the residual cyst cavity.

Postoperative recurrence is considered one of the most important considerations in surgical treatment of hydatid disease. Recurrence may be seen in 8.5%–25% of patients after surgery. The main causes of recurrence include dissemination of protoscolices-rich fluid during surgery and incomplete removal of the germinative membrane from the cyst cavity [7, 9, 10].

8.1.1 Preoperative Medical Therapy

Medical treatment with benzimidazole must be administered to all patients prior to any surgical intervention. The main aim of this medical therapy is to achieve sterilization of the cyst content and prevent dissemination during surgery. It is also associated with decreased risk of anaphylactic reaction during operative intervention [9, 11, 12].

8.1.2 Medical Treatment of Albendazole Include Doses and Duration

Administering albendazole preoperatively at a dosage of 10 mg/kg/day for 1 month results in the sterilization of the hepatic hydatid cyst, eliminating a significant number of protoscolices. However, extended treatment for three uninterrupted months has been reported to enhance the effectiveness of albendazole in liver cyst therapy. The recommended dosage scheme, proposed by Horton (1989) and supported by the WHO, consists of three 28-day courses of 10 mg/kg/day, with doses administered in divided doses and separated by 2-week intervals [13]. The ideal duration of treatment is not firmly established. However, the intake of albendazole before surgery, at a dosage of 10 mg/kg/day for 12 weeks, has shown a 100% success rate in eliminating protoscoleces [14].

8.1.3 Radical vs. Conservative Surgery

Surgical therapy of hepatic hydatid disease can be classified based on the disposal methods of hydatid pericyst. There are two main types of surgical procedures for hepatic hydatid disease, namely, radical surgery which extricates the entire pericyst and conservative surgery that removes the parasite and leave all or part of the pericyst. Options of radical surgery include total pericystectomy and hepatectomy. Conservative surgery includes subtotal pericystectomy, partial pericystectomy, and cystectomy.

8.1.4 Definition of Surgeries

Total pericystectomy involves the complete extraction of the cystic membrane (pericyst) that surrounds the hydatid cyst. The primary goal of this procedure is to remove the cyst without causing rupture, thereby minimizing the risk of spillage and potential complications associated with the dissemination of parasitic material. Opting for total cystopericystectomy eliminates the necessity for hepatic resection [15].

Conversely, hepatectomy is a surgical intervention wherein a section or the entire liver lobe affected by the hydatid cyst is removed. This procedure becomes necessary when the cyst is deeply embedded in the liver, and the associated liver tissue is extensively damaged or nonfunctional [12].

As a conservative surgical option, partial cysto-pericystectomy entails removing both the parasitic foci and the surrounding pericyst. Similar to subtotal pericystectomy, this approach refrains from resecting small pericystic areas in close proximity to vascular and biliary vessels to mitigate the heightened risk of severe complications [16].

The main limitation of conservative procedures is the complications of biliary leakage and fistulas in the postoperative period. Another complication is the septic complications of the residual cavity. The residual pericyst left in situ causes delay liver regeneration which is needed for filling of the residual cavity. This delay leads to accumulation of serum and blood which may cause liver abscess. The residual pericyst may also hide possible biliary communication in the residual cavity leading to biliary fistula, which occurs in up to 50% of patients after conservative procedures. Unlike conservative procedures, radical procedures allow exact detection and ligation of biliary and vascular branches within the hepatic parenchyma and this reduces the risk of biliary leak [1, 4, 6–8, 17–19].

8.1.5 Preoperative Assessment of Cystobiliary Communication

One of the main components of surgical therapy of hepatic hydatid disease is the assessment of communication to the biliary tract. If the assessment is missed during preoperative reparation, it may be a source of significant morbidity in the patients. Patients with recurrent episodes of cholangitis and/or large cysts affecting multiple liver segments are at a higher risk of cystobiliary communications. The search for the communication should be meticulous and multiple reports have shown that MRCP has a high sensitivity and specificity in identifying the communication [7, 8, 20, 21].

8.1.6 Role of Preoperative Endoscopic Retrograde Cholangiopancreatography

While routine endoscopic retrograde cholangiopancreatography (ERCP) is not universally endorsed for uncomplicated hydatid cysts, some clinical centers recommend its utilization to comprehensively assess bile duct anatomy and delineate communication between the cyst and bile ducts. In cases where cystobiliary communication is frequent during liver hydatid cyst surgery (such as with large or multiple cysts, cysts in the hilar or caudate lobe, or cysts accompanied by leukocytosis), the routine application of ERCP in uncomplicated hydatid cystic disease is proposed to mitigate or minimize postoperative biliary complications [22].

There are certain limitations associated with the use of ERCP in diagnosing hydatid disease. Notably, ERCP may fail to detect minor cystobiliary communications due to factors like increased intracystic pressure, limited communication between the cyst and biliary tree, and temporary blockage by daughter cysts at the cystobiliary opening. The higher pressure can prevent complete filling of intrahepatic bile ducts with contrast media, leading to an elevated false-negative rate ranging from 17% to 20%. Additionally, overly intense contrast media may obscure cysts within the main bile ducts, mimicking gallstones with daughter cysts or old hydatid particles. Furthermore, ERCP may overlook hydatid elements exclusively present in intrahepatic ducts. Surgeons should, therefore, maintain vigilance regarding preoperative minor cystobiliary communications, even if they appear to be absent. Some of these limitations can be mitigated by appropriately injecting contrast media from various points or employing occlusion cholangiography. Additionally, ERCP can identify cystobiliary fistulae, filling defects from hydatid elements in bile ducts, and indirect signs of minor cystobiliary communication, such as cystic pressure, dilation, and irregular filling of adjacent bile ducts [22].

8.1.7 Laparoscopic vs. Open Surgical Approach

Surgical treatment of hepatic hydatid disease can be approached via open or laparoscopic approach. Laparoscopic approach is associated with shorter hospital stay, less surgical trauma, lower incidence of surgical site infection, and less postoperative pain. The disadvantages of laparoscopic technique include the following: limited manipulation, difficulty in aspirating the contents of the cyst especially if they are thick and viscous, increased risk spillage of cyst content, and the difficulty in approaching deeply-seated lesions. Additional limitation for the laparoscopic approach is the lack of adequate access to deeply located cysts or cysts located in the posterior segments of the liver. For this reason, laparoscopic approach is limited to superficial cysts on the anterior side of the liver. The choice of open surgery is more approximate in deeply sited cyst, cysts located in the posterior segment of the liver, cysts that are close to large vessels, cysts that contain multiple daughter cells, centrally located cysts, and calcified cysts [7, 8, 12, 23–26].

8.1.8 Robotic Surgery

In recent years, the laparoscopic approach for the treatment of hydatid cysts has become increasingly popular compared to the open approach. Patients undergoing laparoscopic treatment experience benefits such as a shorter hospital stay, minimal blood loss, reduced postoperative pain, and a lower rate of morbidity. However, the laparoscopic approach may face a conversion rate ranging from 4% to 30%, mainly attributed to challenges such as inappropriate exposure of the cyst, the presence of intra-abdominal adhesions, perioperative bleeding, or intraoperative anaphylaxis [27].

To address these limitations, robotic assistance has emerged as a superior alternative to the laparoscopic approach within the realm of minimally invasive surgery. Robotic surgery compensates for challenges encountered in laparoscopic procedures, making it particularly advantageous in intricate spaces where tissue manipulation is restricted by the dexterity and range of motion of the human hand. This is especially crucial when accessing challenging liver segments, such as the caudate lobe or the posterosuperior region, which may be difficult to reach with traditional laparoscopic instruments due to issues like complex anatomy, limited exposure, or proximity to vascular structures.

The robotic approach allows for curved transection lines and maneuvering at difficult angles, enabling the safe excision of cystic lesions in challenging locations. Furthermore, it facilitates the establishment of hemostasis, control of structures in the hepatic hilum, and more comfortable management of vessels. Unlike the laparoscopic approach, robotic surgery overcomes the rigidity of endoscopic instruments and the two-dimensional view of the laparoscopic camera. Instead, it provides a three-dimensional image projection and a filtering system that cancels out human tremor, ensuring precise suture placement for bleeding control and the creation of biliary and enteric anastomoses [27, 28].

8.1.9 The Palanivelu Hydatid System

The Palanivelu hydatid system is designed to prevent spillage of the content of hydatid cyst during laparoscopic interventions which may cause anaphylactic reaction. The system consists of a trocar and cannula along with 5-mm and 3-mm reducers. The trocar is 29 cm in length and it is completely hollow so that it can accommodate a suction cannula. The tip of the trocar is pyramidal shaped and each facet of the pyramid has a fenestration that enables suction of any fluid that leaks when it is inserted. The long shaft of the trocar has two fenestrations opposite each other about 17 cm from the tip. The cannula is 26 cm long with an inner diameter of 12 mm and two side channels. One of the channels is for gas insufflation, while the other channel is for suction. The diameter of the suction channel is about 10 mm and its outer nozzle is designed in such way that the suction tube can be fit onto it in an airtight manner [26, 29].

8.1.10 Cyst Evacuation

This is a safe surgical intervention especially for peripherally located cysts. The approaches can be open or laparoscopic. Open access can be achieved through either abdominal or flank incision and the choice of incision used will depend on cyst location. Once the abdominal cavity is entered, the operation field is lined with hypertonic (20%) saline-soaked gauze to guard against spillage of the fluid in the cyst. The cyst is then opened while the contents are aspirated carefully with a suction machine that can generate high pressure. Once the content has been aspirated, the cyst is then opened completely, which will enable the surgeon to remove any remaining. The remnant cavity may be irrigated with a scolicidal agent [7, 8, 12, 23–26].

Laparoscopic approach is best utilized for cysts located anteriorly and without thick calcified walls. A 11-mm trocar is inserted just above the cyst, and sponges soaked in scolicidal agents (preferably 10% povidone iodine) are placed around the cyst. The cyst is then aspirated with a 14-gauge needle to allow the endocyst to shrink back from the wall to rest at the base of the cyst. The 11-mm trocar is replaced with an 18-mm trocar, and this trocar is used to aspirate the germinal membrane. The laparoscopic camera is then placed directly into the cystic cavity to explore for biliary fistulae or residual daughter cysts. If these are absent, the cavity is then irrigated with a 20% hypertonic saline solution, and the cyst wall is excised. The residual cavity may be closed over a closed suction drain or plugged with omentum [7, 8, 12, 23–26].

8.1.11 The Palanivelu Hydatid System Can Be Used for Laparoscopic Cyst Evacuation

The Palanivelu hydatid system trocar with cannula should be introduced into the peritoneal cavity directly over the hydatid cyst. Once inside the peritoneal cavity, the trocar should be withdrawn gently so that its tip is within the cannula. The cannula should then be advanced till its tip in total contact with the hydatid cyst surface. At this step, strong suction pressure should be applied through the side channel to create a strong negative pressure and maintain airtight contact between the cyst and cannula. Once this is achieved, the trocar containing a 5-mm suction nozzle should be introduced into the cannula and pushed into the cyst along with the cannula by applying gentle steady pressure. Once the Palanivelu hydatid system enters into the hydatid cyst, the trocar should be removed and the cavity irrigated through the main channel while simultaneously maintaining continuous suction.

The deflated cystic wall should then be deroofed using monopolar diathermy. Removal of the laminated membrane and daughter cysts should be done and they are placed into endobag. The telescope should be introduced into the cavity to visualize the interior of the cyst for biliary communication, confirmed by the absence of bile staining gauze inserted into the cystic cavity. The cyst should then be irrigated with 2% savlon for 10 min before it is sucked out [26, 29, 30]. Postoperative morbidity ranges from 8% to 63%, and mortality ranges from 0% to 3%. Major complications like anaphylaxis are more common in laparoscopic interventions as a result of peritoneal spillage during debridement and removal of cyst contents. Cyst evacuation is associated with a recurrence rate of 10% to 30%.

8.1.12 Pericystectomy

Pericystectomy is an improvement over cyst evacuation. It involves complete resection of the cyst wall without entering the cyst cavity. Pericystectomy is done through the plane of the pericyst or along the cyst wall itself. Preoperative localization of the biliary ducts and vasculature is important to reduce complications. If in the course of evaluating the patient the team suspect communication to the bile duct, then preoperative ERCP should be done to confirm the presence of the connection. The procedure is associated with decreased risk of spillage of cyst contents into the peritoneal cavity and this lowers the risk of anaphylaxis and also lowers the risk of recurrence. To reduce risk of excessive intraoperative hemorrhage, ultrasound-guided radiofrequency pericystectomy has been attempted with good results. The use of radiofrequency minimizes blood loss and also results in minimal coagulation on the hepatic parenchyma. It is recommended specifically for cases where the cyst is not located near the liver hilum [1, 4, 19, 31, 32].

Pericystectomy can be total, subtotal, or partial pericystectomy. Total pericystectomy involves complete resection of the cyst wall without entering the cyst cavity. Total pericystectomy can be open or closed. In closed total pericystectomy, the cyst is completely resected by dissecting the potential space between the liver parenchyma and the ectocyst. In open total pericystectomy, the cyst is opened, the cyst contents including echinococcal scolices are removed, and then the pericyst is totally resected. Total pericystectomy is considered the ideal surgical therapy for the treatment of hepatic hydatid disease because it obliterates the residual cystic cavity thereby preventing relapse [7, 21, 33, 34].

In subtotal pericystectomy, there is complete removal of the endocyst with removal of more than 90% of the pericyst. The only part of the capsule left in situ is the small pericystic areas close to the tubular vascular or biliary structures of the liver. Partial pericystectomy can also be termed modified cystectomy, and it involves the resection of the entire endocyst and partial removal of the pericyst (removal of less than 90%). Since pericystectomy eliminates the parasitic foci and all or part of the pericyst, its postoperative recurrence rate is much lower than cystectomy. Partial cystectomy or subtotal pericystectomy are usually performed when the cyst wall abutted the portal vein (peri-hilar) or major hepatic vein and it is considered unsafe for total pericystectomy. These methods can also be utilized if the cyst interfered with the major hepatic vein–inferior vena cava junction superiorly. Lastly, partial or subtotal pericystectomy can be performed, but there is a risk of a low future liver remnant after hepatic resection or total pericystectomy [7, 8, 21, 33, 34].

The approach to pericystectomy can be through open or laparoscopic approach. Open approach can be done through subcostal, transverse, or upper midline incision. Other incisions include inverted J incision and Mercedes Benz incision. Once access to the abdominal cavity is obtained, complete liver mobilization is done. This is followed by identification of the liver hydatid cysts (site and number) and identification of associated cysts (intestine or kidney). Once this has been achieved, gauze towels soaked in hypertonic saline are used to isolate the lesion and safeguard against the risk of spillage of cyst contents into the peritoneal cavity. If the cyst is deeply intraparenchymal, the use of intraoperative ultrasound to identify the location of the cyst has been reported. Pericystectomy is done through the plane of the pericyst or along the cyst wall itself [7, 8, 20, 23, 33].

All patients being planned for laparoscopic approach were positioned in the French position and semi-left lateral position. For lesions located on the right lobe of the liver, four to five ports can be placed under direct visualization. A 10-mm port can be placed 2 cm above and to the right of umbilicus for camera. Two 12-mm ports can be placed about 5 cm to the left and right side of camera port and two 5-mm ports were placed below right and left costal margin for retraction of the liver. For lesions located on the left lobe of the liver, the same trocars can be used in the same positions but 2 cm to the left of the position utilized for right sided lesions. Irrigation of the abdominal cavity is usually performed with hypertonic saline to safeguard against spillage of scolices which may precipitate anaphylactic reaction [23, 25, 30, 35, 36]. Most studies have reported 0.22% mortality and 15% overall complication rate with total pericystectomy [37]. The recurrence rate of 1% was reported and it can occur from 3 to 68 months after the surgery [37].

8.1.13 Hepatic Resection

The role of hepatectomy in the management of hepatic hydatid disease has been a source of controversy. Opponents of hepatectomy as a treatment option believed that resecting a whole part of the liver for a benign parasitic infestation is overkill. They believed that the patient is at a higher risk of intraoperative hemorrhage, and high postoperative complications. But multiple reports have shown that the incidence of intraoperative bleeding and postoperative complications is far less than expected in appropriately selected patients. The long-term results of hepatectomy are also superior to conservative surgeries, so the use of hepatectomy for treatment of hepatic hydatid disease can be justified [12, 21, 38, 39].

The indications for hepatic resection in hydatid disease of the liver include the following

1. Multiple hydatid cysts located in a single liver segment or lobe
2. Coexisting hepatic hydatid cyst with another hepatic lesion affecting the same lobe of the liver
3. Adequate residual liver volume
4. Cysts in a relatively safe location like segment II and III

Hepatectomy for hepatic hydatid disease can be approached via open or laparoscopic approach. Detailed descriptions of the approaches are beyond the scope of this book.

8.1.14 Liver Transplantation as a Treatment Option for Hydatid Cyst

Liver transplantation serves as a viable therapeutic choice for individuals in advanced stages who would not derive benefits from cyst resection. This option becomes particularly relevant when there is involvement of the biliary ducts and blood vessels or the presence of diffuse hepatic disease, resulting in portal hypertension and cirrhosis. In instances where these complications arise, the combination of liver transplantation with antiparasitic drugs stands as the sole potentially curative treatment, offering the prospect of satisfactory long-term results [40].

8.1.15 Postoperative Anthelmintic Treatment Options

8.1.15.1 Duration of Postoperative Albendazole Treatment

In the postoperative context of removing a hydatid cyst to prevent the implantation of protoscolices and potential recurrence, guidelines suggest administering albendazole treatment for a period ranging from 1 to 3 months following the surgery [12, 13].

8.1.16 Management of Postoperative Complications

8.1.16.1 Bile leakage requires keeping the drain in place

Numerous studies indicate that bile leaks can occur in a range of 8.2% to 26% of patients undergoing an open procedure. Such leaks may result from ducts communicating with the residual cavity, leading to cystobiliary communications. Bile leaks indicate that the surgical drain should be kept in place until the amount of leakage drops to less than 50 mL per day [41, 42].

8.1.16.2 Bile Leakage and Role of ERCP

Postoperative bile leakage is a common complication, mainly stemming from substantial fistulas linking the hydatid cavity and the biliary tract. The utilization of ERCP is beneficial in alleviating bile pressure and plays a role in decreasing the likelihood of bile leakage. Additionally, measures such as sphincterotomy or biliary stenting were performed following ERCP to address the bile leaks [43].

Alternative research suggests a preference for a more conservative approach, as it has been observed that the majority of bile leaks tend to resolve spontaneously, especially when characterized by low output. The decision to undergo ERCP and potential interventions such as sphincterotomy or biliary stenting can be influenced by a quantitative assessment of the bile leak. Ahmad et al. suggested that if an efficient drain is left in place during the surgery and is effectively draining bile, the initiation of antibiotics and daily monitoring of drain output may suffice [41].

The likelihood of a cystobiliary fistula is higher in cysts located near the porta hepatis. When dealing with such centrally located cysts, catheter drainage is recommended whenever feasible. The removal of the drainage catheter should only be considered once the daily bile leakage has reduced to approximately 10 cc or below [12]. Radiologic instrument-guided drainage should be considered as the first option in the treatment of collections that develop after hydatid surgery.

References

1. Gomez IGC, López-Andújar R, Belda Ibáñez T, Ramia Ángel JM, Moya Herraiz Á, Orbis Castellanos F, Pareja Ibars E, San Juan Rodríguez F. Review of the treatment of liver hydatid cysts. World J Gastroenterol. 2015;21(1):124–31. https://doi.org/10.3748/wjg.v21.i1.124. PMID: 25574085 PMCID: PMC4284328.
2. Tsaroucha AK, Polychronidis AC, Lyrantzopoulos N, Pitiakoudis MS, A JK, Manolas KJ, Simopoulos CE. Hydatid disease of the abdomen and other locations. World J Surg. 2005;29(9):1161–5. https://doi.org/10.1007/s00268-005-7775-3. PMID: 16088421.
3. Wen H, Vuitton L, Tuxun T, Li J, Vuitton DA, Zhang W, McManus DP. Echinococcosis: advances in the 21st century. Clin Microbiol Rev. 2019;32(2) https://doi.org/10.1128/cmr.00075-18. PMID: 30760475 PMCID: PMC6431127.
4. Sozuer E, Akyuz M, Akbulut S. Open surgery for hepatic hydatid disease. Int Surg. 2014;99(6):764–9. https://doi.org/10.9738/intsurg-d-14-00069.1. PMID: 25437585 PMCID: PMC4254238.
5. Velasco-Tirado V, Alonso-Sardón M, Lopez-Bernus A, Romero-Alegría Á, Burguillo FJ, Muro A, Carpio-Pérez A, Muñoz Bellido JL, Pardo-Lledias J, Cordero M, Belhassen-García M. Medical treatment of cystic echinococcosis: systematic review and meta-analysis. BMC Infect Dis. 2018;18(1):306. https://doi.org/10.1186/s12879-018-3201-y. PMID: 29976137 PMCID: PMC6034244.
6. Dervenis C, Delis S, Avgerinos C, Madariaga J, Milicevic M. Changing concepts in the management of liver hydatid disease. J Gastrointest Surg. 2005;9(6):869–77. https://doi.org/10.1016/j.gassur.2004.10.016. PMID: 15985246.
7. Georgiou GK, Lianos GD, Lazaros A, Harissis HV, Mangano A, Dionigi G, Katsios C. Surgical management of hydatid liver disease. Int J Surg. 2015;20:118–22. https://doi.org/10.1016/j.ijsu.2015.06.058. PMID: 26118608.
8. Vidoura A, Parisidou M, Chatedaki C, Zacharouli D. Surgical management of hydatid disease. In: Inceboz T, editor. Echinococcosis. InTech; 2017.
9. Prousalidis J, Kosmidis C, Anthimidis G, Kapoutzis K, Karamanlis E, Fachantidis E. Postoperative recurrence of cystic hydatidosis. Can J Surg. 2012;55(1):15–20. https://doi.org/10.1503/cjs.013010. PMID: 21939605 PMCID: PMC3270079.
10. Rinaldi F, Brunetti E, Neumayr A, Maestri M, Goblirsch S, Tamarozzi F. Cystic echinococcosis of the liver: a primer for hepatologists. World J Hepatol. 2014;6(5):293–305. https://doi.org/10.4254/wjh.v6.i5.293. PMID: 24868323 PMCID: PMC4033287.
11. Manterola C, Otzen T, Muñoz G, Alanis M, Kruuse E, Figueroa G. Surgery for hepatic hidatidosis. Risk factors and variables associated with postoperative morbidity. Overview of the existing evidence. Cir Esp. 2017;95(10):566–76. https://doi.org/10.1016/j.ciresp.2017.08.009. PMID: 29033069.
12. Öztürk G, Uzun MA, Özkan ÖF, Kayaalp C, Tatlı F, Eren S, Aksungur N, Çoker A, Bostancı EB, Öter V, Kaya E, Taşar P. Turkish HPB Surgery Association consensus report on hepatic cystic echinococcosis (HCE). Turk J Surg. 2022;38(2):101–20. https://doi.org/10.47717/turkjsurg.2022.5757. PMID: 36483170 PMCID: PMC9714645.
13. Arif SH, Shams Ul B, Wani NA, Zargar SA, Wani MA, Tabassum R, Hussain Z, Baba AA, Lone RA. Albendazole as an adjuvant to the standard surgical management of hydatid cyst liver. Int J Surg. 2008;6(6):448–51. https://doi.org/10.1016/j.ijsu.2008.08.003. PMID: 18819855.
14. Vahidirad A, Mansouri M, Shamshirian A, Berenji F, Motie MR. Efficacy of preoperative albendazole on protoscoleces viability in hydatid cyst of the liver. Iran J Parasitol. 2022;17(4):483–7. https://doi.org/10.18502/ijpa.v17i4.11274. PMID: 36694569 PMCID: PMC9825706.
15. Pavlidis ET, Symeonidis NG, Psarras KK, Martzivanou EK, Marneri AG, Stavrati KE, Pavlidis TE. Total cysto-pericystectomy for huge echinococcal cyst located on hepatic segment IVb. Case report and review of the literature. J Surg Case Rep. 2021;2021(1):rjab002. https://doi.org/10.1093/jscr/rjab002. PMID: 33569169 PMCID: PMC7852600.
16. Filippou DK, Kolimpiris C, Anemodouras N, Rizos S. Modified capitonage in partial cystectomy performed for liver hydatid disease: report of 2 cases. BMC Surg. 2004;4:8. https://doi.org/10.1186/1471-2482-4-8. PMID: 15222891 PMCID: PMC446201.
17. Mansy W, Mohamed M, Saber S. Outcomes of radical surgical management in liver hydatid cysts: 7 years center experience. Mini-invasive Surg. 2018;2(10):36. https://doi.org/10.20517/2574-1225.2018.48.
18. Tagliacozzo S, Miccini M, Amore Bonapasta S, Gregori M, Tocchi A. Surgical treatment of hydatid disease of the liver: 25 years of experience. Am J Surg. 2011;201(6):797–804. https://doi.org/10.1016/j.amjsurg.2010.02.011. PMID: 20832053.
19. Wan L, Wang T, Cheng L, Yu Q. Laparoscopic treatment strategies for liver echinococcosis. Infect Dis

Ther. 2022;11(4):1415–26. https://doi.org/10.1007/s40121-022-00664-2. PMID: 35751754 PMCID: PMC9334471.
20. Botezatu C, Mastalier B, Patrascu T. Hepatic hydatid cyst—diagnose and treatment algorithm. J Med Life. 2018;11(3):203–9. https://doi.org/10.25122/jml-2018-0045. PMID: 30364592 PMCID: PMC6197524.
21. Sözen S, Emir S, Tükenmez M, Topuz O. The results of surgical treatment for hepatic hydatid disease. Hippokratia. 2011;15(4):327–9. PMID: 24391414 PMCID: PMC3876848.
22. Dolay K, Akbulut S. Role of endoscopic retrograde cholangiopancreatography in the management of hepatic hydatid disease. World J Gastroenterol. 2014;20(41):15253–61. https://doi.org/10.3748/wjg.v20.i41.15253. PMID: 25386073 PMCID: PMC4223258.
23. Derbel F, Mabrouk MB, Hamida MBH, Mazhoud J, Youssef S, Ali AB, Jemni H, Mama N, Ibtissem H, Nadia A. Hydatid cysts of the liver-diagnosis, complications and treatment. In: Abdominal surgery. InTech; 2012. p. 105–38.
24. Goja S, Saha SK, Yadav SK, Tiwari A, Soin AS. Surgical approaches to hepatic hydatidosis ranging from partial cystectomy to liver transplantation. Ann Hepatobiliary Pancreat Surg. 2018;22(3):208–15. https://doi.org/10.14701/ahbps.2018.22.3.208. PMID: 30215042 PMCID: PMC6125266.
25. Manterola C, Fernández O, Muñoz S, Vial M, Losada H, Carrasco R, Bello N, Barroso M. Laparoscopic pericystectomy for liver hydatid cysts. Surg Endosc. 2002;16(3):521–4. https://doi.org/10.1007/s00464-001-8125-7. PMID: 11928040.
26. Palanivelu C, Jani K, Malladi V, Senthilkumar R, Rajan PS, Sendhilkumar K, Parthasarthi R, Kavalakat A. Laparoscopic management of hepatic hydatid disease. JSLS. 2006;10(1):56–62. PMID: 16709359 PMCID: PMC3015664.
27. Yaghi M, Zorkot M, Kanso M, Faraj W. Robotic resection for hydatid disease of the liver. BMJ Case Rep. 2021;14(6) https://doi.org/10.1136/bcr-2021-241681. PMID: 34155012 PMCID: PMC8217948.
28. Leung U, Fong Y. Robotic liver surgery. Hepatobiliary Surg Nutr. 2014;3(5):288–94. https://doi.org/10.3978/j.issn.2304-3881.2014.09.02. PMID: 25392840 PMCID: PMC4207844.
29. Samala DS, Gedam MC, Gajbhiye R. Laparoscopic management of hydatid cyst of liver with Palanivelu hydatid system over a period of 3 years: a case series of 32 patients. Indian J Surg. 2015;77(Suppl 3):918–22. https://doi.org/10.1007/s12262-014-1064-z. PMID: 27011483 PMCID: PMC4775613.
30. Bakr A, Alsabek MB, Hammadieh AR. Simultaneous laparoscopic partial pericystectomy and cholecystectomy: is it safe? A case report. Int J Surg Case Rep. 2019;57:190–3. https://doi.org/10.1016/j.ijscr.2019.03.028. PMID: 30981074 PMCID: PMC6461566.
31. Kayaalp C, Bzeizi K, Demirbag AE, Akoglu M. Biliary complications after hydatid liver surgery: incidence and risk factors. J Gastrointest Surg. 2002;6(5):706–12. https://doi.org/10.1016/s1091-255x(02)00046-x. PMID: 12399060.
32. Yao P, Morris DL. Radiofrequency ablation-assisted liver resection: review of the literature and our experience. HPB (Oxford). 2006;8(4):248–54. https://doi.org/10.1080/13651820600703272. PMID: 18333135 PMCID: PMC2023895.
33. Stancu B, Andercou O, Pintea D, Mironiuc A, Pop DA. Laparoscopic simultaneous partial pericystectomy and total cystectomy for hydatid liver cysts—case report. Clujul Med. 2015;88(3):415–9. https://doi.org/10.15386/cjmed-416. PMID: 26609279 PMCID: PMC4632905.
34. Yaqub S, Jensenius M, Heieren OE, Drolsum A, Pettersen FO, Labori KJ. Echinococcosis in a non-endemic country—20-years' surgical experience from a Norwegian tertiary referral Centre. Scand J Gastroenterol. 2022;57(8):953–7. https://doi.org/10.1080/00365521.2022.2046846. PMID: 35246008.
35. Ikeda T, Toshima T, Harimoto N, Yamashita Y, Ikegami T, Yoshizumi T, Soejima Y, Shirabe K, Maehara Y. Laparoscopic liver resection in the semi-prone position for tumors in the anterosuperior and posterior segments, using a novel dual-handling technique and bipolar irrigation system. Surg Endosc. 2014;28(8):2484–92. https://doi.org/10.1007/s00464-014-3469-y. PMID: 24622763 PMCID: PMC4077249.
36. Yin Z, Jin H, Ma T, Wang H, Huang B, Jian Z. Laparoscopic hepatectomy versus open hepatectomy in the management of posterosuperior segments of the liver: a systematic review and meta-analysis. Int J Surg. 2018;60:101–10. https://doi.org/10.1016/j.ijsu.2018.10.040. PMID: 30389536.
37. Tuxun T, Zhang JH, Zhao JM, Tai QW, Abudurexti M, Ma HZ, Wen H. World review of laparoscopic treatment of liver cystic echinococcosis—914 patients. Int J Infect Dis. 2014;24:43–50. https://doi.org/10.1016/j.ijid.2014.01.012. PMID: 24747089.
38. Salama IA, Abdelsalam OH, Saleh SM, Shoreem HAM, Houseni M, Amir E-AM. Major hepatic resection in hepatic hydatidosis. Surg Sci. 2013;4(1):27026. https://doi.org/10.4236/ss.2013.41004.
39. Yahya AI, Shwereif HE, Ekheil MA, Thoboot AS, Algader KA, Gyaed FO, Aldarat AS. The role of emergency surgery in hydatid liver disease. World J Emerg Surg. 2014;9:12. https://doi.org/10.1186/1749-7922-9-12. PMID: 24479989 PMCID: PMC4012255.
40. Castro AR, Hyppolito EB, Valença-Júnior JT, Coelho GR, Garcia JHP. Preliminary data on liver transplantation in hydatidosis disease. Arq Bras Cir Dig. 2022;35:e1667. https://doi.org/10.1590/0102-672020210002e1667. PMID: 35766612 PMCID: PMC9254553.
41. Faraj W, Malek E, Itani N, El Naaj A, El Nounou G, El Merhi F, Haidar A, Khalife M. Management of bile leak after laparoscopic hydatid liver surgery. CRSLS.

2014;60:e2014.00060. https://doi.org/10.4293/CRSLS.2014.00060.

42. Kayaalp C, Aydin C, Olmez A, Isik S, Yilmaz S. Leakage tests reduce the frequency of biliary fistulas following hydatid liver cyst surgery. Clinics (Sao Paulo). 2011;66(3):421–4. https://doi.org/10.1590/s1807-59322011000300010. PMID: 21552666 PMCID: PMC3072002.

43. Jiang T, Wang Z, Aji T, Ran B, Guo Q, Zhang R, Tuergan T, Zhong K, Shao Y, Hao W. ERCP management of acute cholangitis caused by rupture of Echinococcus hepaticus into the biliary tract. J Minim Access Surg. 2023;19(4):498–503. https://doi.org/10.4103/jmas.jmas_219_22. PMID: 37282433 PMCID: PMC10695320.

Treatment of Complicated Hydatid Disease

9

Ibrahim Umar Garzali and Ali Aloun

9.1 Introduction

Hydatid disease is a parasitic disease that is distributed globally. It is a zoonotic infection caused by the larval stage of the tapeworm *Echinococcus*. There are two recognized species that are important in human diseases. They are *E. granulosus* and *E. multilocularis*. The infection is acquired through ingestion of the eggs of the parasite released into the feces of the definite host. Once the eggs are ingested by the intermediate host, the eggs hatch and migrate to a broad category of tissues to form multilayered metacestode cysts. The outmost layer of a cyst, the ectocyst or pericyst, is formed by compressed and fibrotic host tissue and may become calcified, whereas inner layers are of parasitic origin and act as a germinal center. Hydatid disease can also be termed hydatid cyst. Up to 75% of all hydatid cysts are found in the liver. Patients may be asymptomatic for years and usually present with nonspecific complaints. However, at any time any particular hydatid cyst can develop complications that can be life threatening unless treated early [1–7].

Complications are seen in about 40% of patients with hepatic hydatid disease. In some cases, complication may be the first manifestation of underlying hepatic hydatid disease, while in other cases, complications may develop in patients being treated medically or patients being prepared for operative intervention after diagnosis. Complications such as perforations are much more common in echinococcal disease, whereas alveolar echinococcal disease invades neighboring tissues like a tumor and therefore causes compression complications rather than perforation [8–11].

The most common complication of liver hydatid disease is superimposed infection, followed by rupture to the biliary tree, rupture to the peritoneal cavity, and rupture to the pleural cavity or the bronchus [8, 10, 12].

Hydatid cyst rupture can be classified into contained, communicating, and direct type of rupture. In contained rupture only the endocyst ruptures, but the pericyst remains intact and contains the spread of cyst content. In cases of communicating rupture, the endocyst ruptures, the pericyst is intact, but cyst contents escape through biliary or bronchial radicles incorporated in the pericyst. In direct rupture, there is rupture of both endocyst and pericyst resulting in the spread of cyst content to the pleura, peritoneum, or pericardium [3, 6, 8, 13].

9.2 Infected Hydatid Cyst

An infected cyst is defined as a symptomatic cyst presenting with signs of infection, pus during surgery, and a positive microbiology and culture.

I. U. Garzali (✉)
Department of Surgery, Aminu Kano Teaching Hospital, Kano, Nigeria

A. Aloun
Department of Hepatobiliary Surgery, Jordanian Royal Medical Services, Amman, Jordan

S. Akbulut (ed.), *Hydatid Disease*, https://doi.org/10.1007/978-3-031-97277-5_9

It is the most common complication associated with liver hydatid cyst. Infection is seen in 5–25% of patients with hydatid disease [14, 15].

9.2.1 Pathogenesis

For infection to occur in a hydatid cyst, leakage of the cyst might have occurred since the laminated membrane is a potent antibacterial barrier. Leakage allows entry of serum or bile into the cyst, and this makes it a perfect medium for bacterial growth. Another mechanism of infection in hepatic hydatid is bacteremia arising from a variety of sites. Some risk factors for bacteremia and superimposed infection among patients with hydatid cyst include cirrhosis, malignancies, diabetes mellitus, and steroid therapy [14–17].

9.2.2 Bacteriology

Analysis and culture of hydatid cystic fluid in cases of infected hydatid cyst results in isolation of the following bacteria: *Escherichia coli*, *Klebsiella pneumonia*, *Proteus mirabilis*, *Enterobacter aerogenes*, coagulase-positive and coagulase-negative *Staphylococci*, and *Pseudomonas aeroginosa*. The most common bacteria were *E. coli* (24–36%) and *K. pneumonia* (23–43%). Superimposed infection by *Aspergillus fumigatus* has also been reported in some series [14, 16, 18].

9.2.3 Clinical Features and Diagnosis

The most frequent symptoms and signs in these patients include fever with chills and localized right upper quadrant pain. Patient may be on follow up for asymptomatic hydatid cyst, or this may be the first manifestation of the disease. Clinical examination may reveal tender hepatomegaly [10, 12, 17, 19].

Laboratory investigations may be unchanged, but in more than 30% of patients, leukocytosis (>12 × 103/μL) and eosinophilia (>450/μL) may be detected. Magnetic resonance imaging and CT scan imaging are used in the evaluation of hydatid cysts and in infected cysts; the following findings may be evident. Infected cyst have poorly defined margins with perilesional inflammation, presence of gas, air-fluid level, and/or communication with hollow viscera or biliary tree [8, 10, 12, 16, 20–22].

9.2.4 Treatment

Treatment of infected hydatid cyst is surgical. The surgical options for treatment include cystopericystectomy, partial cystopericystectomy, cystotomy, deroofing and drainage, percutaneous drainage, endoscopic drainage, laparoscopic drainage, hepatectomy, and marsupialization. Among the options, percutaneous drainage is associated with the least postintervention morbidity seen in 14.8% of patients, while marsupialization has the most morbidity of 33.3%. The commonest morbidity associated with surgical treatment of infected hydatid cyst is surgical site infection (24.8%) followed by biliary fistula (11.6%). Antibiotic therapy is also needed in patients with infected hydatid cysts, and the regimen usually comprises of penicillins, cephalosporins, and aminoglucosides [8, 11, 12, 14, 15, 20, 21, 23–25].

9.3 Intrabiliary Rupture of the Hydatid Cysts

This is one of the most common complications of hepatic hydatid cysts. The incidence of intrabiliary rupture of hydatid cyst ranges from 3% to 17% [26].

9.3.1 Pattern of Rupture

There are two pattern of intrabiliary rupture of hydatid cyst. The first is the occult rupture and it is seen 10–37% of patients. In this type of rupture, only the cystic fluid drains to the biliary tree. In occult rupture, the communication between the biliary tree and the cyst is less than 5 mm in diameter. The second pattern is the frank rupture, which is seen in 3–17% of patients. It is seen when the diameter of communication between the cyst and biliary tree is more than or equals to 5 mm. In this type of rupture, there is passage of intracystic material to the biliary tract. Once the rupture occurs, there is migration of the cystic content into the biliary tree which leads to other biliary complications like: cholangitis, hydatid biliary lithiasis, etc. [10, 26–29].

9.3.2 Risk Factors for Intrabiliary Rupture

Multiple risk factors have been associated with intrabiliary rupture of hepatic hydatid cyst. One of the most commonly encountered risk is the location of the cysts. Intrabiliary rupture is commonly seen in centrally localized cysts. Another risk for intrabiliary rupture is the intracystic pressure. The risk of intrabiliary rupture of hepatic hydatid cyst increases if the intracystic pressure is up to 80 cm of water [10, 26–29].

9.3.3 Sites Affected

Different parts of the biliary tract are affected differently by rupture of hydatid cysts. Right hepatic duct is affected in 60% cases. The left hepatic duct is affected in 30% cases. Other part of the biliary tract such as hepatic duct junction, common bile duct (CBD), or cystic duct are affected in 11%. Rupture into the gall bladder is seen in 5–6% of cases [10, 26, 27].

9.3.4 Clinical Features and Diagnosis

In patients with occult rupture, there may not be change in the symptomatology of the patient. However, in patients with frank rupture, they manifest with biliary obstruction and cholangitis.

Diagnosis of intrabiliary rupture is made using imaging techniques such as ultrasound and CT scan. Imaging reveals dilatation of biliary tracts, daughter vesicles or floating membranes within the biliary radicles, defect on the cyst wall, and cystic lesion in the liver. Magnetic resonance cholangiopancreatography (MRCP) can also be used for the diagnosis of the intrabiliary rupture of hydatid cysts. Endoscopic retrograde cholangiopancreatography (ERCP) can also be used preoperatively in diagnosing intrabiliary rupture of hydatid disease. But because of its role in therapy, it is usually reserved for MRCP diagnosed intrabiliary rupture where it is done for interventions. In addition to diagnosing intrabiliary rupture, the uses of ERCP/MRCP include identification of the type of communication and anatomical location of the communication of the biliary duct with the cyst. The latter information is needed to decide the appropriate operative procedure [10, 15, 26–30]. For patients with occult intrabiliary rupture, diagnosis may not be possible before surgery. It may be identified intraoperatively when bile stained cystic content is observed after initial aspiration to reduce the pressure within the cystic cavity.

9.3.5 Treatment

In patients in which intrabiliary rupture is diagnosed preoperatively, ERCP with endoscopic sphincterotomy can be done. It has multiple applications in patients with intraoperative rupture of hydatid cyst. It serves as an initial treatment for patients with acute presentation like cholangitis and obstruction, allowing time before elective surgery for definitive treatment. It can also be used to evacuate contents from the biliary tract and, by extension, the hydatid cyst. Studies have shown that preoperative ERCP in patients with intrabiliary rupture is associated with reduced complication rates like postoperative biliary fistula and is also associated with reduced hospitalization. In up to 25% of patients with intrabiliary rupture of hydatid disease, ERCP and endoscopic sphincterotomy has resulted in complete resolution with no need for definitive surgery [10, 15, 21, 26–30].

In patients in whom the biliary communication is identified intraoperatively, the anatomy of the communication can be highlighted by intraoperative cholangiography to delineate the point of communication. If the communication is not with major biliary ductules, then the point of communication can be sutured. However, if communication is to the primary or secondary biliary pedicles, primary closure can case biliary obstruction for distal branches. So for this group of patients, radical surgery is recommended [10, 15, 21, 26–30].

Intraoperative identification of biliary rupture requires a high index of suspicion. Small lesion can be seen by applying a dry pad on the cyst's wall while applying pressure on the gallbladder, and methylene blue injection or indocyanine green injection through the cystic duct. Once the point of leak is identified, it can be closed by suture ligation depending on the anatomy [10, 27, 29–31].

9.4 Rupture into the Thoracic Cavity

Intrathoracic rupture of hydatid cyst is a rare complication of liver hydatid cyst. It is seen in approximately 0.6–6% of patients with rupture. The rupture can cause a spectrum of pathologies in the pleura, bronchi, and lung parenchyma. Rupture in to the pericardium has also been reported [10, 32–34].

9.4.1 Pathogenesis of Intrathoracic Rupture

Rupture into the thorax can be precipitated by several factors. One of these factors is the pressure gradient between thoracic and abdominal cavities especially in patients with hydatid cyst of the hepatic dome. This negative pressure results in aspiration of the hepatic hydatid cyst toward the thoracic cavity. Other factors include mechanical compression, ischemia, and subsequent erosion of the diaphragm musculature. Sepsis in the hepatic cyst has also been associated with rupture of the cyst resulting from necrosis of diaphragmatic musculature. The role of biliary communication in the rupture of hydatid cyst has been documented and it is believed to be through chemical erosion of diaphragm, lung, and pleura by bile [10, 32, 33].

9.4.2 Types of Rupture

Type I rupture is the type of rupture that directly communicates with the bronchial tree.

IA: Small bronchial fistulas
IB: Major bronchial fistulas

Type II rupture is the type of rupture that ruptures into the lung parenchyma with formation of a suppurative cavity containing hydatid material and sometimes biliary secretions.

IIA: Small bronchial fistulas
IIB: Major bronchial fistulas

Type III rupture usually result from rupture in to the pleura cavity with formation of encysted pleural pouch that communicates with the hepatic cyst. The pouch can subsequently rupture into the bronchi.

IIIA: No bronchial fistulas
IIIB: With bronchial fistulas
IIIC: With pleuroparietal fistulas

Type IV rupture is the type of rupture that results from acute rupture into the pleural cavity with formation of hydatic and biliary empyema.

IVA: Acute rupture: hydatic and biliary empyema
IVB: Secondary pleural hydatidosis

9.4.3 Clinical Features and Diagnosis

The clinical manifestation of intrathoracic rupture is predominately pulmonary, with less frequent abdominal symptoms.

Coughs productive of bile-stained sputum, hydatic vomica (the expectoration of the clear and salty cyst contents through the mouth and nose), and particularly hydatidoptysis (expectoration with daughter cysts sometimes bile-stained) are pathognomonic. These symptoms are not seen frequently and only one of three patients has all the symptoms. Diagnosis can be confirmed by imaging investigation. Ultrasonography shows the hepatic hydatid cyst usually at the dome of the liver in proximity and in continuity with the thoracic lesions. Continuity is demonstrated by identifying the hole in the diaphragm. Ultrasonography also plays an important role in deciding the approach to treatment by exploring the biliary tract to detect obstruction. Computer tomographic imaging shows the connections between intrathoracic lesions and the hepatic cyst, but the diaphragmatic gap may not be visualized. Sagittal and frontal MRI cross-sections provide better diaphragmatic assessment than CT scan or ultrasound [8, 10, 12, 32, 33, 35, 36].

9.4.4 Treatment

The mainstay of treatment of hepatic hydatid cysts with intrathoracic rupture is surgery. The main goals of surgical treatment include:

1. Treat the liver cyst
2. Ensure free-flowing biliary fluid
3. Hepato-diaphragmatic disconnection
4. Treatment of intrathoracic lesions
5. Diaphragmatic restoration

The approach thus depends on findings on imaging investigation and the type of intrathoracic rupture. For patients with type I rupture, abdominal approach can be used in treatment of the patients. For patients with type II, III, and IV ruptures, thoracoscopic approach is needed. In some patients with type IB rupture, there may be accompanying bronchiectasis which requires parenchymal resection that can only be performed via thoracoscopic approach. Combined abdominothoracic approach is needed for patients with type II, III, and IV rupture, with ultrasound findings of dilated biliary ducts. However, with the advent of endoscopic sphincterotomy, the combined thoracic and abdominal approach can be avoided [8–10, 12, 15, 32, 33, 35–37].

9.5 Rupture into the Peritoneal Cavity

This is one of the few complications of hepatic hydatid diseases. It is a rare complication and it may be complicated by anaphylactic reactions. Most series have reported intraperitoneal rupture of hepatic hydatid cyst to occur in 1% and 8% of patients [5, 6, 38, 39].

9.5.1 Pathogenesis and Risk Factors

Risk factors associated with increased risk of intraperitoneal rupture in hepatic hydatid cyst include the following: young age of patients, patients with superficially located cysts, patients with cyst located at anterior or inferior surface of the liver, and those with cyst of more than 10 cm. Intraperitoneal rupture of hepatic hydatid cyst may occur spontaneously as a result of increase in the pressure within the cysts. Another factor that may precipitate rupture is trauma to the abdomen or in some cases lower part of the chest [7, 40–42].

9.5.2 Clinical Features and Diagnosis

Most cases of intraperitoneal rupture of hepatic hydatid cyst usually present as emergency. Patients usually present with acute onset of abdominal pain. Patients also present with associated nausea, vomiting, and urticaria.

Sudden rupture into the peritoneum may present as acute abdominal pain. Antigenic fluid released into the peritoneal cavity and absorbed into the circulation may present with acute allergic manifestations. In about 25% of patients with intraperitoneal rupture of hydatid cyst, profound hypotension may be observed due to anaphylactic reaction to the peritoneal absorption of the cyst content. If there is biliary communication of the cyst before rupture, then the patients will present with features of peritonitis or hydatid choleperitonitis. Abdominal examination usually reveals guarding, tenderness, and rebound tenderness [5, 6, 8, 40–42].

Some patients may have insidious intraperitoneal rupture of hydatid cyst. In these patients, there is gradual release of scolices and daughter cysts into the peritoneal cavity. These daughter cysts now implant in other parts of the peritoneal cavity which leads to multiple cysts in different parts of the peritoneal cavity. This is what is considered secondary echinococcus [7, 10, 40–42].

The diagnosis of intraperitoneal rupture can be made by imaging. Abdominal ultrasonography or abdominopelvic CT scan is the imaging of choice for diagnosing intraperitoneal rupture of hepatic hydatid cyst with sensitivity of 85% and 100%, respectively. They reveal the collapsing of the hepatic hydatid cyst, multiple peritoneal cyst, and peritoneal effusion [10, 43–46].

9.5.3 Treatment

Immediate resuscitation is required in patients with acute rupture of hepatic hydatid cyst with peritonitis, hypovolemia, and anaphylaxis. Once the patients have been stabilized and a diagnosis of acute intraperitoneal rupture of hepatic hydatid is suspected, then surgical intervention is required [4–6, 13, 24]. The aims of surgical treatment include the following [13, 21, 23–25, 37]:

1. Eliminate local disease
2. Prevent complications
3. Minimize morbidity, mortality, and recurrence rates

The approach to surgical interventions in patients with ruptured hydatid cyst is usually through a laparotomy. The role of laparoscopic approach in intraperitoneal rupture of hydatid cyst is limited. In stable patients, elimination of local disease is achieved by the treatment of hepatic hydatid disease. Surgical treatment of hepatic hydatid disease can be radical or conservative treatment. In patients with intraperitoneal rupture of hydatid cyst, the conservative approach is preferred by most surgeons. These approaches include deroofing and filling of the cavity or external drainage. In nonendemic areas, the radical approach such as hepatectomy or pericystectomy is preferred by some surgeons. Removal of all secondary scolices during the surgery is very important [18, 47–52].

One of the most important interventions during laparotomy for intraperitoneal rupture of hydatid disease is the use of scolicidal agents to adequately irrigate the peritoneal cavity. Some examples of the commonly used scolicidal agents include: hypertonic saline (3–30%), formalin (2%), povidone iodine (10%), chlorhexidine (0.05%), silver nitrate (0.5%), and a combination of cetrimide (0.5%) and chlorhexidine (0.4%). By irrigating the peritoneal cavity with scolicidal agents, there is inactivation of the daughter cyst, and this prevents re-implantation and recurrence. Some series report hypernatremia with the use of hypertonic saline irrigation [21, 53, 54].

Postoperative albendazole therapy for 6 months is indicated for patients that had laparotomy for ruptured hydatid cysts. It is associated with reduced recurrence and reduced risk of secondary hydatidosis. Postoperative mortality of 3% has been reported after surgery for intraperitoneal rupture of hepatic hydatid cysts. The most common causes of death are septic shock and multiorgan failure [3, 5, 6, 10, 13, 21].

9.6 Rupture of Pulmonary Hydatid Cyst

The lung is the second most commonly affected site in adults but in childhood, the lungs are the commonest site. It is usually asymptomatic with diagnosis occurring incidentally during imaging. Symptom usually occurs following the rupture of the cyst. Cyst rupture might be seen spontaneously, following a trauma, or after medical therapy. Pulmonary hydatid cyst may rupture into the pleural cavity or bronchus. Treatment of ruptured pulmonary hydatid cyst may involve decortication, segmentectomy, thoracotomy, and lobectomy [55–60].

9.7 Parietal Complications of Hydatid Cyst

Parietal complications of hydatid cyst occur as a result of cutaneous communication of cyst located in the thoracoabdominal cavities. These complications are categorized into three stages depending on the depth of invasion of the cysts at the abdominal or thoracic wall: stage 1 lesions are lesions in which the hydatid cyst protrudes into the mural muscular layer; stage 2 lesions are lesions in which the hydatid cyst protrudes beyond the muscular layer to invade or protrude into the subcutaneous layer; and stage 3 lesions are lesions in which the hydatid cyst protrudes beyond the subcutaneous tissue to communicate with the external environment thereby forming a fistula. Parietal complications of hydatid cyst can occur spontaneously or it can follow surgical intervention for an organic hydatid cyst. Treatment of the parietal complication is aimed at the treatment of the underlying organ that bears the hydatid cyst, and en bloc resection of the primary hydatid cyst, fistula tract, and involved skin is effective [61–66].

9.8 Vascular Complications of Hepatic Hydatid Cysts

Extrinsic compression of hepatic vasculature and intravascular rupture of hepatic hydatid cysts have been reported but they are very rare. Extrinsic compression of intrahepatic or extrahepatic vascular structures (portal trunk, hepatic vein, and inferior vena cava) is well tolerated and asymptomatic. However, intravascular rupture (venous and/arterial system) is a very serious complication. These complications are said to occur in 0.8% of hepatic hydatid cysts, but with significant operative mortality. This intravascular rupture remains one of the most serious complications of the hydatid cyst, which can lead to death from anaphylactic shock. Extrinsic compression will present as portal hypertension (when there is compression of the portal bifurcation and the portal trunk) or Budd-Chiari syndrome (when there is compression of the cavo-hepatic vein junction). Intravascular rupture will result in release of vesicles into the circulation. This will cause two things. The first is the obstruction of the vessel with stasis, and this will result in an attempt to bypass the point of obstruction which will manifest as portal cavernoma. The second and the most serious response is anaphylactic shock occurring as a result of release of the daughter vesicles into the circulation [67–76].

References

1. Derici H, Tansug T, Reyhan E, Bozdag AD, Nazli O. Acute intraperitoneal rupture of hydatid cysts. World J Surg. 2006;30(10):1879–83; discussion 1884–5. https://doi.org/10.1007/s00268-005-0699-0. PMID: 16847712.
2. Koc C, Akbulut S, Sahin TT, Tuncer A, Yilmaz S. Intraperitoneal rupture of the hydatid cyst disease: single-center experience and literature review. Ulus Travma Acil Cerrahi Derg. 2020;26(5):789–97. https://doi.org/10.14744/tjtes.2020.32223. PMID: 32946087.
3. Lencioni R, Franchini C. Hydatid disease, abdominal. In: Baert AL, editor. Encyclopedia of diagnostic imaging. Berlin/Heidelberg: Springer; 2008. p. 894–8.
4. Mejri A, Arfaoui K, Ayadi MF, Aloui B, Yaakoubi J. Primitive isolated hydatid cyst of the spleen: total splenectomy versus spleen saving surgical modalities. BMC Surg. 2021;21(1):46. https://doi.org/10.1186/s12893-020-01036-8. PMID: 33472623 PMCID: PMC7819254.
5. Mejri A, Arfaoui K, Omry A, Yaakoubi J, Mseddi MA, Rchidi J, Saad S, Ellouze MM. Acute intraperitoneal rupture of hydatid cysts of the liver: case series. Medicine (Baltimore). 2021;100(44):e27552. https://doi.org/10.1097/md.0000000000027552. PMID: 34871219 PMCID: PMC8568461.
6. Mouaqit O, Hibatallah A, Oussaden A, Maazaz K, Taleb KA. Acute intraperitoneal rupture of hydatid cysts: a surgical experience with 14 cases. World J Emerg Surg. 2013;8:28. https://doi.

org/10.1186/1749-7922-8-28. PMID: 23885766 PMCID: PMC3725171.
7. Tinsley B, Abbara A, Kadaba R, Sheth H, Sandhu G. Spontaneous intraperitoneal rupture of a hepatic hydatid cyst with subsequent anaphylaxis: a case report. Case Reports Hepatol. 2013;2013:320418. https://doi.org/10.1155/2013/320418. PMID: 25431702 PMCID: PMC4238233.
8. Alexiou K, Mitsos S, Fotopoulos A, Karanikas I, Tavernaraki K, Konstantinidis F, Antonopoulos P, Ekonomou N. Complications of hydatid cysts of the liver: spiral computed tomography findings. Gastroenterology Res. 2012;5(4):139–43. https://doi.org/10.4021/gr460e. PMID: 27785194 PMCID: PMC5051081.
9. Aribas OK, Kanat F, Gormus N, Turk E. Pleural complications of hydatid disease. J Thorac Cardiovasc Surg. 2002;123(3):492–7. https://doi.org/10.1067/mtc.2002.119341. PMID: 11882820.
10. Derbel F, Mabrouk MB, Hamida MBH, Mazhoud J, Youssef S, Ali AB, Jemni H, Mama N, Ibtissem H, Nadia A. Hydatid cysts of the liver-diagnosis, complications and treatment. In: Derbel F, editor. Abdominal surgery. IntechOpen; 2012. p. 105–38.
11. Sozuer E, Akyuz M, Akbulut S. Open surgery for hepatic hydatid disease. Int Surg. 2014;99(6):764–9. https://doi.org/10.9738/intsurg-d-14-00069.1. PMID: 25437585 PMCID: PMC4254238.
12. Brunetti E, Kern P, Vuitton DA. Expert consensus for the diagnosis and treatment of cystic and alveolar echinococcosis in humans. Acta Trop. 2010;114(1):1–16. https://doi.org/10.1016/j.actatropica.2009.11.001. PMID: 19931502.
13. Altintop I, Yilmaz M, Alincak Z, Vural A. Fatal rupture of a giant hydatid cyst in an elderly patient. Open J Emerg Med. 2016;4(1):1–5. https://doi.org/10.4236/ojem.2016.41001.
14. García MB, Lledías JP, Pérez IG, Tirado VV, Pardo LF, Bellvís LM, Varela G, Sánchez MC. Primary super-infection of hydatid cyst—clinical setting and microbiology in 37 cases. Am J Trop Med Hyg. 2010;82(3):376–8. https://doi.org/10.4269/ajtmh.2010.09-0375. PMID: 20207859 PMCID: PMC2829895.
15. Malik AA, Bari SU, Amin R, Jan M. Surgical management of complicated hydatid cysts of the liver. World J Gastrointest Surg. 2010;2(3):78–84. https://doi.org/10.4240/wjgs.v2.i3.78. PMID: 21160854 PMCID: PMC2999219.
16. Belhassen-García M, Romero-Alegria A, Velasco-Tirado V, Alonso-Sardón M, Lopez-Bernus A, Alvela-Suarez L, del Villar LP, Carpio-Perez A, Galindo-Perez I, Cordero-Sanchez M, Pardo-Lledias J. Study of hydatidosis-attributed mortality in endemic area. PLoS One. 2014;9(3):e91342. https://doi.org/10.1371/journal.pone.0091342. PMID: 24632824 PMCID: PMC3954695.
17. Khazaei S, Rezaeian S, Khazaei Z, Goodarzi E, Khazaei S, Mohammadian M, Salehiniya H, Ayubi E, Mohammadian-Hafshejani A. Epidemiological and clinical characteristics of patients with hydatid cysts in Khorasan Razavi Province, from 2011 to 2014. Iran J Parasitol. 2016;11(3):364–70. PMID: 28127342 PMCID: PMC5256053.
18. Yahya MQ, Mohi-Aldeen ZK-A. Secondary bacterial infection of hydatid cysts infected livestock animals (in vitro study). Iraqi J Vet Sci. 2022;36(3):761–8. https://doi.org/10.33899/ijvs.2022.131873.2016.
19. Siracusano A, Teggi A, Ortona E. Human cystic echinococcosis: old problems and new perspectives. Interdiscip Perspect Infect Dis. 2009;2009:474368. https://doi.org/10.1155/2009/474368. PMID: 19888428 PMCID: PMC2771156.
20. Baruah A, Sarma K, Barman B, Phukan P, Nath C, Boruah P, Rajkhowa P, Baruah M, Dutta A, Naku N. Clinical and laboratory presentation of hydatid disease: a study from Northeast India. Cureus. 2020;12(9):e10260. https://doi.org/10.7759/cureus.10260. PMID: 33042698 PMCID: PMC7537570.
21. Öztürk G, Uzun MA, Özkan ÖF, Kayaalp C, Tatlı F, Eren S, Aksungur N, Çoker A, Bostancı EB, Öter V, Kaya E, Taşar P. Turkish HPB Surgery Association consensus report on hepatic cystic echinococcosis (HCE). Turk J Surg. 2022;38(2):101–20. https://doi.org/10.47717/turkjsurg.2022.5757. PMID: 36483170 PMCID: PMC9714645.
22. Rasheed K, Zargar SA, Telwani AA. Hydatid cyst of spleen: a diagnostic challenge. N Am J Med Sci. 2013;5(1):10–20. https://doi.org/10.4103/1947-2714.106184. PMID: 23378949 PMCID: PMC3560132.
23. Deo KB, Kumar R, Tiwari G, Kumar H, Verma GR, Singh H. Surgical management of hepatic hydatid cysts—conservative versus radical surgery. HPB (Oxford). 2020;22(10):1457–62. https://doi.org/10.1016/j.hpb.2020.03.003. PMID: 32229090.
24. Marom G, Khoury T, Gazla SA, Merhav H, Padawer D, Benson AA, Zamir G, Luques L, Safadi R, Khalaileh A. Operative treatment of hepatic hydatid cysts: a single center experience. Asian J Surg. 2019;42(6):702–7. https://doi.org/10.1016/j.asjsur.2018.09.013. PMID: 30446425.
25. Stoot JH, Jongsma CK, Limantoro I, Terpstra OT, Breslau PJ. More than 25 years of surgical treatment of hydatid cysts in a nonendemic area using the "frozen seal" method. World J Surg. 2010;34(1):106–13. https://doi.org/10.1007/s00268-009-0267-0. PMID: 19898895 PMCID: PMC2795857.
26. Avcu S, Ünal Ö, Arslan H. Intrabiliary rupture of liver hydatid cyst: a case report and review of the literature. Cases J. 2009;2:6455. https://doi.org/10.1186/1757-1626-2-6455. PMID: 19829807 PMCID: PMC2709968.
27. Atli M, Kama NA, Yuksek YN, Doganay M, Gozalan U, Kologlu M, Daglar G. Intrabiliary rupture of a hepatic hydatid cyst: associated clinical factors and proper management. Arch Surg.

2001;136(11):1249–55. https://doi.org/10.1001/archsurg.136.11.1249. PMID: 11695968.
28. Greco S, Cannella R, Giambelluca D, Pecoraro G, Battaglia E, Midiri M, Brancatelli G, Vernuccio F. Complications of hepatic echinococcosis: multimodality imaging approach. Insights Imaging. 2019;10(1):113. https://doi.org/10.1186/s13244-019-0805-8. PMID: 31792750 PMCID: PMC6889260.
29. Manouras A, Genetzakis M, Antonakis PT, Lagoudianakis E, Pattas M, Papadima A, Giannopoulos P, Menenakos E. Endoscopic management of a relapsing hepatic hydatid cyst with intrabiliary rupture: a case report and review of the literature. Can J Gastroenterol. 2007;21(4):249–53. https://doi.org/10.1155/2007/410308. PMID: 17431515 PMCID: PMC2657701.
30. Ahire P, Iyer N, Gada PB. Complication of hepatic hydatid cyst surgery presenting as obstructive jaundice. Cureus. 2023;15(2):e35410. https://doi.org/10.7759/cureus.35410. PMID: 36994267 PMCID: PMC10042515.
31. Buddingh KT, Nieuwenhuijs VB, van Buuren L, Hulscher JB, de Jong JS, van Dam GM. Intraoperative assessment of biliary anatomy for prevention of bile duct injury: a review of current and future patient safety interventions. Surg Endosc. 2011;25(8):2449–61. https://doi.org/10.1007/s00464-011-1639-8. PMID: 21487883 PMCID: PMC3142332.
32. Kabiri EH, El Maslout A, Benosman A. Thoracic rupture of hepatic hydatidosis (123 cases). Ann Thorac Surg. 2001;72(6):1883–6. https://doi.org/10.1016/s0003-4975(01)03204-0. PMID: 11789764.
33. Laamiri R, Sellami S, Kechiche N, Abdelaali M, Mekki M, Belghith M. Intrathoracic rupture of hydatid cyst of the liver in children: a report of two cases. Egypt Liver J. 2021;11(1):1–4. https://doi.org/10.1186/s43066-021-00116-5.
34. Nagarajan M, Vydianathan P, Muralidharan S. Hydatid cyst of the heart—a case report. Indian J Thorac Cardiovasc Surg. 2002;18:118–20. https://doi.org/10.1007/s12055-002-0019-3.
35. El Hammoumi M, Kabiri EH. Bilio-bronchial and bilio-pleuro-bronchial fistulas of hydatic origin. Kardiochir Torakochirurgia Pol. 2021;18(4):239–46. https://doi.org/10.5114/kitp.2021.112192. PMID: 35079267 PMCID: PMC8768848.
36. Morar R, Feldman C. Pulmonary echinococcosis. Eur Respir J. 2003;21(6):1069–77. https://doi.org/10.1183/09031936.03.00108403. PMID: 12797504.
37. Gomez IGC, Lopez-Andujar R, Belda Ibanez T, Ramia Angel J, Moya Herraiz A, Orbis Castellanos F, Pareja Ibars E, Rodriguez SJ, F. Review of the treatment of liver hydatid cysts. World J Gastroenterol. 2015;21(1):124–31. https://doi.org/10.3748/wjg.v21.i1.124. PMID: 25574085 PMCID: PMC4284328.
38. Imam A, Khoury T, Weis D, Khalayleh H, Adeleh M, Khalaileh A. Laparoscopic cystectomy for pancreatic echinococcosis: a case report and literature review. Ann Hepatobiliary Pancreat Surg. 2019;23(1):87–90. https://doi.org/10.14701/ahbps.2019.23.1.87. PMID: 30863816 PMCID: PMC6405372.
39. Wu Y, Gong J, Xiong W, Yu X, Lu X. Primary pancreatic hydatid cyst: a case report and literature review. BMC Gastroenterol. 2021;21(1):164. https://doi.org/10.1186/s12876-021-01753-1. PMID: 33849455 PMCID: PMC8045313.
40. Akbulut S, Ozdemir F. Intraperitoneal rupture of the hydatid cyst: four case reports and literature review. World J Hepatol. 2019;11(3):318–29. https://doi.org/10.4254/wjh.v11.i3.318. PMID: 30967909 PMCID: PMC6447420.
41. Arikanoglu Z, Taskesen F, Aliosmanoğlu İ, Gul M, Cetincakmak MG, Onder A, Kapan M. Spontaneous intraperitoneal rupture of a hepatic hydatid cyst. Int Surg. 2012;97(3):245–8. https://doi.org/10.9738/cc96.1. PMID: 23113854 PMCID: PMC3723219.
42. Dhua AK, Sharma A, Sarin YK. Intraperitoneal rupture of hepatic hydatid cyst following blunt abdominal trauma. APSP J Case Rep. 2012;3(2):10. PMID: 22953304 PMCID: PMC3418044.
43. Botezatu C, Mastalier B, Patrascu T. Hepatic hydatid cyst—diagnose and treatment algorithm. J Med Life. 2018;11(3):203–9. https://doi.org/10.25122/jml-2018-0045. PMID: 30364592 PMCID: PMC6197524.
44. Kantarci M, Pirimoglu B. Diffusion-weighted MR imaging findings in a growing problem: hepatic alveolar echinococcosis. Eur J Radiol. 2014;83(10):1991–2. https://doi.org/10.1016/j.ejrad.2014.07.013. PMID: 25112672.
45. Kantarci M, Pirimoglu B, Kizrak Y. Diagnostic imaging and interventional procedures in a growing problem: hepatic alveolar echinococcosis. World J Surg Proced. 2014;4(1):13–20. https://doi.org/10.5412/wjsp.v4.i1.13.
46. Kern P, Grüner B, Wahlers K. Diagnosis and course of echinococcocal diseases in the transplant setting. Transpl Infect Dis. 2011;13(3):217–21. https://doi.org/10.1111/j.1399-3062.2011.00643.x. PMID: 21518162.
47. Georgiou GK, Lianos GD, Lazaros A, Harissis HV, Mangano A, Dionigi G, Katsios C. Surgical management of hydatid liver disease. Int J Surg. 2015;20:118–22. https://doi.org/10.1016/j.ijsu.2015.06.058. PMID: 26118608.
48. Manterola C, Fernández O, Muñoz S, Vial M, Losada H, Carrasco R, Bello N, Barroso M. Laparoscopic pericystectomy for liver hydatid cysts. Surg Endosc. 2002;16(3):521–4. https://doi.org/10.1007/s00464-001-8125-7. PMID: 11928040.
49. Palanivelu C, Jani K, Malladi V, Senthilkumar R, Rajan PS, Sendhilkumar K, Parthasarthi R, Kavalakat A. Laparoscopic management of hepatic hydatid disease. JSLS. 2006;10(1):56–62. PMID: 16709359 PMCID: PMC3015664.
50. Stancu B, Andercou O, Pintea D, Mironiuc A, Pop DA. Laparoscopic simultaneous partial pericystectomy and total cystectomy for hydatid liver cysts—case report. Clujul Med. 2015;88(3):415–9. https://

doi.org/10.15386/cjmed-416. PMID: 26609279 PMCID: PMC4632905.
51. Vidoura A, Parisidou M, Chatedaki C, Zacharouli D. Surgical management of hydatid disease. In: Inceboz T, editor. Echinococcosis. IntechOpen; 2017.
52. Wan L, Wang T, Cheng L, Yu Q. Laparoscopic treatment strategies for liver echinococcosis. Infect Dis Ther. 2022;11(4):1415–26. https://doi.org/10.1007/s40121-022-00664-2. PMID: 35751754 PMCID: PMC9334471.
53. Sahin M, Eryilmaz R, Bulbuloglu E. The effect of scolicidal agents on liver and biliary tree (experimental study). J Investig Surg. 2004;17(6):323–6. https://doi.org/10.1080/08941930490524363. PMID: 15764499.
54. Topcu O, Kuzu I, Karayalcin K. Effects of peritoneal lavage with scolicidal agents on survival and adhesion formation in rats. World J Surg. 2006;30(1):127–33. https://doi.org/10.1007/s00268-005-7960-4. PMID: 16369703.
55. Karimi M, Rostami A, Spotin A, Rouhani S. Ruptured pulmonary hydatid cyst: a case report. J Parasit Dis. 2017;41(3):899–902. https://doi.org/10.1007/s12639-017-0880-z. PMID: 28848299 PMCID: PMC5555925.
56. Keramidas D, Mavridis G, Soutis M, Passalidis A. Medical treatment of pulmonary hydatidosis: complications and surgical management. Pediatr Surg Int. 2004;19(12):774–6. https://doi.org/10.1007/s00383-003-1031-4. PMID: 14714132.
57. Koul P, Kanjwal M. Albendazole induced rupture of pulmonary hydatid cysts. Internet J Pulm Med. 2006;6(1)
58. Louhaichi S, Zairi S, Hamdi B, Abdennadher M, Zribi H, Dridi A, Ouerghi S, Marghli A. Management of intact and ruptured pulmonary hydatid cysts. Eur Respir J. 2018:PA1356.
59. Pangeni RP, Regmi PR, Khadka S, Timilsina B. Ruptured pulmonary hydatid cyst presenting as hemoptysis in TB endemic country: a case report. Ann Med Surg (Lond). 2022;78:103836. https://doi.org/10.1016/j.amsu.2022.103836. PMID: 35734680 PMCID: PMC9207051.
60. Sheikhy K, Abbasi Dezfuli A, Pejhan S, Sadegh BF. Fatal outcome of ruptured pulmonary hydatid cyst. Tanaffos. 2018;17(2):138–41. PMID: 30627187 PMCID: PMC6320564.
61. Akbulut S. Parietal complication of the hydatid disease: comprehensive literature review. Medicine (Baltimore). 2018;97(21):e10671. https://doi.org/10.1097/md.0000000000010671. PMID: 29794743 PMCID: PMC6392988.
62. Bouassida M, Sassi S, Mighri MM, Laajili A, Chebbi F, Chtourou MF, Bouzaidi K, Touinsi H, Sassi S. [Parietal complications of hydatid cyst of the liver. Report of two cases in Tunisia]. Bull Soc Pathol Exot. 2012;105(4):259–61. https://doi.org/10.1007/s13149-012-0224-2. PMID: 23086495.
63. Jayant K, Agrawal S, Agarwal R, Dayama KG. Spontaneous external fistula: the rarest presentation of hydatid cyst. BMJ Case Rep. 2014;2014 https://doi.org/10.1136/bcr-2014-203784. PMID: 24827660 PMCID: PMC4025379.
64. Korwar V, Subhas G, Gaddikeri P, Shivaswamy BS. Hydatid disease presenting as cutaneous fistula: review of a rare clinical presentation. Int Surg. 2011;96(1):69–73. https://doi.org/10.9738/1364.1. PMID: 21675624.
65. Yakan S, Yildirim M, Coker A. Spontaneous cutaneous fistula of infected liver hydatid cyst. Turk J Gastroenterol. 2009;20(4):299–300. https://doi.org/10.4318/tjg.2009.0033. PMID: 20084579.
66. Gupta BB, Schangole S, Rnandagawali V, Kjadhav A. Spontaneous cutaneous fistulization, eventration of right hemidiaphragm and invasion of the pericardial cavity by a liver hydatid cyst. Trop Gastroenterol. 2013;34(4):287–9. https://doi.org/10.7869/tg.155. PMID: 25046901.
67. Akbulut S, Yilmaz M, Kahraman A, Yilmaz S. Budd-Chiari syndrome due to giant hydatid cyst: a case report and brief literature review. J Infect Dev Ctries. 2013;7(6):489–93. https://doi.org/10.3855/jidc.2712. PMID: 23771293.
68. Berkane S, Belkherchi S, Messaoudi N, Hassani D, Mahmoudi Y. Hydatid cyst of liver ruptured in the portal vein. J Clin Med Res. 2020;2(4):1–7. https://doi.org/10.37191/Mapsci-2582-4333-2(4)-043.
69. Ertan G, Ulus S, Kilicarslan R, Yilmaz M, Paksoy Y, Erol C. Hydatid disease of the liver with portal vein invasion mimicking portal vein thrombosis. Arab J Gastroenterol. 2019;20(1):50–2. https://doi.org/10.1016/j.ajg.2019.01.010. PMID: 30853256.
70. Hassine HB. Hydatid disease of the liver with portal vein invasion and cavernous transformation: a case report and literature review. IDCases. 2021;23:e01006. https://doi.org/10.1016/j.idcr.2020.e01006. PMID: 33251112 PMCID: PMC7677698.
71. Kirmizi S, Kayaalp C, Yilmaz S. Hydatid liver cyst causing portal vein thrombosis and cavernous transformation: a case report and literature review. Gastroenterol Hepatol Bed Bench. 2016;9(4):331–5. PMID: 27895860 PMCID: PMC5118859.
72. Neelakantan S, Babu AA, Anandarajan R, Philip B. Hepatic hydatid disease presenting as secondary Budd-Chiari syndrome. BMJ Case Rep. 2016;2016 https://doi.org/10.1136/bcr-2016-217118. PMID: 27681355 PMCID: PMC5051358.
73. Ramia J, De La Plaza R, Kuhnhardt A, Gonzalez J, Valenzuela J. Severe vascular complications due to liver hydatid cyst relapse: a case report and review of the literature. J Surg Surgical Res. 2015;1(1):1–3. https://doi.org/10.17352/2454.
74. San Juan López C, Lázaro Sáez M, Barrientos Delgado A, Casado Martín M, Hallouch Toutouh S, Vega Sáenz JL. Portal hypertension as a complication of hepatic hydatidosis. Gastroenterol Hepatol. 2018;41(10):648–9. https://doi.org/10.1016/j.gastrohep.2018.01.002. PMID: 29496233.
75. Sarawagi R, Keshava SN, Surendrababu NR, Zachariah UG, Eapen EC. Budd-Chiari syndrome

complicating hydatid cyst of the liver managed by venoplasty and stenting. Cardiovasc Intervent Radiol. 2011;34(Suppl 2):S202–5. https://doi.org/10.1007/s00270-010-9833-y. PMID: 20333383.
76. Yilmaz C, Erkan N, Arslan M, Yildirim Y, Kalaycioglu S. Budd-Chiari syndrome associated with liver hydatid disease: retrospective evaluation of color Doppler US findings with emphasis on intrahepatic venous collateralization. Radiol Oncol. 2009;43(4):225–32. https://doi.org/10.2478/v10019-009-0036-6.

Role of Endoscopic Retrograde Cholangiopancreatography in the Management of Hepatic Hydatid Disease

10

Kemal Dolay

10.1 Introduction

Liver hydatidosis, also known as hepatic hydatid cysts, is a disease that occurs worldwide and is caused by the *Echinococcus granulosus* parasite. Symptoms of hydatid cysts may occur due to the mass effect of the cyst and compression on organs, obstruction of blood or lymph flow, or complications such as rupture and secondary bacterial infections. The liver is the most commonly affected organ, with a rate of 55–80%, and the right lobe is the most common. Hepatic hydatid cyst disease is often asymptomatic or has mild symptoms. After the cyst reaches 10 cm in diameter, the symptoms become evident. The symptoms of hepatic hydatid cysts vary depending on the size, growth rate, type of cyst, location of the cyst in the liver, and complications. The most common symptom in symptomatic patients is pain in the right upper quadrant or epigastric area, and the most common finding during examination is an enlarged liver and a palpable mass. Abdominal fullness, nausea, and vomiting may also be present [1–5].

Most patients with hepatic hydatid cyst present an uncomplicated clinical course at diagnosis, and treatment planning is not complex. Although complications related to hepatic hydatid cysts have been reported as high as 60% in some series, they are generally seen in one-third of patients. Complicated hepatic hydatid cysts may result in mortality if not diagnosed and treated correctly. Complications include rupture of the bile ducts and associated biliary colic, obstructive jaundice, cholangitis, and acute pancreatitis. In addition, complications of systemic disease, anaphylaxis, and organ failure due to free rupture in the peritoneum, pleura, and pericardium may also occur. Infection of the cyst is another important complication. The diagnosis and management of complicated courses of hepatic hydatid cyst disease depend on the experience levels of the surgeon, interventional radiologist, and therapeutic endoscopist [2, 4–9].

Intrabiliary rupture, also known as cystobiliary fistula, is the most common and serious complication of hepatic hydatid cyst disease, reported in 2%–42% of cases. Cystobiliary fistula usually occurs in the biliary ducts of the right lobe (55%–60%), less commonly in the left lobe biliary ducts (30%–35%), and rarely in the common bile duct. The clinical presentation of cystobiliary fistula can range from asymptomatic to jaundice, cholecystitis, cholangitis, liver abscess, pancreatitis, and septicemia, depending on the size of the cystobiliary communication. However, undiagnosed ruptures can result in biliary leakage, biloma, cavitary infection, and obstructive jaundice after hepatic hydatid cysts surgery [2, 3, 5, 10–12].

In the past, and currently in endemic areas where endoscopic retrograde cholangiopancreatography (ERCP) is not available, conventional surgery for major cystobiliary fistula complica-

K. Dolay (✉)
Hepatopancreatobiliary Surgery, Istanbul, Türkiye

S. Akbulut (ed.), *Hydatid Disease*, https://doi.org/10.1007/978-3-031-97277-5_10

tions has up to 55% morbidity and up to 7% death rates. Recently, ERCP has been suggested as a valid alternative for hepatic hydatid cyst patients with biliary complications, both for the pre- and postoperative periods. In the clinical practice, it is generally agreed that ERCP is indicated for patients with biliary adverse events after surgery, such as persistent biliary fistulae and jaundice, and preoperative frank cystobiliary fistula that is suspected clinically (because of jaundice), biochemically (because of overt cholestasis), or radiologically (as a dilated main biliary ductal system or hydatid elements evident in the bile ducts). However, the use of routine preoperative ERCP and prophylactic endoscopic sphincterotomy in patients with suspected minor cystobiliary communications and the role of ERCP as a definitive treatment of frank cystobiliary fistula of hepatic hydatid cysts still remains unclear. Furthermore, the increasing use of magnetic resonance cholangiopancreatography fuels the debate on the routine use of preoperative diagnostic ERCP [2, 3, 5, 10, 13]. The role of ERCP for the diagnosis and treatment of cystobiliary fistula of hepatic hydatid cysts during the pre- and postoperative periods will be reviewed here.

10.2 Intrabiliary Rupture of Hepatic Hydatid Cysts

The parasite that causes the hydatid cyst becomes visible after 3 weeks and can reach 3 cm in diameter by the third month. When the cyst continues to grow, it causes atrophy and fibrosis as it compresses the surrounding liver parenchyma. During this growing period, the intracystic pressure may rise to 80 cm H_2O within a live cyst and can cause compression and obstruction of the biliary tree if centrally located. Even if there is no rupture into the bile ducts, cholestasis and partial dilatation of the proximal part of the affected ducts can be detected biochemically and radiologically. Continued growth can result in rupture into the bile ducts due to the increased intracystic pressure and fragility of the cyst wall [2, 3, 5, 11, 14].

Rupture into the bile ducts is the most common complication of hepatic hydatid cysts. Although its incidence is reported as 13–17% in radiological series, this rate varies between 1.3–42% in surgical series [2, 3, 5, 11, 12]. Several authors report some form of association with the biliary tract in up to 90% of hydatid cysts, but this variability in frequency is due to the lack of a clear internationally accepted definition [12].

Two theories have been postulated about the pathogenesis of cystobiliary fistula: First, the enlarged hydatid cyst causes necrosis and rupture of the bile duct wall due to persistent compression. The second hypothesis states that small bile radix are trapped in the pericystic wall and high intracystic pressure leads to atrophy of the bile radix and eventually rupture of the bile ducts [2, 3, 5, 12].

The communication between the ruptured cyst and bile ducts is divided into minor/simple and major/frank rupture according to its size. Simple communications (intrabiliary rupture <5 mm) are seen in 10%–37% of patients with hepatic hydatid cysts. In these cases, bile ducts are in communication with the cyst, but neither daughter cysts nor germinative membrane is found within the bile ducts. Cystic fluid, scolices, or very tiny hydatid elements may enter the bile ducts, but do not cause obstruction due to their size. Minor communications are difficult to demonstrate preoperatively as they do not exhibit prominent symptoms and radiologic findings, such as dilatation of bile ducts, seldom cause biliary colic, and patients are generally asymptomatic. Although minor communications can be demonstrated intraoperatively, they are most often recognized as biliary fistulas after surgery or percutaneous drainage, resulting in prolonged hospitalization and increased morbidity [2, 3, 5, 7, 12, 15–19].

Major intrabiliary ruptures are seen in 3%–17% of patients with hepatic hydatid cysts and can be easily recognized preoperatively. In this type of cystobiliary communication (intrabiliary rupture >5 mm), cystic content empties into the biliary tree, causing intermittent or total obstruction [2, 3, 5, 12]. The classic cystobiliary fistula triad of biliary colic, jaundice, and urticaria in hydatid disease, and finally fever (in the case of cholangitis) are most common in patients with major cystobiliary fistula. The clinical course may worsen and progress to liver abscess, septicemia, or anaphylaxis. Although rare, complications such as acute cholecystitis and pancreatitis can also occur due to hydatid remnants within bile ducts. In some patients, due to the decrease in intracystic pressure after rupture of the biliary tract, the discharge of cystic contents into the bile is stopped and jaundice is temporarily resolved. Minor leakage from the cyst to the biliary tract may cause flushing and urticaria, and major rupture may result in a potentially fatal anaphylactic reaction. In major cystobiliary fistula, the parasite usually dies as a result of destruction of the germinative layer in contact with the bile, but there is a risk of secondary infection in the cavity and biliary tract [2, 4, 5, 8, 12, 15, 17–21].

Cystobiliary communication occurs more frequently in cysts that are located in the center of the liver and close to the hilum (segments I, III, IV, V, and VI), advanced stage and/or multiple cysts, and large cysts. The size of the cyst has always been reported as an important marker for cystobiliary fistula. A threshold value between 7.5 cm and 14.5 cm has been shown in different studies, but the most widely accepted is 10 cm. Other identified markers are Gharbi type IV cysts with a fibrotic or calcified pericystic wall, bile duct dilation on imaging, clinical and preoperative laboratory findings such as leukocytosis, eosinophilia, hyperbilirubinemia, high alkaline phosphatase (ALP), and gamma-glutamyl transpeptidase (GGT) [1, 2, 5, 12, 16, 17, 19, 21–23].

10.3 Preoperative ERCP

In clinical practice, preoperative ERCP for the diagnosis and treatment of cystobiliary communication is performed for the following four purposes: (i) to perform elective surgery after overcoming acute conditions such as cholangitis and biliary obstruction; (ii) to reduce the risk of persistent biliary fistula after surgery and hospital stay with preoperative endoscopic sphincterotomy in patients with minor communications; (iii) to perform surgery on the hydatid cyst only without interfering with the bile ducts by removing the hydatid elements in the bile ducts; and (iv) to be able to apply a definitive therapeutic approach in cases of major rupture without the need for surgery by evacuating the hydatid contents from the cyst and bile ducts [2, 3, 5, 8, 11, 14, 18, 20, 24].

10.3.1 Diagnostic ERCP

Clinical history, detailed physical examination, routine blood tests (complete blood count, liver biochemistry), serology, and radiological imaging methods such as ultrasonography and computed tomography, if necessary, magnetic resonance imaging evaluation of liver hydatid cyst disease constitutes the first step in management. Based on these evaluations, hydatid disease is classified as either complicated or uncomplicated. Management is planned based on the patient's performance status, the characteristics of the cysts, the expertise of the physician and team, available medical resources, and patient compliance with long-term follow-up [2–5, 14, 18, 19].

Routine use of preoperative ERCP for the diagnosis of intrabiliary rupture is not yet accepted for uncomplicated hydatid cysts, although occult cystobiliary fistula may cause postoperative biliary fistula complications and prolonged hospitalization in patients who are considered uncomplicated hepatic hydatid cysts and have undergone surgery. Some authors have recommended routine preoperative ERCP in uncomplicated hepatic hydatid cysts to prevent or reduce postoperative biliary complications in patients where cystobiliary communication is frequently reported in surgical series (e.g., large or multiple cysts, hilar or caudate lobe cysts, or cysts with leukocytosis) [20, 25, 26]. However, postoperative bile leakage has also been reported in patients who underwent preoperative ERCP and endoscopic sphincterotomy to prevent fistula formation; not all communications are detectable with diagnostic ERCP, and most of the patients with postoperative bile leakage recover spontaneously without a need for ERCP [20, 21, 25–27]. Moreover, cyst compression on the bile duct alone was observed in up to 30% of patients who underwent ERCP (i.e., presenting with jaundice and/or dilated bile ducts) with suspected frank cystobiliary fistula [11, 20, 28].

ERCP has some limitations in the diagnosis of hepatic hydatid cysts complications. Minor cystobiliary communications may not be detected on ERCP primarily due to increased intracystic pressure, minimal connection between the cyst and the biliary tree, or daughter cysts that temporarily obstruct the cystobiliary opening. Due to the increased pressure, intrahepatic bile ducts may not be fully filled with contrast media and the false-negative rate may be up to 17–20%. Also, if contrast media is too concentrated, tiny hydatid materials within the extrahepatic bile ducts may not be apparent, and daughter cysts or old hydatid particles may mimic gallstones. Finally, hydatid elements found only in the intrahepatic bile ducts may be overlooked during ERCP. Therefore, surgeons should be alert intraoperatively for minor cystobiliary communications despite their apparent absence [2, 3, 5, 8, 14, 20, 27, 29, 30]. Some of these limitations can be minimized by applying accurate pressure injection of contrast media from different locations or occlusion cholangiography during ERCP. Direct findings of cystobiliary rupture in ERCP are as follows: hydatid elements within the duodenum or protruding from the papilla, some types of filling defects due to hydatid elements in the bile ducts on cholangiography, and filling of the cyst cavity with contrast media. There may be indirect signs of minor cystobiliary communication, such as cystic compression on the biliary tract, dilatation of the intrahepatic bile ducts, and irregular contour of the biliary tract adjacent to the cyst.

Studies have shown that magnetic resonance cholangiopancreatography is comparable to ERCP as a diagnostic tool in bile duct pathologies due to its high sensitivity and recommends its use instead of ERCP as it is a noninvasive method [2, 5, 20, 31, 32]. In cases of cystobiliary communication, ERCP can then be used as a therapeutic means. In our clinical practice, we do not perform preoperative diagnostic ERCP and prophylactic endoscopic sphincterotomy unless there is a history of cystobiliary pancreatic symptoms such as acute cholangitis and acute pancreatitis, or significant cholestasis or obstructive jaundice on laboratory testing, or dilatation of the bile ducts distal to the cyst on ultrasonography and computed tomography images. In cases of minor cholestasis or partial dilatation of intrahepatic bile ducts proximal to the cyst, magnetic resonance cholangiopancreatography in addition to ultrasonography and computed tomography prove to be more useful than ERCP for diagnosing minor communications in uncomplicated cases [20, 31–33]. In cases of suspected hepatic hydatid cysts with minor communication, we perform a careful examination for bile leakage intraoperatively and routinely perform bile leakage testing, and if detected, we apply primary sutures to the bile leakage orifices.

Frank cystobiliary fistula can be easily diagnosed with ultrasonography, computed tomography, or magnetic resonance cholangiopancreatography, and ERCP is often the first modality of choice in these complicated cases [2, 3, 5, 8, 14, 18, 21, 29, 34]. Hence, in such cases, preoperative ERCP is more commonly used therapeutically than diagnostically [2, 3, 8, 27, 35].

In cases of frank cystobiliary fistula, white, shiny hydatid membranes are mostly seen within the duodenum or protruding from the papilla of Vater during ERCP. On cholangiography, linear wavy filling defects of laminated hydatid membranes can be easily seen within the main bile duct. Three types of filling defects can be seen on cholangiography. Membranes appear filiform, daughter cysts round, and hydatid sand as debris. In late phases, membrane particles lose their shiny characteristic and mimic bile stones as they become colorized by bile acids (Fig. 10.1a–d).

If bile is seen in cyst puncture or if a biliary fistula develops after PAIR (puncture, aspiration, injection, and re-aspiration) protocol, early ERCP is not planned in these patients because most of the fistulas will spontaneously close if there is no hydatid element in the biliary tract on fistulography. If persistent biliary fistula develops, therapeutic ERCP is planned [2, 5, 36, 37].

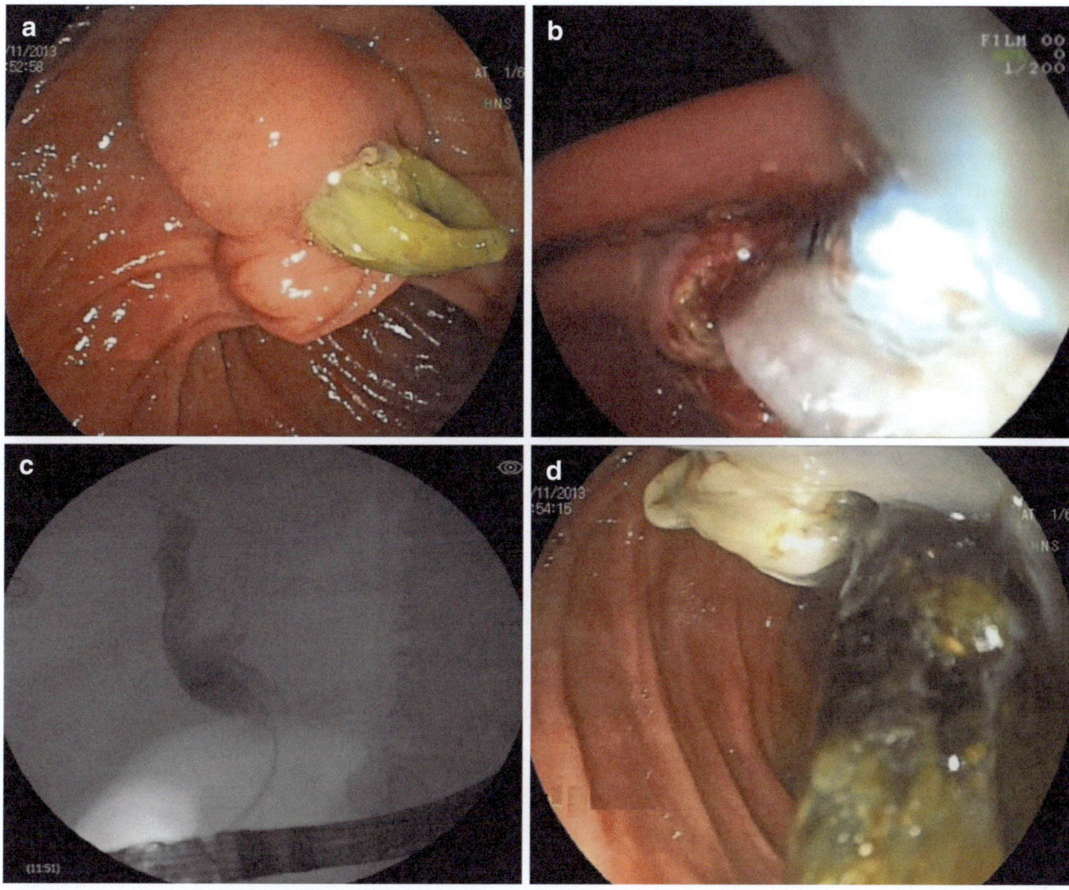

Fig. 10.1 At ERCP: In cases of frank cystobiliary fistula, hydatid membranes are mostly seen as protruding from the papilla of Vater (**a**) or as white, shiny hydatid membranes in the duodenum (**b**). Cholangiography shows oval and linear filling defects due to hydatid elements in the common bile duct (**c**). Hydatid elements can sometimes mimic gallstones (**d**)

10.3.2 Therapeutic ERCP

Preoperative therapeutic ERCP is an important tool in cases where a major rupture of the biliary tree has occurred, with a reported success rate of 80%-100, and it allows both the diagnosis of major biliary communication and the clearing of the common bile duct of hydatid materials [2, 3, 5, 8, 37–47]. Endoscopic sphincterotomy, hydatid cyst material extraction with balloon or Dormia basket catheter, biliary stenting, and nasobiliary drainage are some of the therapeutic interventions that can be performed thorough an ERCP scope. In our clinic, although cystobiliary communication cannot be demonstrated in cholangiography in patients who underwent ERCP with the suspicion of frank cystobiliary fistula, we perform endoscopic sphincterotomy and explore for hydatid sand and materials in the biliary tract with the extraction.

Even today, the treatment of hepatic hydatid cyst patients with major cystobiliary fistula presents difficulties. According to the series, postoperative mortality rates vary between 1.25% and 7% in hepatic hydatid cyst patients with major cystobiliary fistula, and the main causes are hepatic failure due to sepsis and cholangitis. Postoperative morbidity ranges from 16.7% to 55%, with the common complications including biliary fistula, wound infection, lung complications, biloma, bleeding, and abscess formation [2, 5, 19–21]. In a recent study, morbidity of 50.9% and mortality of 14.5% were reported in patients who were directly operated on due to major cystobiliary fistula and cholangitis, as routine ERCP was not able [13]. Emergency ERCP is vital in such patients.

In major ruptures, hydatid membranes or daughter cysts detected in bile ducts can be evacuated to the duodenum using a Dormia basket and balloon catheter after endoscopic sphincterotomy. If large amounts of hydatid material fill the duodenum, the sphincterotomized papillary orifice may not be visible endoscopically. In our ERCP procedure, a guidewire is placed routinely before endoscopic sphincterotomy when large amounts of hydatid elements are detected in bile ducts, to avoid such trouble. Hydatid material extraction with a balloon catheter over the guidewire can be performed successively under C-arm fluoroscopy, even when hydatid material around the papilla makes endoscopic manipulation difficult. After removal of the hydatid membranes and large daughter vesicles, saline irrigation of the biliary tract facilitates the removal of tiny hydatid elements and small daughter cysts.

Biliary stenting or nasobiliary drainage may be added to endoscopic sphincterotomy when the bile ducts cannot be completely cleared of hydatid elements to prevent cholangitis. Nasobiliary drainage may be the first choice of temporary treatment, especially when life-threatening sepsis due to acute cholangitis occurs. After the critical phase, ERCP can be repeated to completely clean out bile ducts of remaining hydatid elements. Endoscopic balloon dilation and biliary stenting may be required if there is a stricture in the bile tree. Elective surgery for the hydatid cyst can be performed after the hydatid elements in the bile ducts are completely cleared with ERCP [2, 11, 18, 20, 46, 48–50].

In previous studies, it was observed that the cyst volume was significantly reduced after complete removal of cyst materials in the bile ducts with ERCP in patients with major intrabiliary rupture, and even up to 25% cure was achieved without the need for surgical treatment [14, 29, 37, 38, 40, 41, 43, 45, 46, 51–53]. In 1987, Al Karawi et al. [54] published the first case report in which nonsurgical endoscopic transpapillary treatment of the ruptured hepatic hydatid cysts with hypertonic saline lavage via nasobiliary drainage tube was successfully performed [54]. Five of the six patients treated by these authors fully recovered without the need for further surgical treatment [43]. Other studies and case reports have also achieved similar results with the same method, and some have even reported a 100% success rate. They considered that ERCP and nasobiliary drainage safe and effective definitive treatment option and should be the first-line treatment in hepatic hydatid cyst patients with major cystobiliary fistula [2, 38, 55–59].

In our clinic, some modifications were made to this nasobiliary drainage technique. After standard endoscopic sphincterotomy and emptying of the hydatid material into the duodenum, a hydrophilic guidewire is negotiated into the cyst cavity. The basket and balloon catheter over the guidewire are used to evacuate as much of the cyst content as possible, and a redesigned 10F nasobiliary drainage catheter is inserted into the cystic cavity to facilitate complete evacuation of the cyst contents. The cystic cavity and common bile duct are irrigated periodically for 7 days using saline solution through the nasobiliary catheter to ensure complete clearance of the germinal layers and remaining small cysts. Small additional holes are placed in the common bile duct and duodenal portion of the nasobiliary drainage catheter to prevent the risk of cholangitis due to the continuous drainage and obstruction of hydatid material into the bile ducts (Fig. 10.2).

This allowed the nasobiliary drainage catheter to act as a stent between the common bile duct and duodenum, and also allowed for intermittent irrigation of the cyst cavity with saline. Another benefit is reduction in external biliary loss [2].

When looking at some studies and case reports, it is seen that there is no need for surgery at a high rate after the bile ducts are cleared of hydatid elements with ERCP, endoscopic sphincterotomy, and/or stenting in hepatic hydatid cyst patients with major ruptures. However, it is seen that repeated ERCP due to increased cholangitis is required [37, 40, 43, 45, 51, 55, 56, 58, 60–62].

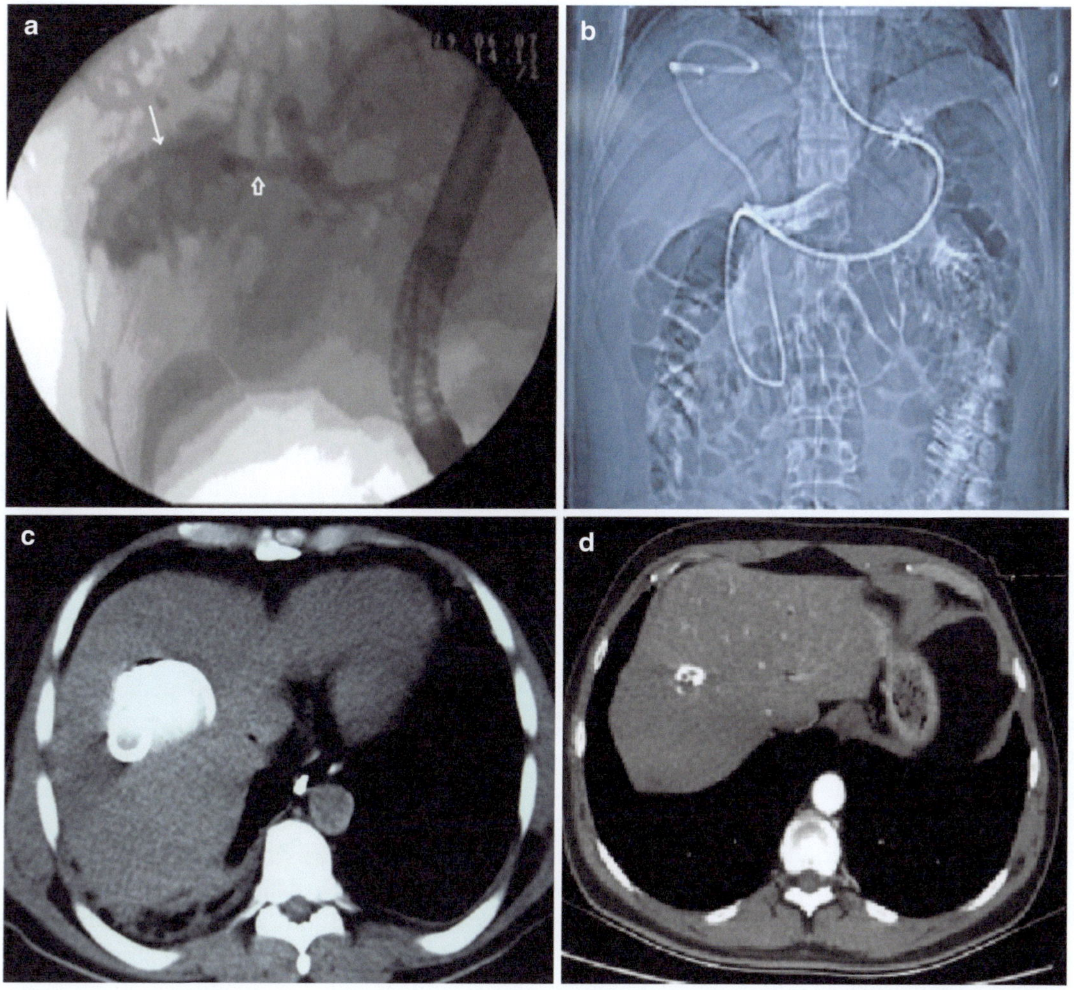

Fig. 10.2 (**a**) Thick arrow in cholangiography indicates major intrabiliary rupture and thin arrow indicates hydatid cyst cavity. (**b**) Topographic image after placement of a nasobiliary drainage catheter into the cyst cavity. (**c**) Computed tomography image taken before removal of the nasobiliary drainage catheter. (**d**) Computed tomography image of the same patient taken at 14 years follow-up

Therefore, we also suggest that ERCP and nasobiliary drainage should be the first choice of treatment in hepatic hydatid cyst patients with major cystobiliary fistula in whom common bile duct is dilated.

10.4 Postoperative ERCP

Although several studies have demonstrated fewer complications following radical surgical approaches, conservative surgical procedures for hepatic hydatid cysts are still the most widely performed techniques. The overall postoperative morbidity of hepatic hydatid cyst surgery ranges between 9% and 42%. Surgical complications are usually associated with complicated hydatid cysts. Therapeutic ERCP is indicated for early (persistent biliary fistula and obstructive jaundice) and late (sclerosing cholangitis, biliary strictures, and sphincter of Oddi stenosis) postoperative biliary adverse events. Therapeutic procedures such as endoscopic sphincterotomy, stenting, and nasobiliary drainage are preferred based on the type of biliary complication identified during ERCP [2, 3, 5, 8, 10, 14, 18, 28, 48, 51, 63].

10.4.1 Biliary Fistulae

Biliary fistulas are the most common postoperative complication. Although the risk of bile leakage is higher in the early postoperative period, biliary fistulas usually close spontaneously and may persist in 4–27.5% of patients, of which 11% may develop ascending cholangitis. Persistent biliary fistulas are defined by a high amount of biliary drainage lasting more than 10 days after surgery, and ERCP is indicated in such cases. It is difficult to predict whether postoperative biliary fistulas will resolve spontaneously or persist. However, the presence of excessive bile leakage in the early postoperative period or the presence of drainage that does not decrease day by day suggests that the fistula will not resolve spontaneously. ERCP is also indicated if fistula output is high, or rising each day, and in such cases, ERCP can be performed in the early postoperative period without waiting for 10 days. Any delay in ERCP may result in fistulas becoming chronic and an increase in infective complications. Conversely, bile leaks with low-output or decreasing amounts can be followed conservatively [2, 3, 5, 8, 10, 11, 13, 64].

When bile leakage is observed through the abdominal drain, the priority in management is to convert the bile leakage into a controlled fistula rather than treating the bile fistula with ERCP. If intra-abdominal bile collection is not drained effectively, it becomes infected and can lead to life-threatening generalized peritonitis and sepsis [2, 49].

The goal of endoscopic treatment for postoperative bile leaks is to reduce the transpapillary pressure gradient that helps bile flow through the bile ducts into the duodenum rather than leaking into the cystic cavity. This can be achieved by performing the endoscopic sphincterotomy, biliary stenting, and nasobiliary drainage procedures alone or in combination. Although the different advantages of each of these techniques are debated, the overall endoscopic treatment success rate is over 90% (range 83.3–100%) and mortality has been reported very low [2, 3, 8, 10, 14, 28, 37, 46, 48, 50, 51, 65–68]. Reports in the literature suggest that postoperative biliary fistula closes with endoscopic sphincterotomy within 3–43 days, yet the hospital stays for ERCP procedures last typically 3–9 days [2, 8, 10, 46, 51, 59, 68]. Only endoscopic sphincterotomy success rate is lower than biliary stenting, but it is an advantage that it requires a single session of ERCP. Nasobiliary drainage is more effective in reducing intrabiliary pressure and allows follow-up with cholangiography; however, it has some disadvantages, such as low tolerability and long hospital stay [2, 3, 8, 10, 28].

Our current practice is to perform early ERCP in cases of high biliary fistula outflow or lack of significant daily flow reduction in the first week following hydatid cyst surgery, and endoscopic sphincterotomy is the preferred method due to its safety and effectiveness. Since hydatid sand and small hydatid remnants may not be easily visualized on cholangiography, endoscopic sphincterotomy is performed in all patients, balloon and basket catheters are used to routinely examine the bile ducts, and if hydatid elements are detected, they are removed into the duodenum. If biliary fistulas accompanied by bile duct strictures, biliary stenting is added. Nasobiliary drainage is preferred in cases of biliary sepsis, especially when the bile ducts are filled with hydatid elements [2, 3, 8, 10].

10.4.2 Biliary Stricture

Biliary strictures are rare but complex postoperative complications of hydatid cyst disease. Approximately 27% of all postoperative biliary strictures are reported to be secondary to hydatid cyst surgery [30]. Several factors can predispose a patient to develop biliary strictures, including the use of scolicidal agents, cystobiliary communications, hydatid cystic elements in bile ducts, secondary biliary infections, distal bile duct obstruction due to gallstones, papillary stenosis, and centrally located hydatid cysts [19]. The most dramatic form is caustic sclerosing cholangitis due to scolicidal agents, and diffuse caustic sclerosing cholangitis may result in secondary biliary cirrhosis, portal hypertension, and liver decompensation with ascites and bleeding esophageal varices, which may ultimately require liver transplantation [69, 70].

A few studies claim that endoscopic or percutaneous methods should be used first for the management of postoperative strictures. In the case of biliary strictures secondary to hepatic hydatid disease, endoscopic stenting can help to alleviate symptoms such as jaundice, abdominal pain, and fever. It can also help to prevent further complications such as cholangitis. However, biliary strictures affecting long and multiple segments, particularly those located proximally (Bismuth type IV), and with cholangiographic signs showing diffuse involvement, are not suitable for surgical repair. Biliary stenting should be preferred for biliary strictures located distally and involving single short segments, whereas percutaneous transhepatic cholangiography and internal biliary drainage are preferred for proximally located, multiple and long-segment biliary strictures [2, 14, 30, 37, 42].

References

1. Akcan A, Sozuer E, Akyildiz H, Ozturk A, Atalay A, Yilmaz Z. Predisposing factors and surgical outcome of complicated liver hydatid cysts. World J Gastroenterol. 2010;16(24):3040–8. https://doi.org/10.3748/wjg.v16.i24.3040. PMID: 20572308 PMCID: PMC2890945.
2. Dolay K, Akbulut S. Role of endoscopic retrograde cholangiopancreatography in the management of hepatic hydatid disease. World J Gastroenterol. 2014;20(41):15253–61. https://doi.org/10.3748/wjg.v20.i41.15253. PMID: 25386073 PMCID: PMC4223258.
3. Reddy DN, Rao GV, Banerjee R. Tropical parasitic infestations. In: Baron T, Kozarek R, Carr-Locke D, editors. ERCP. Elsevier; 2019. p. 464–468. e462.
4. Schulick R, Ibrahim-zada I. Management of echinococcal cyst disease of the liver. In: Cameron J, Cameron A, editors. Current surgical therapy. 13th ed. Elsevier; 2020. p. 361–5.
5. Lendoire J. Hydatid disease of the liver. In: Jarnagin WR, editor. Blumgart's surgery of the liver, biliary tract and pancreas. 7th ed. Philadelphia: Elsevier; 2022. p. 978–88.
6. Bektasoglu HK, Hasbahceci M, Tasci Y, Aydogdu I, Malya FU, Kunduz E, Dolay K. Comparison of laparoscopic and conventional cystotomy/partial cystectomy in treatment of liver hydatidosis. Biomed Res Int. 2019;2019:1212404. https://doi.org/10.1155/2019/1212404. PMID: 30868067 PMCID: PMC6379859.
7. Brunetti E, Kern P, Vuitton DA. Expert consensus for the diagnosis and treatment of cystic and alveolar echinococcosis in humans. Acta Trop. 2010;114(1):1–16. https://doi.org/10.1016/j.actatropica.2009.11.001. PMID: 19931502.
8. Nabi Z, Korman A, Reddy N, Carr-Locke D. Infections of the biliary tract. In: Chandrasekhara VE, Joseph B, Khashab MA, Muthusamy R, editors. Clinical gastrointestinal endoscopy. Elsevier; 2019. p. 636–651. e633.
9. Symeonidis N, Pavlidis T, Baltatzis M, Ballas K, Psarras K, Marakis G, Sakantamis A. Complicated liver echinococcosis: 30 years of experience from an endemic area. Scand J Surg. 2013;102(3):171–7. https://doi.org/10.1177/1457496913491877. PMID: 23963031.
10. Dolay K, Akçakaya A, Soybir G, Cabioğlu N, Müslümanoğlu M, Iğci A, Topuzlu C. Endoscopic sphincterotomy in the management of postoperative biliary fistula A complication of hepatic hydatid disease. Surg Endosc. 2002;16(6):985–8. https://doi.org/10.1007/s00464-001-9020-y. PMID: 12163969.
11. Kayaalp C, Bostanci B, Yol S, Akoglu M. Distribution of hydatid cysts into the liver with reference to cystobiliary communications and cavity-related complications. Am J Surg. 2003;185(2):175–9. https://doi.org/10.1016/s0002-9610(02)01202-3. PMID: 12559452.
12. Ramia JM, Figueras J, De la Plaza R, García-Parreño J. Cysto-biliary communication in liver hydatidosis. Langenbeck's Arch Surg. 2012;397(6):881–7. https://doi.org/10.1007/s00423-012-0926-8. PMID: 22374106.
13. Toumi O, Ammar H, Gupta R, Ben Jabra S, Hamida B, Noomen F, Zouari K, Golli M. Management of liver hydatid cyst with cystobiliary communication and acute cholangitis: a 27-year experience. Eur J Trauma Emerg Surg. 2019;45(6):1115–9. https://doi.org/10.1007/s00068-018-0995-7. PMID: 30191292.
14. El Malki HO, El Mejdoubi Y, Souadka A, Mohsine R, Ifrine L, Abouqal R, Belkouchi A. Predictive model of biliocystic communication in liver hydatid cysts using classification and regression tree analysis. BMC Surg. 2010;10:16. https://doi.org/10.1186/1471-2482-10-16. PMID: 20398342 PMCID: PMC2867769.
15. Bhutani N, Kajal P. Hepatic echinococcosis: a review. Ann Med Surg (Lond). 2018;36:99–105. https://doi.org/10.1016/j.amsu.2018.10.032. PMID: 30450204 PMCID: PMC6226561.
16. Demircan O, Baymus M, Seydaoglu G, Akinoglu A, Sakman G. Occult cystobiliary communication presenting as postoperative biliary leakage after hydatid liver surgery: are there significant preoperative

clinical predictors? Can J Surg. 2006;49(3):177–84. PMID: 16749978 PMCID: PMC3207592.
17. El Nakeeb A, Salem A, El Sorogy M, Mahdy Y, Ellatif MA, Moneer A, Said R, El Ghawalby A, Ezzat H. Cystobiliary communication in hepatic hydatid cyst: predictors and outcome. Turk J Gastroenterol. 2017;28(2):125–30. https://doi.org/10.5152/tjg.2017.17553. PMID: 28119271.
18. Lewall DB, McCorkell SJ. Rupture of echinococcal cysts: diagnosis, classification, and clinical implications. AJR Am J Roentgenol. 1986;146(2):391–4. https://doi.org/10.2214/ajr.146.2.391. PMID: 3510519.
19. Unalp HR, Baydar B, Kamer E, Yilmaz Y, Issever H, Tarcan E. Asymptomatic occult cysto-biliary communication without bile into cavity of the liver hydatid cyst: a pitfall in conservative surgery. Int J Surg. 2009;7(4):387–91. https://doi.org/10.1016/j.ijsu.2009.06.012. PMID: 19573629.
20. Murty TV, Sood KC, Rakas FS. Biliary obstruction due to ruptured hydatid cyst. J Pediatr Surg. 1989;24(4):401–3. https://doi.org/10.1016/s0022-3468(89)80281-7. PMID: 2732886.
21. Zaouche A, Haouet K, Jouini M, El Hachaichi A, Dziri C. Management of liver hydatid cysts with a large biliocystic fistula: multicenter retrospective study. Tunisian Surgical Association. World J Surg. 2001;25(1):28–39. https://doi.org/10.1007/s002680020005. PMID: 11213153.
22. Kayaalp C, Bzeizi K, Demirbag AE, Akoglu M. Biliary complications after hydatid liver surgery: incidence and risk factors. J Gastrointest Surg. 2002;6(5):706–12. https://doi.org/10.1016/s1091-255x(02)00046-x. PMID: 12399060.
23. Ozdemir Y, Ilter O, Surmelioglu A, Dalgic T, Aksoy E, Caparlar MA, Murat U, Ozoğul YB, Bostanci EB, Akoglu M. Risk factors determining bile leakage after multiple liver hydatid cyst surgery. J Surg Med. 2022;6(3):306–12. https://doi.org/10.28982/josam.1072054.
24. Ozaslan E, Bayraktar Y. Endoscopic therapy in the management of hepatobiliary hydatid disease. J Clin Gastroenterol. 2002;35(2):160–74. https://doi.org/10.1097/00004836-200208000-00009. PMID: 12172363.
25. El-Gendi AM, El-Shafei M, Bedewy E. The role of prophylactic endoscopic sphincterotomy for prevention of postoperative bile leak in hydatid liver disease: a randomized controlled study. J Laparoendosc Adv Surg Tech A. 2018;28(8):990–6. https://doi.org/10.1089/lap.2017.0674. PMID: 29641366.
26. Pitt HA, Korzelius J, Tompkins RK. Management of hepatic echinococcosis in Southern California. Am J Surg. 1986;152(1):110–5. https://doi.org/10.1016/0002-9610(86)90159-5. PMID: 3728803.
27. Ponchon T, Bory R, Chavaillon A. Endoscopic retrograde cholangiography and sphincterotomy for complicated hepatic hydatid cyst. Endoscopy. 1987;19(4):174–7. https://doi.org/10.1055/s--2007-1018274. PMID: 3622400.
28. Cicek B, Parlak E, Disibeyaz S, Oguz D, Cengiz C, Sahin B. Endoscopic therapy of hepatic hydatid cyst disease in preoperative and postoperative settings. Dig Dis Sci. 2007;52(4):931–5. https://doi.org/10.1007/s10620-006-9426-4. PMID: 17333353.
29. Magistrelli P, Masetti R, Coppola R, Costamagna G, Durastante V, Nuzzo G, Picciocchi A. Value of ERCP in the diagnosis and management of pre- and postoperative biliary complications in hydatid disease of the liver. Gastrointest Radiol. 1989;14(4):315–20. https://doi.org/10.1007/bf01889226. PMID: 2680737.
30. Yilmaz U, Sakin B, Boyacioglu S, Saritas U, Cumhar T, Akoglu M. Management of postoperative biliary strictures secondary to hepatic hydatid disease by endoscopic stenting. Hepato-Gastroenterology. 1998;45(19):65–9. PMID: 9496489.
31. Kaltenthaler EC, Walters SJ, Chilcott J, Blakeborough A, Vergel YB, Thomas S. MRCP compared to diagnostic ERCP for diagnosis when biliary obstruction is suspected: a systematic review. BMC Med Imaging. 2006;6:9. https://doi.org/10.1186/1471-2342-6-9. PMID: 16907974 PMCID: PMC1579209.
32. Kumar R, Reddy SN, Thulkar S. Intrabiliary rupture of hydatid cyst: diagnosis with MRI and hepatobiliary isotope study. Br J Radiol. 2002;75(891):271–4. https://doi.org/10.1259/bjr.75.891.750271. PMID: 11932222.
33. Marrone G, Crino F, Caruso S, Mamone G, Carollo V, Milazzo M, Gruttadauria S, Luca A, Gridelli B. Multidisciplinary imaging of liver hydatidosis. World J Gastroenterol. 2012;18(13):1438–47. https://doi.org/10.3748/wjg.v18.i13.1438. PMID: 22509075 PMCID: PMC3319939.
34. Zargar SA, Khuroo MS, Khan BA, Dar MY, Alai MS, Koul P. Intrabiliary rupture of hepatic hydatid cyst: sonographic and cholangiographic appearances. Gastrointest Radiol. 1992;17(1):41–5. https://doi.org/10.1007/bf01888506. PMID: 1544557.
35. Galati G, Sterpetti AV, Caputo M, Adduci M, Lucandri G, Brozzetti S, Bolognese A, Cavallaro A. Endoscopic retrograde cholangiography for intrabiliary rupture of hydatid cyst. Am J Surg. 2006;191(2):206–10. https://doi.org/10.1016/j.amjsurg.2005.09.014. PMID: 16442947.
36. Akhan O, Salik AE, Ciftci T, Akinci D, Islim F, Akpinar B. Comparison of long-term results of percutaneous treatment techniques for hepatic cystic echinococcosis types 2 and 3b. AJR Am J Roentgenol. 2017;208(4):878–84. https://doi.org/10.2214/ajr.16.16131. PMID: 28199132.
37. Rodriguez AN, Sánchez del Río AL, Alguacil LV, De Dios Vega JF, Fugarolas GM. Effectiveness of endoscopic sphincterotomy in complicated hepatic hydatid disease. Gastrointest Endosc. 1998;48(6):593–7. https://doi.org/10.1016/s0016-5107(98)70041-0. PMID: 9852449.
38. Akkiz H, Akinoglu A, Colakoglu S, Demiryürek H, Yagmur O. Endoscopic management of biliary hydatid disease. Can J Surg. 1996;39(4):287–92. PMID: 8697318 PMCID: PMC3950134.

39. Akoglu M, Hilmioglu F, Balay AR, Sahin B, Davidson BR. Endoscopic sphincterotomy in hepatic hydatid disease open to the biliary tree. Br J Surg. 1990;77(9):1073. https://doi.org/10.1002/bjs.1800770940. PMID: 2207579.
40. Becker K, Frieling T, Saleh A, Häussinger D. Resolution of hydatid liver cyst by spontaneous rupture into the biliary tract. J Hepatol. 1997;26(6):1408–12. https://doi.org/10.1016/s0168-8278(97)80479-5. PMID: 9210631.
41. de Aretxabala X, Perez OL. The use of endoprostheses in biliary fistula of hydatid cyst. Gastrointest Endosc. 1999;49(6):797–9. https://doi.org/10.1016/s0016-5107(99)70306-8. PMID: 10343233.
42. Dumas R, Le Gall P, Hastier P, Buckley MJ, Conio M, Delmont JP. The role of endoscopic retrograde cholangiopancreatography in the management of hepatic hydatid disease. Endoscopy. 1999;31(3):242–7. https://doi.org/10.1055/s-1999-14209. PMID: 10344429.
43. Hilmioglu F, Karincaoglu M, Yilmaz S, Yildirim B, Kirimlioglu V, Aladag M, Onmus H. Complete treatment of ruptured hepatic cyst into biliary tree by ERCP. Dig Dis Sci. 2001;46(3):463–7. https://doi.org/10.1023/a:1005670325782. PMID: 11318516.
44. Kattan YB. Intrabiliary rupture of hydatid cyst of the liver. Br J Surg. 1975;62(11):885–90. https://doi.org/10.1002/bjs.1800621108. PMID: 1191949.
45. Sáez-Royuela F, Yuguero L, López-Morante A, Pérez-Alvarez JC, Martín-Lorente JL, Ojeda C. Acute pancreatitis caused by hydatid membranes in the biliary tract: treatment with endoscopic sphincterotomy. Gastrointest Endosc. 1999;49(6):793–6. https://doi.org/10.1016/s0016-5107(99)70305-6. PMID: 10343232.
46. Spiliadis C, Georgopoulos S, Dailianas A, Konstantinidis A, Rimikis M, Skandalis N. The use of ERCP in the study of patients with hepatic echinococcosis before and after surgical intervention. Gastrointest Endosc. 1996;43(6):575–9. https://doi.org/10.1016/s0016-5107(96)70194-3. PMID: 8781936.
47. Ulualp KM, Aydemir I, Senturk H, Eyuboğlu E, Cebeci H, Unal G, Unal H. Management of intrabiliary rupture of hydatid cyst of the liver. World J Surg. 1995;19(5):720–4; discussion 728. https://doi.org/10.1007/bf00295913. PMID: 7571669.
48. Bilsel Y, Bulut T, Yamaner S, Buyukuncu Y, Bugra D, Akyuz A, Sokucu N. ERCP in the diagnosis and management of complications after surgery for hepatic echinococcosis. Gastrointest Endosc. 2003;57(2):210–3. https://doi.org/10.1067/mge.2003.64. PMID: 12556786.
49. Kornaros SE, Aboul-Nour TA. Frank intrabiliary rupture of hydatid hepatic cyst: diagnosis and treatment. J Am Coll Surg. 1996;183(5):466–70. PMID: 8912615.
50. Saritas U, Parlak E, Akoglu M, Sahin B. Effectiveness of endoscopic treatment modalities in complicated hepatic hydatid disease after surgical intervention. Endoscopy. 2001;33(10):858–63. https://doi.org/10.1055/s-2001-17342. PMID: 11571682.
51. Giouleme O, Nikolaidis N, Zezos P, Budas K, Katsinelos P, Vasiliadis T, Eugenidis N. Treatment of complications of hepatic hydatid disease by ERCP. Gastrointest Endosc. 2001;54(4):508–10. https://doi.org/10.1067/mge.2001.118256. PMID: 11577320.
52. Shemesh E, Klein E, Abramowich D, Pines A. Common bile duct obstruction caused by hydatid daughter cysts—management by endoscopic retrograde sphincterotomy. Am J Gastroenterol. 1986;81(4):280–2. PMID: 3962953.
53. Somani S, Srivastava A. Resolution of hepatic hydatid cyst with biliary communication with ERCP. J Gastroint Dig Syst. 2012;2(114):2.
54. Al Karawi MA, Mohamed AR, Yasawy I, Haleem A. Non-surgical endoscopic trans-papillary treatment of ruptured echinococcus liver cyst obstructing the biliary tree. Endoscopy. 1987;19(2):81–3. https://doi.org/10.1055/s-2007-1013021. PMID: 3569154.
55. Biswas SN, Kulkarni A, Jain M, Chaudhary N, Ghoshal UC. Intra-biliary rupture of hepatic hydatid cyst: an uncommon presentation of a common disease. Indian J Gastroenterol. 2021;40(4):445–7. https://doi.org/10.1007/s12664-020-01130-y. PMID: 33387305.
56. Borahma M, Afifi R, Benelbarhdadi I, Ajana FZ, Essamri W, Essaid A. Endoscopic retrograde cholangiopancreatography in ruptured liver hydatid cyst. Indian J Gastroenterol. 2015;34(4):330–4. https://doi.org/10.1007/s12664-015-0585-0. PMID: 26345677.
57. Manouras A, Genetzakis M, Antonakis PT, Lagoudianakis E, Pattas M, Papadima A, Giannopoulos P, Menenakos E. Endoscopic management of a relapsing hepatic hydatid cyst with intrabiliary rupture: a case report and review of the literature. Can J Gastroenterol. 2007;21(4):249–53. https://doi.org/10.1155/2007/410308. PMID: 17431515 PMCID: PMC2657701.
58. Sharma BC, Reddy RS, Garg V. Endoscopic management of hepatic hydatid cyst with biliary communication. Dig Endosc. 2012;24(4):267–70. https://doi.org/10.1111/j.1443-1661.2011.01225.x. PMID: 22725113.
59. Singh V, Reddy DC, Verma GR, Singh G. Endoscopic management of intrabiliary-ruptured hepatic hydatid cyst. Liver Int. 2006;26(5):621–4. https://doi.org/10.1111/j.1478-3231.2006.01259.x. PMID: 16762008.
60. Abid H, Ousseur G, Lahmidani N, El Yousfi M, Aqodad N, El Abkari M, Ibrahimi A, Benajah DA. What is the role of Endoscopic Retrograde Cholangio-Pancreato-Grahy in the management of hepatic hydatid disease complications? Open J Gastroenterol. 2020;10(05):107. https://doi.org/10.4236/ojgas.2020.105011.
61. Bektaş M, Dökmeci A, Cinar K, Halici I, Oztas E, Karayalcin S, Idilman R, Sarioglu M, Ustun Y, Nazligul Y, Ormeci N, Ozkan H, Bozkaya H, Yurdaydin C. Endoscopic management of biliary parasitic diseases. Dig Dis Sci. 2010;55(5):1472–8. PMID: 19513838 DOI:10.1007/s10620-009-0850-0.

62. Govindaraj S, Bhadra V, Prakash C, Govindaraj S. Unusual case of disappearing hepatic hydatid cyst: COVID-19 times. BMJ Case Rep. 2021;14(8) https://doi.org/10.1136/bcr-2021-243533. PMID: 34353833 PMCID: PMC8344285.
63. Agarwal S, Sikora SS, Kumar A, Saxena R, Kapoor VK. Bile leaks following surgery for hepatic hydatid disease. Indian J Gastroenterol. 2005;24(2):55–8. PMID: 15879650.
64. Zeybek N, Dede H, Balci D, Coskun AK, Ozerhan IH, Peker S, Peker Y. Biliary fistula after treatment for hydatid disease of the liver: when to intervene. World J Gastroenterol. 2013;19(3):355–61. https://doi.org/10.3748/wjg.v19.i3.355. PMID: 23372357 PMCID: PMC3554819.
65. Adas G, Arikan S, Gurbuz E, Karahan S, Eryasar B, Karatepe O, Tekant Y. Comparison of endoscopic therapeutic modalities for postoperative biliary fistula of liver hydatid cyst: a retrospective multicentric study. Surg Laparosc Endosc Percutan Tech. 2010;20(4):223–7. https://doi.org/10.1097/SLE.0b013e3181e12ee6. PMID: 20729689.
66. Akcakaya A, Sahin M, Karakelleoglu A, Okan I. Endoscopic stenting for selected cases of biliary fistula after hepatic hydatid surgery. Surg Endosc. 2006;20(9):1415–8. https://doi.org/10.1007/s00464-005-0572-0. PMID: 16736309.
67. Tekant Y, Bilge O, Acarli K, Alper A, Emre A, Arioğul O. Endoscopic sphincterotomy in the treatment of postoperative biliary fistulas of hepatic hydatid disease. Surg Endosc. 1996;10(9):909–11. https://doi.org/10.1007/bf00188481. PMID: 8703149.
68. Vignote ML, Miño G, de la Mata M, de Dios JF, Gomez F. Endoscopic sphincterotomy in hepatic hydatid disease open to the biliary tree. Br J Surg. 1990;77(1):30–1. https://doi.org/10.1002/bjs.1800770110. PMID: 2302509.
69. Belghiti J, Benhamou JP, Houry S, Grenier P, Huguier M, Fékété F. Caustic sclerosing cholangitis. A complication of the surgical treatment of hydatid disease of the liver. Arch Surg. 1986;121(10):1162–5. https://doi.org/10.1001/archsurg.1986.01400100070014. PMID: 3767649.
70. Loinaz C, González EM, Jiménez C, García I, Gómez R, González-Pinto I, Colina F, Gimeno A. Long-term biliary complications after liver surgery leading to liver transplantation. World J Surg. 2001;25(10):1260–3. https://doi.org/10.1007/s00268-001-0106-4. PMID: 11596886.

11 Liver Transplantation for Hepatic Echinococcosis

Ez El Din Abu Zeid and Sami Akbulut

11.1 Introduction

Echinococcosis, also known as hydatidosis or hydatid disease, is caused by the larval stages of cestodes (tapeworms) from the genus *Echinococcus*. The disease is distributed globally and it constitutes a major public health problem. The two main types of these parasitic infestations include: cystic echinococcosis and alveolar echinococcosis caused by *Echinococcus granulosus* and *Echinococcus multilocularis*, respectively. It is endemic in countries where sheep are commonly reared. Such countries include Mediterranean countries including North Africa, Spain, and Portugal. The common site affected by the disease is the liver and it usually manifests as cystic lesion that may not be symptomatic [1].

Liver echinococcosis is the most common type of the disease because of the pattern of life cycle of the parasite. This disease is ignored initially because of its asymptomatic early-middle stages. This results in delayed presentation and high mortality. With the advent of improvements in diagnostic techniques, more cases are diagnosed at early, asymptomatic phase [2].

Though the initial infection is asymptomatic, progressive enlargement of the hydatid liver cyst can manifest as abdominal pain, nausea, hepatomegaly, or a palpable mass. There may be associated upper right quadrant pain, urticaria, and pruritus [3].

Treatment of hydatid cyst is multidisciplinary involving multiple specialties of medical practice. Some of the specialties include internal medicine, parasitology, interventional radiologist, hepatobiliary surgeons, and others [4, 5].

The role of medical therapy in the treatment of hydatid cyst has been the subject of controversy. Some selected role of medical therapy in hydatid disease includes reduction of the size and infectivity of the cysts and prevention of recurrence. In patients with disseminated or inoperable hydatid cyst affecting multiple systems, medical therapy may be the only treatment [5, 6]. The medical therapy of hydatid disease involves the administration of medications such as albendazole and mebendazole. In recent times, praziquantel has also been adopted for the treatment of hydatid disease. For patients with disseminated hydatid disease an additional use of nitazoxanide have also been reported [7, 8].

Surgical therapy constitutes the mainstay of treatment of hydatid disease. It can be an image guided intervention, laparoscopic surgery, or open surgery. Surgical therapy of hepatic hydatid disease can be classified as radical surgery, which

E. E. D. Abu Zeid (✉)
Department of Surgery, Soroka Medical Center, Beer Sheva, Israel

S. Akbulut
Department of Surgery and Liver Transplantation Institute, Inonu University Faculty of Medicine, Malatya, Türkiye

Department Biostatistics and Medical Informatics, Inonu University Faculty of Medicine, Malatya, Türkiye

S. Akbulut (ed.), *Hydatid Disease*, https://doi.org/10.1007/978-3-031-97277-5_11

extricates the entire pericyst or conservative surgery that removes the parasite and leave all or part of the pericyst. Radical surgery includes procedures such as total pericystectomy and hepatectomy. Conservative surgical techniques used for hydatid disease include subtotal pericystectomy, partial pericystectomy, and cystectomy [9–11]. Conservative surgical procedures are associated with higher complication rates compared to radical surgeries. Some of the complications are biliary leakage and fistulas in the postoperative period. Additional complication is the septic complications of the residual cavity. The residual pericyst left in situ causes delayed liver regeneration, which is needed for filling of the residual cavity. This delay leads to accumulation of serum and blood which may cause liver abscess. The residual pericyst may also hide possible biliary communication in the residual cavity leading to biliary fistula, which occurs in up to 50% of patients after conservative procedures. Unlike conservative procedures, radical procedures allow exact detection and ligation of biliary and vascular branches within the hepatic parenchyma, and this reduces the risk of biliary leak [12–15].

11.1.1 Hydatid History of Liver Transplantation for Hepatic Hydatid Disease

Liver transplantation for hepatic hydatid disease was first reported in 1989 and then a follow report was published in 1994. Moreno-Gonzalez et al. [16] reported a series of six patients that had hepatic hydatid cyst that was treated surgically, but patients developed recurrence which necessitated liver transplantation [16, 17]. Some of the patients had repeated surgery, but hydatid cyst still recurred. The indication for liver transplant was acute Budd–Chiari syndrome due to obstruction of hepatic veins in one patient. The other five patients received liver transplants for end stage liver disease caused by secondary sclerosing cholangitis in two patients, secondary biliary cirrhosis in two patients, and one case of post-necrotic cirrhosis of the liver [16, 17]. Since this initial reports of liver transplantation for hepatic hydatid disease, there were no other cases reported. This may be related to improved treatment of hydatid cyst with the increased use of adjunctive medical therapy for patients treated with surgery there by reducing recurrence, complication, and the need for transplantation.

11.1.2 The Evolution of PNM Classification in Alveolar Echinococcosis

The World Health Organization—Informal Working Group on Echinococcosis (WHO-IWGE) PNM classification system was developed specifically for alveolar echinococcosis as a response to the challenge of standardizing diagnostic and therapeutic strategies. The evolution of the PNM system was driven by the need to establish a common language among clinicians dealing with alveolar echinococcosis, which is often compared to malignant tumors due to its invasive growth pattern.

The idea for the PNM classification originated from the TNM system used in oncology, which categorizes cancers based on Tumor size (T), Node involvement (N), and Metastasis (M). Similarly, the PNM system for alveolar echinococcosis was designed to evaluate three crucial components: the extent of the parasitic mass within the liver (P), the involvement of adjacent organs (N), and the presence of metastases (M).

The initial formulation of the PNM classification involved a retrospective analysis of 97 patient records from two treatment centers, Besançon (France) and Ulm (Germany) [18]. This effort aimed to develop a classification that would not only reflect disease progression but also be practical for use in diverse clinical settings. Later, the classification was tested and validated with a larger cohort of 222 patients from four centers worldwide (Besançon, Urumqi, Sapporo, and Ulm), demonstrating its robustness and applicability across various healthcare environments [18].

Once the P, N, and M components are determined, they are combined to assign an overall stage ranging from I to IV. To further specify disease severity, Stage III is divided into subcategories IIIA and IIIB. This structured approach ensures consistency in disease assessment and helps guide

Table 11.1 PNM system for classification of hepatic alveolar echinococcosis

P	Hepatic localization of the primary lesion (parasitic lesion)
PX	Primary lesion cannot be assessed
P0	No detectable lesion in the liver
P1	Peripheral lesions without proximal vascular and/or biliar involvement
P2	Central lesions with proximal vascular and/or biliar involvement of one lobe[a]
P3	Central lesions with hilar vascular and biliar involvement of both lobes and/or with involvement of two hepatic veins
P4	Any liver lesion with extension along the vessels[b] and the biliary tree
N	Extrahepatic involvement of neighboring organs or tissues[c]
NX	Cannot be evaluated
N0	No regional involvement
N1	Regional involvement of contiguous organs or tissues
M	Absence or presence of distant metastases
MX	Not completely evaluated
M0	No metastasis[d]
M1	Metastasis present

[a] for classification, the plane projecting between the bed of the gallbladder and the inferior vena cava divides the liver in two lobes; [b] vessels means inferior vena cava, portal vein and arteries; [c] diaphragm, lung, pleura, pericardium, heart, gastric and duodenal wall, adrenal glands, peritoneum, retroperitoneum, parietal wall (muscles, skin, bone), pancreas, regional lymph nodes, liver ligaments; [d] chest X-ray and cerebral CT negative. This table has been widely used in various studies and is adapted from Kern et al. (2006) [18]

clinical management. The PNM classification system has since become the internationally accepted standard for alveolar echinococcosis staging, enabling more precise treatment planning and outcome prediction. PNM classification and staging of alveolar echinococcosis are summarized in Tables 11.1 and 11.2, respectively [18, 19].

11.1.3 Liver Transplantation for Alveolar Echinococcosis

Unlike hydatid disease that is caused by *E. granulosus*, hepatic alveolar echinococcosis is caused by *E. multilocularis*. Hepatic alveolar echinococcosis is considered a more serious disease compared to hydatid disease because one of the main character of hepatic alveolar echinococcosis is the tumor-like infiltrative growth in the liver parenchyma with associated progressive destruction of the hepatic parenchyma, invasion of adjacent organs, and metastasis to distant organs. This parasitic infestation of the liver shares similarities to liver tumors in terms of biological behavior. It has the ability of outward budding and the ability to invade neighboring structures and distant structures like brain and lungs, among others [20, 21]. The outcome of the disease depends on the stage of the disease. The PNM staging is used to stage the disease. In the classification, P denotes primary lesion in the liver, N is for lymph nodes and neighboring organs, and M stands for metastases [22, 23]. The mortality rate of hepatic alveolar echinococcosis in untreated patients is reported at 70% in 5 years and 94% in 10 years. The most important method of improving prognosis of the disease is early diagnosis and prompts treatment with complete surgical resection. However, because of the long clinical latency period of the disease, some cases cannot be diagnosed early enough, and at the time of diagnosis the lesion is advanced and cannot be resected completely. Advanced hepatic alveolar echinococcosis is characterized by extensive intrahepatic lesions, invasion of main hepatic veins, inferior vena cava or hepatic hilum, invasion of the diaphragm, retroperitoneal space, and distant metastatic lesions.

Table 11.2 Staging of alveolar echinococcosis based on PNM classification

Stage of alveolar echinococcosis	P	N	M
Stage I	P1	N0	M0
Stage II	P2	N0	M0
Stage IIIa	P3	N0	M0
Stage IIIb	P1–3	N1	M0
Stage IIIb	P4	N0	M0
Stage IV	P4	N1	M0
Stage IV	Any P	Any N	M1

This table has been widely used in various studies and is adapted from Kern et al. (2006) [18]

Treatment of early localized hepatic alveolar echinococcosis is surgical resection and postoperative albendazole therapy for 2 years to reduce recurrence. However, for patients with advanced disease, the treatment options are tenuous and the outcome has been a subject of discussion among specialists [21, 24–26].

Options for treatment of advanced hepatic alveolar echinococcosis include long albendazole treatment, palliative resections, palliative non-resective surgeries, and liver transplantation [27]. The survival of patients with albendazole therapy alone is reported to be around 55%, 36%, and 9% at 1 year, 3 years, and 5 years, respectively [25]. Palliative resections have a superior survival compared to isolated albendazole therapy. The survival in these patients was reported at 87%, 53%, and 40% at 1 year, 3 years, and 5 years, respectively. Options for palliative resection included partial liver resection with drainage of the residual cysts, partial liver resection with percutaneous transhepatic drainage, or partial resection of the liver and other structure invaded like lungs and diaphragm. Palliative non-resective surgeries have a better short-term survival compared to both albendazole therapy alone and palliative resective surgeries. The survival rate at 1 year was reported to be about 92.3%. However, the long-term survival is inferior to the other options with survival rate at 3 years and 5 years reported to be around 31% and 8%, respectively. Options for palliative non-resective procedures include percutaneous transhepatic biliary drainage (PTCD) guided by ultrasound, cholangiojejunostomy, and modified Longmire procedure [25–27].

11.1.4 Indications for Liver Transplantation

Liver transplantation for advanced hepatic alveolar echinococcosis is associated with the best short-term and long-term survival compared to other treatment options. The survival is reported to be 100%, 100%, and 67%, at 1 year, 3 years, and 5 years, respectively [25].

In 2007, expert consensus for the diagnosis and treatment of cystic and alveolar echinococcosis in humans conclude that the indications for liver transplantation in patents with hepatic alveolar echinococcosis include the following [19]:

1. Severe liver insufficiency or recurrent life threatening cholangitis
2. PNM IIIa, IIIb, or IV
3. Inability to perform a radical liver resection
4. Absence of extrahepatic alveolar echinococcosis location

Other additional indications reported include recurrent cholangitis, bleeding caused by portal hypertension, and chronic Budd–Chiari syndrome. Options for liver transplant in hepatic alveolar echinococcosis include liver transplantation and ex vivo liver resection and autotransplantation.

11.1.5 Liver Transplantation

The most common form of liver transplantation performed for hydatid disease is the liver transplantation which entails the implantation of a liver allograft from a donor. Liver transplantation is technically challenging in patients with hepatic alveolar echinococcosis. This is due to fact that the patients might have undergone liver resection in the past. Some patients might have had more than one hepatic resection before they came for liver transplant. Other reason for technical difficulty may be seen in patients with advanced disease with invasion of the hepatic artery, hepatic vein, inferior vena cava, or diaphragm, and these will not be readily accessible for mobilization during liver transplant. Aliakbarian et al. [28] reported that patients going for liver transplant for hepatic alveolar echinococcosis had previous hepatectomy in 17% of the cohorts and that 12% of them had invasion of one or more major hepatic vasculature [27–30].

Liver transplantation from allograft donors was the most common form of liver transplantation done for patients with hepatic alveolar echinococcosis needing liver transplant. This form of transplant is associated with certain limitations. The main limitation is the fact that patients with hepatic alveolar echinococcosis do not have enough derangement of their liver function to warrant a high model for end-stage liver disease (MELD) score and so they are almost always given a low priority, and this will result in prolonged waiting period during which they may develop metastatic hepatic alveolar echinococcosis which precludes liver transplant. Another limitation is the fact that with allograft donor, there is need for post-transplant immunosuppression to reduce the risk of graft rejection. In this category of patients, immunosuppression is associated with an increased risk of recurrence of the disease which may cause graft loss. To reduce the risk of recurrence, post-transplant albendazole therapy has been advocated by some surgeons, but recurrence rate of 8.9% to 60% has been reported despite the therapy [24, 28–30].

There is an ongoing effort to reduce recurrence of hepatic alveolar echinococcosis after liver transplantation. One such effort is the modification of post-transplant immunosuppression by using minimum doses of immunosuppressive medications needed. Another modification of immunosuppression is the exclusion of antilymphocyte antibodies from the regimen. Studies have also shown that early albendazole therapy within the first month of liver transplant is associated with reduced incidence of recurrence of hepatic alveolar echinococcosis. The therapy should be continued for minimum of 2 years. During the follow-up period, regular radiological (abdominal, thoracic, and cerebral computed tomography scans once a year) and serological (anti-Eg antibodies) must be done. If at the end of the second year post transplant all test are negative, then albendazole therapy can be suspended. However, if the tests are positive at the end of second year, lifelong albendazole therapy is needed [31–33]. Another effort at preventing recurrence among few selected patients is ex vivo liver resection and autotransplantation.

11.1.6 Ex Vivo Liver Resection and Autotransplantation

Ex vivo liver resection and autotransplantation was first reported as an option for patients with hepatic alveolar echinococcosis by Wen et al. [34]. It involves explantation of pathologic liver, resection of all the affected liver parenchyma and re implantation of the native liver. It was first used for advanced hepatic malignancy not amenable to conventional treatment. It has multiple advantages in patients with end stage hepatic alveolar echinococcosis. The first advantage is that there is no need for donor, which is one of the main limitations of liver transplantation for hepatic alveolar echinococcosis. The other advantage is there is no need for post-transplant immunosuppression since this is a form of autotransplantation. The fear of recurrence is reduced in this group of patients because there is no need for immunosuppression. Studies have shown that overall survival of patients that had ex vivo liver resection and autotransplantation is preferable to those that received allograft liver transplantation. Currently, more than a hundred cases of hepatic alveolar echinococcosis were treated with ex vivo liver resection and autotransplantation and the reports are encouraging [34–37].

11.1.7 Comparison of Liver Transplantation and Ex Vivo Liver Resection and Autotransplantation Procedures

However, patients who underwent ex vivo liver resection and autotransplantation experienced a significantly longer anhepatic phase and overall operation time [37, 38]. Postoperative parameters such as intensive care unit (ICU) stay, total hospital stay, and complication rates showed no significant difference between the two groups. Importantly, overall survival was higher in patients treated with ex vivo liver resection and autotransplantation, and these patients also exhibited lower recurrence rates of the parasite compared to those who received allograft liver transplantation [37, 38]. Despite these promising findings, the current evidence remains limited. The study by Shen et al [38]. was a retrospective, single-center analysis, which restricts the generalizability and strength of the conclusions. Therefore, further multicenter prospective studies and comprehensive meta-analyses are needed to more definitively compare these surgical strategies and refine their respective indications. 11.1.8. Conclusion Hepatic alveolar echinococcosis is a rare but aggressive parasitic disease that may require radical treatment in advanced stages. While complete surgical resection remains the optimal therapy in early disease, it is often unachievable due to extensive intrahepatic invasion and involvement of major vascular and biliary structures. Liver transplantation has shown satisfactory outcomes in selected patients; however, the need for lifelong immunosuppression poses a significant risk for disease recurrence. Additionally, donor organ scarcity and prior surgical interventions can complicate the transplant procedure. In contrast, ex vivo liver resection and autotransplantation has emerged as a promising alternative for carefully selected cases, offering comparable or even superior long-term results without the burden of immunosuppressive therapy. Comparative analyses suggest that this technique may lead to lower recurrence rates and better overall survival. Nevertheless, its widespread application is limited by technical complexity, prolonged operative duration, and the requirement for highly specialized surgical expertise. In conclusion, the management of advanced hepatic alveolar echinococcosis demands meticulous multidisciplinary planning. Treatment strategies should be individualized based on the extent of disease, feasibility of resection, and institutional capabilities.

References

1. Otero-Abad B, Torgerson PR. A systematic review of the epidemiology of echinococcosis in domestic and wild animals. PLoS Negl Trop Dis. 2013;7(6):e2249. https://doi.org/10.1371/journal.pntd.0002249. PMID: 23755310 PMCID: PMC3674998.
2. Wen H, Vuitton L, Tuxun T, Li J, Vuitton DA, Zhang W, McManus DP. Echinococcosis: advances in the 21st century. Clin Microbiol Rev. 2019;32(2) https://doi.org/10.1128/cmr.00075-18. PMID: 30760475 PMCID: PMC6431127.
3. Tsaroucha AK, Polychronidis AC, Lyrantzopoulos N, Pitiakoudis MS, Karayiannakis AJ, Manolas KJ, Simopoulos CE. Hydatid disease of the abdomen and other locations. World J Surg. 2005;29(9):1161–5. https://doi.org/10.1007/s00268-005-7775-3. PMID: 16088421.
4. Calame P, Weck M, Busse-Cote A, Brumpt E, Richou C, Turco C, Doussot A, Bresson-Hadni S, Delabrousse E. Role of the radiologist in the diagnosis and management of the two forms of hepatic echinococcosis. Insights Imaging. 2022;13(1):68. https://doi.org/10.1186/s13244-022-01190-y. PMID: 35394226 PMCID: PMC8994011.
5. Gomez IGC, Lopez-Andujar R, Belda Ibanez T, Ramia Angel J, Moya Herraiz A, Orbis Castellanos F, Pareja Ibars E, San Juan Rodriguez F. Review of the treatment of liver hydatid cysts. World J Gastroenterol. 2015;21(1):124–31. https://doi.org/10.3748/wjg.v21.i1.124. PMID: 25574085 PMCID: PMC4284328.
6. Velasco-Tirado V, Alonso-Sardón M, Lopez-Bernus A, Romero-Alegría Á, Burguillo FJ, Muro A, Carpio-Pérez A, Muñoz Bellido JL, Pardo-Lledias J, Cordero M, Belhassen-García M. Medical treatment of cystic echinococcosis: systematic review and meta-analysis. BMC Infect Dis. 2018;18(1):306. https://doi.org/10.1186/s12879-018-3201-y. PMID: 29976137 PMCID: PMC6034244.
7. Alvela-Suárez L, Velasco-Tirado V, Belhassen-Garcia M, Novo-Veleiro I, Pardo-Lledías J, Romero-Alegría A, Pérez del Villar L, Valverde-Merino MP, Cordero-Sánchez M. Safety of the combined use of praziquantel and albendazole in the treatment of human hydatid disease. Am J Trop Med Hyg. 2014;90(5):819–22. https://doi.org/10.4269/ajtmh.13-0059. PMID: 24615131 PMCID: PMC4015571.
8. Siles-Lucas M, Casulli A, Cirilli R, Carmena D. Progress in the pharmacological treatment of human cystic and alveolar echinococcosis: compounds and therapeutic targets. PLoS Negl Trop Dis. 2018;12(4):e0006422. https://doi.org/10.1371/journal.pntd.0006422. PMID: 29677189 PMCID: PMC5931691.
9. Dervenis C, Delis S, Avgerinos C, Madariaga J, Milicevic M. Changing concepts in the management of liver hydatid disease. J Gastrointest Surg. 2005;9(6):869–77. https://doi.org/10.1016/j.gassur.2004.10.016. PMID: 15985246.
10. Sozuer E, Akyuz M, Akbulut S. Open surgery for hepatic hydatid disease. Int Surg. 2014;99(6):764–9. https://doi.org/10.9738/intsurg-d-14-00069.1. PMID: 25437585 PMCID: PMC4254238.
11. Wan L, Wang T, Cheng L, Yu Q. Laparoscopic treatment strategies for liver echinococcosis. Infect Dis Ther. 2022;11(4):1415–26. https://doi.org/10.1007/s40121-022-00664-2. PMID: 35751754 PMCID: PMC9334471.
12. Kayaalp C, Bzeizi K, Demirbag AE, Akoglu M. Biliary complications after hydatid liver surgery: incidence and risk factors. J Gastrointest Surg. 2002;6(5):706–12. https://doi.org/10.1016/s1091-255x(02)00046-x. PMID: 12399060.
13. Mansy W, Mohamed M, Saber S. Outcomes of radical surgical management in liver hydatid cysts: 7 years center experience. Mini-invasive Surg. 2018;2(10):36. https://doi.org/10.20517/2574-1225.2018.48.
14. Shalayiadang P, Jiang T, Yimiti Y, Ran B, Aini A, Zhang R, Guo Q, Ahan A, Abulizi A, Wen H, Shao Y, Aji T. Double versus single T-tube drainage for frank cysto-biliary communication in patients with hepatic cystic echinococcosis: a retrospective cohort study with median 11 years follow-up. BMC Surg.

2021;21(1):12. https://doi.org/10.1186/s12893-020-01028-8. PMID: 33407348 PMCID: PMC7789643.

15. Tagliacozzo S, Miccini M, Amore Bonapasta S, Gregori M, Tocchi A. Surgical treatment of hydatid disease of the liver: 25 years of experience. Am J Surg. 2011;201(6):797–804. https://doi.org/10.1016/j.amjsurg.2010.02.011. PMID: 20832053.
16. Moreno González E, Landa García JI, Calleja Kempin J, Gómez Gutiérrez M, Jover Navalón JM, Arias Díaz J, Boundrandi S. [Role of liver transplant in the treatment of hepatic hydatidosis]. Rev Esp Enferm Apar Dig. 1989;76(1):1–3. PMID: 2799031.
17. Moreno-González E, Loinaz Segurola C, García Ureña MA, García García I, Gómez Sanz R, Jiménez Romero C, González Pinto I, Corral Sanchez MA, Palma CF. Liver transplantation for Echinococcus granulosus hydatid disease. Transplantation. 1994;58(7):797–800. PMID: 7940713.
18. Kern P, Wen H, Sato N, Vuitton DA, Gruener B, Shao Y, Delabrousse E, Kratzer W, Bresson-Hadni S. WHO classification of alveolar echinococcosis: principles and application. Parasitol Int. 2006;55(Suppl):S283–7. https://doi.org/10.1016/j.parint.2005.11.041. PMID: 16343985.
19. Brunetti E, Kern P, Vuitton DA. Expert consensus for the diagnosis and treatment of cystic and alveolar echinococcosis in humans. Acta Trop. 2010;114(1):1–16. https://doi.org/10.1016/j.actatropica.2009.11.001. PMID: 19931502.
20. Kantarci M, Pirimoglu B. Diffusion-weighted MR imaging findings in a growing problem: hepatic alveolar echinococcosis. Eur J Radiol. 2014;83(10):1991–2. https://doi.org/10.1016/j.ejrad.2014.07.013. PMID: 25112672.
21. Strohaeker J, Sulyok M, Koenigsrainer A, Nadalin S. Alveolar echinococcosis-a challenging task for the hepatobiliary surgeon. Pathogens. 2021;11(1) https://doi.org/10.3390/pathogens11010040. PMID: 35055988 PMCID: PMC8778716.
22. Bulakci M, Kartal M, Yilmaz S, Yilmaz E, Yilmaz R, Sahin D, Asik M, Erol OB. Multimodality imaging in diagnosis and management of alveolar echinococcosis: an update. Diagn Interv Radiol. 2016;22(3):247–56. https://doi.org/10.5152/dir.2015.15456. PMID: 27082120 PMCID: PMC4859741.
23. WHO/OIE manual on echinococcosis in humans and animals: a public health problem of global concern (2002).
24. Kern P, Grüner B, Wahlers K. Diagnosis and course of echinococcocal diseases in the transplant setting. Transpl Infect Dis. 2011;13(3):217–21. https://doi.org/10.1111/j.1399-3062.2011.00643.x. PMID: 21518162.
25. Qu B, Guo L, Sheng G, Yu F, Chen G, Wang Y, Shi Y, Zhan H, Yang Y, Du X. Management of Advanced Hepatic Alveolar Echinococcosis: report of 42 cases. Am J Trop Med Hyg. 2017;96(3):680–5. https://doi.org/10.4269/ajtmh.16-0557. PMID: 28070011 PMCID: PMC5361545.
26. Xu X, Gao C, Qian X, Liu H, Wang Z, Zhou H, Zhou Y, Wang H, Hou L, He S, Feng X, Fan H. Treatment of complicated hepatic alveolar echinococcosis disease with suspicious lymph node remote metastasis near the inferior vena cava-abdominal aorta: a case report and literature review. Front Oncol. 2022;12:849047. https://doi.org/10.3389/fonc.2022.849047. PMID: 35402272 PMCID: PMC8987535.
27. Salm LA, Lachenmayer A, Perrodin SF, Candinas D, Beldi G. Surgical treatment strategies for hepatic alveolar echinococcosis. Food Waterborne Parasitol. 2019;15:e00050. https://doi.org/10.1016/j.fawpar.2019.e00050. PMID: 32095621 PMCID: PMC7034045.
28. Aliakbarian M, Tohidinezhad F, Eslami S, Akhavan-Rezayat K. Liver transplantation for hepatic alveolar echinococcosis: literature review and three new cases. Infect Dis (Lond). 2018;50(6):452–9. https://doi.org/10.1080/23744235.2018.1428823. PMID: 29363377.
29. Aydinli B, Ozturk G, Arslan S, Kantarci M, Tan O, Ahıskalioglu A, Özden K, Colak A. Liver transplantation for alveolar echinococcosis in an endemic region. Liver Transpl. 2015;21(8):1096–102. https://doi.org/10.1002/lt.24195. PMID: 26074280.
30. Bresson-Hadni S, Blagosklonov O, Knapp J, Grenouillet F, Sako Y, Delabrousse E, Brientini MP, Richou C, Minello A, Antonino AT, Gillet M, Ito A, Mantion GA, Vuitton DA. Should possible recurrence of disease contraindicate liver transplantation in patients with end-stage alveolar echinococcosis? A 20-year follow-up study. Liver Transpl. 2011;17(7):855–65. https://doi.org/10.1002/lt.22299. PMID: 21455928.
31. Bresson-Hadni S, Koch S, Beurton I, Vuitton DA, Bartholomot B, Hrusovsky S, Heyd B, Lenys D, Minello A, Becker MC, Vanlemmens C, Mantion GA, Miguet JP. Primary disease recurrence after liver transplantation for alveolar echinococcosis: long-term evaluation in 15 patients. Hepatology. 1999;30(4):857–64. https://doi.org/10.1002/hep.510300426. PMID: 10498634.
32. Castro A, Hyppolito E, Valença-Júnior J, Coelho G, Garcia J. Preliminary data on liver transplantation in hydatidosis disease. Arq Bras Cir Dig. 2022;35:e1667. https://doi.org/10.1590/0102-672020210002e1667. PMID: 35766612 PMCID: PMC9254553.
33. Zavoikin VD, Zelya OP, Tumolskaya NI. The importance of uninterrupted albendazole treatment in patients with unresectable alveolar echinococcosis undergoing liver transplantation. Transpl Infect Dis. 2020;22(4):e13291. https://doi.org/10.1111/tid.13291. PMID: 32277859.
34. Wen H, Dong JH, Zhang JH, Duan WD, Zhao JM, Liang YR, Shao YM, Ji XW, Tai QW, Li T, Gu H, Tuxun T, He YB, Huang JF. Ex vivo liver resection and autotransplantation for end-stage alveolar echinococcosis: a case series. Am J Transplant. 2016;16(2):615–24. https://doi.org/10.1111/ajt.13465. PMID: 26460900.
35. Aji T, Dong JH, Shao YM, Zhao JM, Li T, Tuxun T, Shalayiadang P, Ran B, Jiang TM, Zhang RQ, He YB, Huang JF, Wen H. Ex vivo liver resection and autotransplantation as alternative to allotransplantation for end-stage hepatic alveolar echinococcosis. J Hepatol.

2018;69(5):1037–46. https://doi.org/10.1016/j.jhep.2018.07.006. PMID: 30031886.
36. Yang X, Qiu Y, Huang B, Wang W, Shen S, Feng X, Wei Y, Lei J, Zhao J, Li B, Wen T, Yan L. Novel techniques and preliminary results of ex vivo liver resection and autotransplantation for end-stage hepatic alveolar echinococcosis: a study of 31 cases. Am J Transplant. 2018;18(7):1668–79. https://doi.org/10.1111/ajt.14621. PMID: 29232038 PMCID: PMC6055796.
37. Yuan J, Chen X, Hou L, Wang H, Zhou Y, Pang M, YangDan C, Wang Z, Fan H. Single-center experience of ex vivo liver resection and autotransplantation for complex hepatic alveolar echinoccosis. Front Surg. 2023;10:1089788. https://doi.org/10.3389/fsurg.2023.1089788. PMID: 36874451 PMCID: PMC9975350.
38. Shen S, Kong J, Qiu Y, Zhang S, Qin Y, Wang W. Ex vivo liver resection and autotransplantation versus allotransplantation for end-stage hepatic alveolar echinococcosis. Int J Infect Dis. 2019;79:87–93. https://doi.org/10.1016/j.ijid.2018.11.016. PMID: 30496849.

Treatment of Hydatid Disease in Unusual Locations

12

Ibrahim Umar Garzali
and Abdurrahman Abba Sheshe

12.1 Introduction

Hydatid cyst is a zoonotic caused by adult or larval stages of *Echinococcus granulosus*. In normal situations, the worm is harbored in the intestine of carnivores such as dogs and wolves, and these are the definitive hosts. The eggs are usually passed in the feces of the infected carnivores before ingestion by herbivores such as sheep, and these are considered the intermediate host. Humans are the accidental intermediate host and they became infected after ingestion of feeds contaminated by the eggs of *Echinococcus granulosus*. After ingestion, the larvae emerge from the eggs in the human intestine, and after invasion to the blood vessels, they can migrate into almost every part of the body. The commonest path of invasion is through the portal vein where the liver is commonly affected. In some occasions, the larvae can pass through the liver barrier and reach the lungs where they can cause pulmonary hydatid disease. The larvae may also go beyond the lungs into the systemic circulation where they can cause hydatid disease in unusual locations [1–8].

The commonest site affected by hydatid disease is the liver and hepatic hydatid disease constitutes 70% to 80% of all adult hydatid disease. This is followed by pulmonary hydatid disease which constitute about 10% to 15% of all adult hydatid disease. In pediatric patients, the distribution of hydatid disease showed significant variation from adults. It has been reported that in children, the incidence rates of 41–70% were observed for pulmonary hydatid diseases followed by the incidence rate of 43–48% for hepatic hydatid disease. Simultaneous hepatopulmonary hydatid disease is reported to be around 5–13%. Apart from the lungs and the liver manifestations of hydatid disease, there are reports of more unusual locations such as the spleen, heart, orbit, brain, muscle, salivary gland, bone, urinary tract, and pancreas [1–8]. The clinical manifestation and treatment of hydatid disease have been discussed in other chapters; the main focus of this chapter is to discuss the manifestation and treatment of non-hepatic hydatid disease.

12.2 Pulmonary Hydatid Disease

Pulmonary hydatid disease constitutes 10–15% of all adult hydatid diseases and 41–70% of all pediatric hydatid disease. Pulmonary hydatid disease may occur in isolation or it may coexist with hepatic hydatid disease. On rare occasions, pulmonary hydatid disease may be as a result of complication of hepatic hydatid disease. The parasite obtains access to the lungs through various means. The first pathway of access is through the liver. Most of the embryos of *Echinococcus granulosus* are trapped in the hepatic sinusoids.

I. U. Garzali (✉) · A. A. Sheshe
Department of Surgery, Aminu Kano Teaching Hospital, Kano, Nigeria

S. Akbulut (ed.), *Hydatid Disease*, https://doi.org/10.1007/978-3-031-97277-5_12

However, for embryos with diameters <0.3 mm, they may pass through the hepatic sinusoids, hepatic vein, and the inferior vena cava (IVC), the right side of heart, pulmonary arteries, and finally settle in the lungs. Another pathway that embryos reach the lungs is through the thoracic duct. The embryo enters the thoracic duct via lymphatics of the small intestine. The thoracic duct enters the internal jugular vein and then the right side of the heart, pulmonary arteries, and finally the lungs. In this pathway, the hydatid larva bypasses the liver and involves the lungs. Additional pathway includes the lymphatics of the dome of the liver and the diaphragm. These lymphatics ascend to the chest through the parasternal and intercostal lymph nodes. The third possible route is a venal-venous anastomosis in the liver and the space of Retzius. Another possibility is direct pulmonary exposure through the inhalation of air contaminated with *Echinococcus* eggs [1, 2, 4, 9–12].

12.2.1 Synchronous Hepatopulmonary Hydatid Cysts

Synchronous hepatopulmonary hydatid cysts are an uncommon manifestation of hydatid disease that are observed in less than 10% of case. They represent a distinct clinical entity named hepatopulmonary hydatidosis (HPH). The condition is more commonly seen in female patients, particularly in those older than 40 years of age. The pathogenesis of hepatopulmonary hydatidosis is believed to be as a result of trans-diaphragmatic dissemination of the parasite through a bronchobiliary fistula resulting from trans-diaphragmatic rupture of liver hydatid. Synchronous hepatopulmonary hydatid cysts are more commonly seen in adult patients compared to children. Concomitant hepatic cysts are seen in 79% of adults as compared to 33% of children. Pulmonary hydatid disease is characterized by unilateral, solitary cyst in about 82% of cases. The right lung is more commonly affected compare to the left. The cysts are commonly found in the lower lobe of the lungs with predilection for the posterior segments. Bilateral involvement occurs in 20% of cases, and multiple cysts in 30% of cases [4, 9, 11–15].

Clinical presentation of pulmonary hydatid cyst may vary, but most patients present with symptoms of cough followed by chest pain, fever, hemoptysis, breathlessness, expectoration, and anaphylactic reactions. Hemoptysis is most common symptom among adults with pulmonary hydatid disease. Complications associated with pulmonary hydatid disease include: bronchial compression especially for cysts greater that 5 cm, cyst rupture, secondary infection, suppuration, and pneumothorax. Cyst rupture often leads to sudden onset of chest pain, hemoptysis, cough, and fever. Ruptured pulmonary hydatid cyst can also cause anaphylactic reaction which may present with urticaria and wheezing. All patients with pulmonary hydatid should be screened for the presence of hepatic hydatid cysts to rule out synchronous hepatopulmonary hydatid disease [9, 11, 12].

12.2.2 Diagnosis

Diagnosis of pulmonary hydatid disease can be achieved with chest imaging combined with serology. Peripheral blood smear may show leukocytosis, eosinophilia, and elevated erythrocyte sedimentation rate, but these findings are nonspecific.

12.2.3 Imaging

Chest imaging is the main investigation for diagnosis of pulmonary hydatid cyst. Conventional X-ray, computed tomography, and magnetic resonance imaging of the lungs are the various modalities useful in the diagnosis of thoracic hydatid cyst. Radiological presentation depends on whether the cysts are complicated or not. The radiological features of a pulmonary hydatid disease during X-ray evaluation include sharply defined, round-to-oval homogenous opacity. The size of the cyst varies and peripherally located cysts are larger than those located at the major bronchovascular structures. The cysts are usually solitary but multiple cysts may also be found occasionally. The findings of calcification and daughter cyst formation are a rare occurrence in pulmonary hydatid disease. The findings on pulmonary hydatid disease in computed tomography scan include smooth walls of variable thickness and homogenous internal contents of water or near-water density. The appearance of a single wall in computed tomography scan is because the adventitious (pericyst), laminated, and germinal layers of an intact cyst are clubbed together. Features of complication may also be seen during computed tomography imaging. Detachment of the parasitic membrane from the pericyst will manifest as "the crescent" or the "meniscus sign". This will appear as air into the potential space between the pericyst and ectocyst. Ruptured pulmonary hydatid cyst manifest as collapsed of the endocyst with air fluid level within the cyst. This sign has different names and it is termed double-arch or cumbo sign or iceberg sign. Superimposed infection within the cyst manifest as the presence of air bubbles within the cyst and ring enhancement of the pericyst on contrast-enhanced computed tomography scan. Similar findings are also seen in cyst-bronchial tree communication [4, 12–18].

12.2.4 Treatment

Pulmonary hydatid disease may be treated pharmacologically, surgically, or with a combination of the two. Surgical intervention is the treatment of choice.

Indications for pharmacotherapy include small cysts less than 6 cm, patients with contraindication for surgery, disseminated disease, multiple cysts, recurrent cysts, and patients with intraoperative spillage of hydatid fluid. Pharmacological treatment entails oral administration of benzimidazoles group of medications mainly albendazole. Albendazole is the drug of choice because it has a high bioavailability. The recommended dose is 10–15 mg/kg/day, taken twice daily, for at least 3–6 months. Pharmacotherapy is contraindicated in patients with cysts larger than 6 cm in diameter because there is a risk of rupture; patients with inactive or calcified cysts; patients who are prone to bone marrow depression; and pregnancy [1, 4, 11, 12, 14, 18].

Surgery is the gold standard treatment of choice for pulmonary hydatid disease of any size. Surgery is done for patients with large cysts that are superficially located because they are at risk of rupture. Surgery is also preferred in patients with infected cysts, cysts in close proximity to vital anatomical structures, and mass effect resulting from the cyst [1, 4, 11, 12, 14, 18].

Surgical approach is achieved via posterolateral thoracotomy in most cases. However, for patients with bilateral cysts, median sternotomy or two-stage thoracotomy is done. In patients with bilateral cysts and if one of the cysts ruptures, most surgeons recommend operating on the un-ruptured cyst first because it is also at risk of rupture [18–22].

Surgical treatment options include enucleation, pericystectomy, cystotomy, segmental resection, and lobectomy. Enucleation entails removal of the cyst with its intact germinative membrane. It is applicable for the small PHCs and associated with little risk of rupture. This procedure should be avoided in patients with large cysts because of the risk of rupture. There is low incidence of postoperative complications like air leak and infection. Pericystectomy involves removal of hydatid cyst along with the pericyst. This procedure allows complete removal of the parasite. Pericystectomy may be associated with increased risk of airway leak. Cystotomy involves aspirating fluid from the cyst along with removal of the germinative membrane. After the cystotomy, the residual cavity can be managed with capitonnage (Barrett's method) or if there is bronchial opening within the cyst, the bronchial opening is closed before capitonnage (Posadas' method). Capitonnage reduces the risk of infection of the residual cavity, airway leak, and empyema formation, but there is a risk of disfigurement of the lung parenchyma which may lead to atelectasis. Cystotomy can also be done with closure of the bronchial opening but without capitonnage procedure. This technique is associated with minimal parenchymal disfigurement, but the risk of air leak and infection is increased. Open aspiration (Figuera technique) of pulmonary hydatid cyst has been described. This technique is similar to percutaneous aspiration, instillation of scolicidal agents, and re-aspiration (PAIR) used for hepatic hydatid cyst. It is less invasive but there is an increased risk of infection, empyema formation, and air leak. Segmental resection of the lung segments can be done in ruptured pulmonary hydatid disease. The resection can be anatomic or nonanatomic. It is associated with reduced infection and recurrence, but it also causes reduction of lung volume with resulting decrease in lung compliance. Lobectomy involves the anatomic resection of one or more of the lobes of the lung affected by the cysts. It is indicated in the following group of patients: patients with a cyst involving greater than 50% of the lobe; patients with multiple unilobar cysts; and patients with sequelae of hydatid disease such as bronchiectasis, pulmonary fibrosis, or severe hemorrhage [18–24].

All patients that had surgery for hydatid cyst should receive albendazole (10 mg/kg/day) for 6 months to prevent recurrence of the disease. Recurrence rate of 11% has been reported if post-surgical chemoprophylaxis was withheld. Follow-up should be done monthly for the first 3 months and then every 3 months for the first year after surgery.

12.3 Splenic Hydatid Disease

Spleen is the third most common site for *Echinococcus* infestations. The liver is the most common involved organ (75%), followed by the lung (15.4%), and then the spleen (0.5–8%). In endemic areas, hydatid cysts constitute up to 50–80% of splenic cysts. Splenic hydatid cysts are endemic to parts of South America, Middle East, Mediterranean, India, and Australia. Global incidence of splenic hydatid is 0.5–4% with the highest incidence seen in Iran where it constitutes up to 4% of hydatid disease [25–28].

Seedling of the parasites in the spleen occurs through any of the following routes. The first route is the arterial route and it usually occurs after the organism passes through the liver and the lung, and ends up in the spleen. The second route is the venous route and it occurs through the portal circulation, avoiding passage through the liver. A third route is through secondary infestation of the spleen after systemic dissemination or intraperitoneal seeding following the rupture of a hepatic cyst. Once the parasite infests the spleen, it starts to grow at a rate of 0.3 cm to 1 cm per year. Because the rate of growth is slow, symptoms may take several years to appear. In fact, up to 25% to 35% of patients are asymptomatic, and splenic hydatid is diagnosed as an incidental finding during evaluation for other pathologies [25, 26, 29–31].

The most common presenting symptom in patients with splenic hydatid cyst is left upper quadrant or epigastric pain. The pain is usually a dull dragging ache. Abdominal evaluation may reveal mass in the abdomen mostly in left hypochondrium and less frequently in epigastrium. Other symptoms related to mass effect of the cyst compressing neighboring structures include constipation due to pressure on colon and dyspnea due to pushing up of the left diaphragm. Symptoms related to rupture in to neighboring viscera include fistula of the colon or perforation into the diaphragm or bronchial tree. There have been reports of acute abdominal pain and anaphylactic reaction due to intraabdominal rupture of splenic hydatid cysts [25, 26, 29–35].

12.3.1 Diagnosis

Diagnosis of splenic hydatid disease can be achieved by abdominal imaging combined with serology. Peripheral blood smear may show leukocytosis, eosinophilia, and elevated erythrocyte sedimentation rate, but these findings are nonspecific.

12.3.2 Imaging

Plain abdominal radiograph may reveal marginal egg-shell-like calcification in the splenic area. Other findings include elevated left hemidiaphragm, displacement of stomach and/or left colonic flexure. The classic appearance of splenic hydatid disease on ultrasonography is an anechoic smooth, round cyst, which can be difficult to distinguish from splenic cysts. To differentiate hydatid cysts from nonparasitic splenic cysts can be achieved by axial sonogram, which reveals parallel stripes, called "double-line" signs that delineate the space between parasite's ectocyst and host's pericyst. Ultrasonography also plays a role in diagnosing complicated splenic hydatid disease. If the cyst contains membranes, mixed internal echoes will appear in the cyst. The presence of daughter cysts is characterized by internal septations and hydatid sand. Ultrasonography is also used for the classification of the biologic activity of the cyst as this may influence the choice of treatment; splenic hydatid cyst can be active and the characteristics of active cyst are mentioned above. Inactive splenic hydatid cysts are characterized as collapsing, flattened, elliptical cysts with detachment of the germinal layer from the cyst wall ("water lily sign") and calcification of the cyst wall [25, 26, 36, 37].

The sensitivity of abdominopelvic computed tomography scan in diagnosing splenic hydatid disease is between 95% and 100%. It is the best option in diagnosis and characterizing the cyst. It can also be used for monitoring patients for non-operative therapy and to identify recurrence cases.

The main findings on abdominal computed tomography scan are a cystic lesion in the cyst that does not enhance after administration of intravenous contrast. The cyst has a homogenous content that has water attenuation values. High computed tomography attenuation values may be seen if the cyst contains debris, hydatid sand, or inflammatory cells. Other findings include calcification of the cyst wall, and this calcification involves the pericyst. One of the advantages of computed tomography scan is the ability to readily identify the presence of daughter cysts which may present with findings of multivesicular hydatid cyst with peripheral daughter cysts attached to the membrane of mother cyst in the center [25, 26, 29, 36, 37].

The role of magnetic resonance imaging in diagnosing splenic hydatid is increasing and it is especially important to patients with negative serology and equivocal ultrasonography and computed tomography scan evaluation. It can differentiate parasitic and nonparasitic splenic cysts. The features of splenic hydatid on magnetic resonance imaging include hypointense rim sign identified in the T2-weighted sequence, hyperintense septated mass with smaller peripheral masses on echo T2-weighted image [25, 26, 29, 36, 37].

12.3.3 Serology

Immunoelectrophoresis of patients with hydatid cysts has 95% sensitivity, and it is still positive up to 1 year after eradication of the parasite. It plays a significant role in screening for patients living in endemic areas. Indirect hemagglutination is another serological test for diagnosis of splenic hydatid and its sensitivity is around 85% [25, 26].

12.3.4 Treatment

Surgical therapy is the most important intervention for splenic hydatid.

Isolated pharmacotherapy is indicated in selected group of patients and they include those with multiple cysts and recurrence in multiple organs. The medication of choice is albendazole administered continuously at a daily dosage of 15 mg/kg/day. The follow-up of patients on pharmacotherapy includes weekly complete blood count and liver transaminases for the first month, and subsequently these tests are done monthly [31, 38–40].

Percutaneous aspiration irrigation and re-aspiration (PAIR) under image guidance can be used in the treatment of splenic hydatid cysts. The procedure can be used in patients with small cysts of less than 5 cm, splenic hydatid cysts in patients who refused surgery, or in patients in whom the risk of surgical intervention is high.

Splenectomy is considered the main treatment of splenic hydatid cyst because it has a low morbidity and mortality rate. Total splenectomy is encouraged for larger cysts, because of the reduction in splenic parenchyma from pressure. This makes the spleen thin and fragile and there is increased risk of intraoperative rupture during conservative approach. Total splenectomy can be performed via open or laparoscopic approach. The laparoscopic approach is preferred in solitary, small cysts that are located superficially. For multiple cysts, large cysts, or deep-seated cysts, open approach is preferred. Laparoscopic approach is associated with lower morbidity compared to open technique. Splenectomy is associated with overwhelming postsplenectomy infection (OPSI), especially among pediatric patients, and it has a mortality of up to 50%. Recurrence rate after splenectomy is around 14% [29–31, 41].

Due the complication of OPSI, splenic preservation has been recommended for certain group of patients, like patients with superficial cysts confined to one of the poles of spleen, patients with cysts with extensive adhesions, and in children to avoid OPSI. Some of the options for splenic preserving surgery include partial splenectomy, cyst enucleation, and deroofing of cyst with omentoplasty. Recurrence after splenic preservation procedure is around 12% [26, 29–31, 41].

12.4 Intracranial Hydatid Disease

Echinococcal infestation of the brain is the most common brain parasitic infestation in the world. It is responsible for only 1–2% of all hydatid disease. About 50–70% of all patients affected by brain hydatid disease are children and young adults, and the disease shows significant preponderance to the male gender [42–44].

Brain hydatid cyst also can be classified as primary or secondary (multiple). The primary cysts are usually solitary and occur when there is direct infestation of the brain without involvement of other organs. In secondary brain hydatid disease, there are multiple cysts and this usually occurs when there is rupture of a solitary cranial cyst. The rupture may be spontaneous, traumatic, or surgically induced. Brain hydatid disease can affect any part of the brain, but it commonly occurs in the supratentorial region, specifically in the middle cerebral artery territory. The parietal lobe is the most commonly affected lobe [42–46]. Mass effect is the main manifestation of brain hydatid cyst, and patients usually present with headache, vomiting, and convulsions due to raised intracranial pressure. On physical examination, papilledema and focal neurological deficit may be identified in the patients [42–46].

12.4.1 Diagnosis

Imaging diagnosis is the gold standard for the diagnosis of brain hydatid disease. Computed tomography scan usually reveals a spherical, well-defined, thin-walled, homogeneous, and nonenhancing cystic lesion without peripheral edema. The fluid is usually isodense to cerebrospinal fluid. Magnetic resonance imaging usually reveals a low signal intensity rim of the cyst wall while the content shows signal intensity similar to CSF [42, 44–46].

12.4.2 Treatment

Surgical management of brain hydatid cyst is the main stay of treatment. Dowling procedure is the most effective method of delivering intact cyst without rupture. It involves saline irrigation between cyst wall–brain interfaces. Pharmacotherapy with albendazole 10–15 mg/kg/day can be used in younger patients or patients with multivisceral hydatid disease. Pharmacotherapy can also be used to prevent secondary hydatid disease of the brain in cases of intraoperative rupture [47, 48].

12.5 Renal Hydatid Disease

Hydatid disease of the urinary tract is rare. The kidney is the commonest organ affected. It is usually part of disseminated disease but isolated renal hydatid has been reported at 2–4% of all cases. Infestation of the kidney occurs via hematological dissemination. The left kidney is more commonly affected compared to the right kidney and the reason postulated for this is that the left renal artery is shorter so the spread to the left kidney is easy [49–54].

Patients with isolated renal hydatid disease can remain asymptomatic for years. When the patients finally become symptomatic, the symptoms are usually nonspecific. Patients may present with vague pain of lumbar or flank region. They may also have fever, malaise, and/or palpable mass. As the disease progresses, the cyst may rupture into the collecting system of the urinary tract and then patients may have hematuria and hydatiduria which is diagnostic of renal hydatidosis. Hydatiduria is seen in only 10–20% of patients with renal hydatidosis [49–54].

The main modality of diagnosing renal hydatidosis is the use of imaging. The imaging will characterize the cyst as single or multiple, and uni- or multilocular. Ultrasonography is the most common imaging used for hydatid disease, and it shows the floating membranes, daughter cysts, and hydatid sand characteristically seen in purely cystic lesions. Another imaging used is the computed tomography scan. It is superior to ultrasonography because it demonstrates intracystic gas and minute calcifications. The role of serology in the diagnosis of isolated renal hydatidosis is minimal. Eosinophilia is seen in 50% of these patients [51–56].

The main therapeutic option in patients with isolated renal hydatid cyst is surgical intervention. Simple excision of the cyst is an acceptable treatment modality. Another option is pericystectomy. In patients with extensive kidney damage, nephrectomy should be considered. Puncture, aspiration, injection and re-aspiration (PAIR) can be done in some patients [49–54, 57].

Albendazole therapy for a week before surgery may reduce tension within the cyst, reduce the activity of *Echinococcus* larvae, and thereby reduce the risk of rupture and spread during surgery. Postsurgical therapy with albendazole for 3 months also reduces postoperative recurrence, especially in cases with inadvertent rupture of the cyst at the time of surgery [49, 52, 54, 57–60].

12.6 Hydatid Cyst of the Adrenal Glands

Adrenal glands are among the organs that are rarely involved by hydatid disease even in endemic regions. Adrenal hydatidosis accounts for less than 1% of all hydatid diseases. It is usually secondary to hematogenous dissemination of hydatid disease and it is almost always part of multiorgan hydatid disease. Isolated adrenal hydatidosis is relatively rare. Adrenal hydatid cyst constitutes 7% of adrenal gland cysts. Autopsy reports showed an incidence between 0.06% and 0.18% with more than 90% being unilateral [61–65].

Adrenal hydatidosis is usually asymptomatic. The most common form of detection is incidental during radiological evaluation for other indications. Some patients may have nonspecific symptoms such as vague hypochondrial or flank pain, nausea and vomiting, palpable mass, and loss of appetite. In some rare instances, it is discovered by arterial hypertension due to compression of renal parenchyma by large adrenal hydatid cysts [61–67].

Most of the cases of adrenal hydatidosis were diagnosed intraoperatively. However, if the diagnosis is suspected preoperatively, imaging is the most important diagnostic method. Ultrasonography is usually the first investigation of choice, but because of the location of the adrenal deep in the retroperitoneum, the use of ultrasonography is not reliable. Abdominal computed tomography scan is more reliable as it confirms the diagnosis and assesses the relationship of the cyst with the surrounding structures. The role of serology in diagnosis is negligible; positive serological tests do not confirm the diagnosis and negative serological tests do not exclude it. However it can be used to detect recurrence disease during follow-up [56, 61–64, 66, 67].

The treatment of adrenal hydatidosis is mostly surgical. Pericystectomy with preservation of the nonaffected part of the gland can be considered in small cyst. For patient with large cysts, adrenalectomy should be considered. Puncture, aspiration, injection, and re-aspiration (PAIR) treatment of hydatidosis is avoided in patients with adrenal hydatidosis except in inoperable cases [58, 60–64, 66, 67].

12.7 Pancreatic Hydatid Disease

Isolated involvement of the pancreas by hydatid disease is rare, and secondary acute pancreatitis caused by primary pancreatic hydatid cyst is even rarer. Pancreatic hydatid cyst (PHC) has reported incidence of 0.14%–2%. The most common part of the pancreas affected is the pancreatic head (57%), followed by the body (24%) and the tail (19%) [65, 68–70].

Multiple pathways for pancreatic invasion by the parasite have been described. One of the pathways is hematogenous dissemination, and this is believed to be the commonest mode of spread to the pancreas. Other proposed pathways of spread include passage through the biliary system or lymphatic spread from the intestinal mucosa. Additional pathways include direct passage via the pancreatic veins and retroperitoneal dissemination [65, 68–72].

The mode of clinical presentation depends on part of the pancreas involved by the disease. If the cysts are located in the head, patient will develop obstructive jaundice due to extrinsic compression of the common bile duct. For cysts located in the body and tail of pancreas, they may remain asymptomatic until the cyst has attained a large enough size to present compressive symptoms like epigastric pain, nausea, and vomiting. Rare cases of sinistral hypertension with splenomegaly have been reported in cysts located in the pancreatic tail with compression on the splenic vein. Some selected patients present with complications such as cholangitis, pancreatic fistula, and recurrent pancreatitis [65, 68–75].

Diagnosis of pancreatic hydatidosis is usually attained with imaging. The imaging modalities commonly utilized include ultrasonography, computed tomography, and magnetic resonance imaging. During abdominal ultrasonography, findings like floating membranes, hydatid sand, and daughter cysts may be diagnostic. However, the sensitivity of ultrasonography is reduced due to retroperitoneal location of the pancreas. Evaluation by computed tomography scan or magnetic resonance imaging may reveal the presence of an undulating membrane and multiple daughter cysts within a mother cyst. For lesions located in the pancreatic head, magnetic resonance cholangiopancreatography is important because it plays a role in delineating the biliary tree and pancreatic duct. Serology plays minimal role in diagnosis but it is important during follow-up [65, 68–75].

For cysts located in the pancreatic head without communication to the pancreatic or biliary duct can be treated with pericystectomy or partial pericystectomy with external drainage or omentopexy. Pancreatoduodenectomy is also an acceptable procedure for these patients. If there is communication between the cyst and the pancreatic duct, cystoenteric anastomosis is the procedure of choice. For cysts located in the body and tail of pancreas, distal pancreatectomy can be performed. Endoscopic retrograde cholangiopancreatography and palliative stenting can be done in cases of cholangitis or pancreatitis secondary to biliary/pancreatic ductal compression [68, 71–76].

12.8 Hydatid Disease of the Breast

The mammary gland is a rare site of involvement by hydatid disease, and it accounts for only 0.27% of all cases of hydatid disease. The main source of breast affectation is via systemic dissemination. The presentation is usually insidious with patients typically presenting with a long history of a painless breast lump that slowly increases in size. The regional lymph nodes are usually disease free [14, 22, 77–80].

It is very difficult to differentiate the lesion from other breast diseases. And the use of mammography, magnetic resonance imaging, and breast ultrasonography scans are not helpful because it may be indistinguishable from benign cysts. Serologic tests can be used to confirm the hydatid origin of a breast cyst diagnosed by imaging. Fine-needle aspiration cytology may assist in confirming the diagnosis if hooklets are found within the sample. However, if hydatid cyst of the breast is suspected, fine-needle aspiration cytology should be avoided [77, 78, 80–83]. Treatment is usually by complete excision of the cyst without spillage. Recurrence rate of 10% has been reported after excision. The rate of recurrence can be reduced by perioperative administration of albendazole [77, 78, 80–83].

12.9 Tubo-Ovarian Hydatid Cyst

Pelvic involvement with hydatidosis is rare and it is usually as a result of secondary seeding. Reports have shown that approximately 80% of all pelvic hydatid disease involves fallopian tubes and ovaries. Tubo-ovarian disease can be unilateral or bilateral. Symptoms and signs of tubo-ovarian hydatidosis are related to local compression of the genital organs, urinary tracts, vascular structures, and bony structures. Other forms of presentation may be due to complications like infection, and it may present like ovarian abscess [84–89].

Ultrasonography is usually enough to make diagnosis. The finding of cysts with daughter cells which may appear as honeycomb is highly suggestive of tubo-ovarian hydatidosis. Magnetic resonance imaging is used to assess the relationship between the cyst and neighboring structures. Surgical excision is considered the ideal treatment for tubo-ovarian hydatid cyst. The cyst should be excised completely and care must be taken to avoid rupture of the cyst during surgery. For nonsurgical candidates, other alternatives like puncture, aspiration, injection and re-aspiration (PAIR) can be used. The recurrence rate after surgical excision is 2% with a survival rate of 95% [84–89].

12.10 Hydatid Disease of Soft Tissues

Soft tissue hydatidosis are rare even in endemic areas. When they occur, they are usually secondary, and tend to have a predisposition for the lower limbs. Soft tissue hydatidosis usually present as painless, slow-growing mass with normal overlying skin. High index of suspicion is required for diagnosis. Imaging investigations are predominantly used to diagnose these lesions [90–95]. Plain radiography of the region shows an unruptured cyst appearing as a spherical, well-circumscribed, homogenous opacity. The sensitivity of ultrasonography and computed tomography in diagnosing soft tissue hydatidosis is reported at 85% and 100%, respectively. Computed tomography scan and ultrasonography show hydatid sands in purely cystic lesions. Other findings include floating membranes, daughter cysts, and vesicles [36, 37, 90–95]. En bloc excision of the cyst is considered the best treatment option. In cases of spillage during removal, postsurgical albendazole therapy is indicated [90–95].

12.11 Other Unusual Locations

The recorded incidence of hydatid cyst in some unusual sites is about 2%. These sites include the peritoneal cavity, muscles, heart, brain, vertebral column, gallbladder, thyroid gland, and bones [47, 96–98].

Primary retroperitoneal hydatid disease should be considered in patients with retroperitoneal mass in endemic regions. It usually presents with abdominal mass, flank pains, and nonspecific symptoms of nausea and vomiting. Diaphragmatic hydatid cyst is very rare and it is usually associated with liver disease. Pelvic hydatid disease has also been reported, and the patients usually present with pressure effect. Pressure on the bladder may manifest as frequency, urine retention, obstructive uropathy and renal failure. Hydatid cyst has been reported in the musculoskeletal system. There are reports of involvement of the right psoas muscle which may obstruct the ureter and cause hydronephrosis [47, 96–103].

Surgical therapy is the most effective therapy for hydatid cyst in all locations. Pharmacotherapy is indicated in selected group of patients and they include those with multiple cysts in multiple locations, recurrence in multiple organs and spillage during intraoperative manipulation. The medication of choice is albendazole administered continuously at a daily dosage of 15 mg/kg/day. The follow-up of patients on pharmacotherapy includes weekly complete blood count and liver transaminases for the first month, and subsequently these tests are done monthly.

References

1. Derbel F, Mabrouk MB, Hamida MBH, Mazhoud J, Youssef S, Ali AB, Jemni H, Mama N, Ibtissem H, Nadia A. Hydatid cysts of the liver-diagnosis, complications and treatment. In: Derbel F, editor. Abdominal surgery. IntechOpen; 2012. p. 105–38.
2. Kabiri EH, El Maslout A, Benosman A. Thoracic rupture of hepatic hydatidosis (123 cases). Ann Thorac Surg. 2001;72(6):1883–6. https://doi.org/10.1016/s0003-4975(01)03204-0. PMID: 11789764.
3. Khasawneh RA, Mohaidat ZM, Khasawneh RA, Zoghoul SB, Henawi YM. Unusual intramuscular locations as a first presentation of hydatid cyst disease in children: a report of two cases. BMC Pediatr. 2021;21(1):371. https://doi.org/10.1186/s12887-021-02843-5. PMID: 34465295 PMCID: PMC8406844.
4. Morar R, Feldman C. Pulmonary echinococcosis. Eur Respir J. 2003;21(6):1069–77. https://doi.org/10.1183/09031936.03.00108403. PMID: 12797504.
5. Moshref L, Malaekah H. A rare presentation of hydatid cyst in abdominal wall: case report. Pan Afr Med J. 2021;40:246. https://doi.org/10.11604/pamj.2021.40.246.32301. PMID: 35233266 PMCID: PMC8831219.
6. Nagarajan M, Vydianathan P, Muralidharan S. Hydatid cyst of the heart—a case report. Indian J Thorac Cardiovasc Surg. 2022;18:118–20. https://doi.org/10.1007/s12055-002-0019-3.
7. Peña Huertas M, Zafra Martín J, García Á, de Quesada I, Díaz Gavela AA, Guerrero Gómez LL, Sánchez García S, Pardo Pérez E, Couñago F, Del Cerro PE. IJRadiation therapy for recurrent hydatid cyst of the pelvic bone: a case report. Int J Infect Dis. 2022;115:168–70. https://doi.org/10.1016/j.ijid.2021.12.313. PMID: 34883235.
8. Thapa S, Ghosh A, Ghartimagar D, Shrestha S, Lalchan S, Talwar OP. Hydatidosis of infratemporal fossa with proptosis—an unusual presentation: a case report and review of the literature. J Med Case Rep. 2018;12(1):309. https://doi.org/10.1186/s13256-018-1812-y. PMID: 30326941 PMCID: PMC6192125.
9. Emlik D, Ödev K, Poyraz N, Kaya HE. Radiological characteristics of pulmonary hydatid cysts. In: Rodriguez-Morales AJ, editor. Current topics in echinococcosis. IntechOpen; 2015. p. 148–52.
10. Laamiri R, Sellami S, Kechiche N, Abdelaali M, Mekki M, Belghith M. Intrathoracic rupture of hydatid cyst of the liver in children: a report of two cases. Egypt Liver J. 2021;11(1):1–4. https://doi.org/10.1186/s43066-021-00116-5.
11. Punia RS, Kundu R, Dalal U, Handa U, Mohan H. Pulmonary hydatidosis in a tertiary care hospital. Lung India. 2015;32(3):246–9. https://doi.org/10.4103/0970-2113.156241. PMID: 25983410 PMCID: PMC4429386.
12. Sarkar M, Pathania R, Jhobta A, Thakur BR, Chopra R. Cystic pulmonary hydatidosis. Lung India. 2016;33(2):179–91. https://doi.org/10.4103/0970-2113.177449. PMID: 27051107 PMCID: PMC4797438.
13. Anyfantakis D, Blevrakis E, Vlachakis I, Arbiros I. Hepatopulmonary hydatidosis in a ten-year-old girl: a case report. J Med Case Rep. 2010;4:205. https://doi.org/10.1186/1752-1947-4-205. PMID: 20598122 PMCID: PMC2909991.
14. Hammami A, Hellara O, Mnari W, Loussaief C, Bedioui F, Safer L, Golli M, Chakroun M, Saffar H. Unusual presentation of severely disseminated and rapidly progressive hydatic cyst: malignant hydatidosis. World J Hepatol. 2015;7(3):633–7. https://doi.org/10.4254/wjh.v7.i3.633. PMID: 25848488 PMCID: PMC4381187.
15. Marjanovic Z, Slavkovic A, Djordjevic I. Advanced form of hepatopulmonary hydatidosis in a child: case report. West Indian Med J. 2021;69(6):435–7. https://doi.org/10.7727/wimj.2015.064.
16. Aribas OK, Kanat F, Gormus N, Turk E. Pleural complications of hydatid disease. J Thorac Cardiovasc Surg. 2002;123(3):492–7. https://doi.org/10.1067/mtc.2002.119341. PMID: 11882820.
17. Garg MK, Sharma M, Gulati A, Gorsi U, Aggarwal AN, Agarwal R, Khandelwal N. Imaging in pulmonary hydatid cysts. World J Radiol. 2016;8(6):581–7. https://doi.org/10.4329/wjr.v8.i6.581. PMID: 27358685 PMCID: PMC4919757.
18. Rawat S, Kumar R, Raja J, Singh RS, Thingnam SKS. Pulmonary hydatid cyst: review of literature. J Family Med Prim Care. 2019;8(9):2774–8. https://doi.org/10.4103/jfmpc.jfmpc_624_19. PMID: 31681642 PMCID: PMC6820383.
19. Alloubi I. Thoracic hydatid cyst: clinical presentation, radiological features and surgical treatment. In: Firstenberg M, editor. Principles and practice of cardiothoracic surgery. IntechOpen; 2013. p. 195–217.
20. Burgos R, Varela A, Castedo E, Roda J, Montero CG, Serrano S, Téllez G, Ugarte J. Pulmonary hydatidosis: surgical treatment and follow-up of 240 cases. Eur J Cardiothorac Surg. 1999;16(6):628–34; discussion 634–5. https://doi.org/10.1016/s1010-7940(99)00304-8. PMID: 10647832.
21. Ferreres AR. Management of thoracic hydatid disease and its complications. In: Tsoulfas G, Hoballah J, Velmahos G, Ho Y, editors. The surgical management of parasitic diseases. Cham: Springer; 2020. p. 157–77.

22. Novick RJ, Tchervenkov CI, Wilson JA, Munro DD, Mulder DS. Surgery for thoracic hydatid disease: a North American experience. Ann Thorac Surg. 1987;43(6):681–6. https://doi.org/10.1016/s0003-4975(10)60253-6. PMID: 3296977.
23. Manterola C, Otzen T. Surgical alternatives used in the treatment of liver hydatid cyst. A systematized approach based on evidence (an overview). Int J Morphol. 2016;34(2):699–707.
24. Nabi MS, Waseem T. Pulmonary hydatid disease: what is the optimal surgical strategy? Int J Surg. 2010;8(8):612–6. https://doi.org/10.1016/j.ijsu.2010.08.002. PMID: 20732461.
25. Karabicak I, Yurtseven I, Yuruker SS, Ozen N, Kesim M. Splenic hydatid cyst. Can J Surg. 2009;52(5):E209–10. PMID: 19865563 PMCID: PMC2769128.
26. Rasheed K, Zargar SA, Telwani AA. Hydatid cyst of spleen: a diagnostic challenge. N Am J Med Sci. 2013;5(1):10–20. https://doi.org/10.4103/1947-2714.106184. PMID: 23378949 PMCID: PMC3560132.
27. Akbulut S, Sogutcu N, Eris C. Hydatid disease of the spleen: single-center experience and a brief literature review. J Gastrointest Surg. 2013;17(10):1784–95. https://doi.org/10.1007/s11605-013-2303-5. PMID: 23949423.
28. Eris C, Akbulut S, Yildiz MK, Abuoglu H, Odabasi M, Ozkan E, Atalay S, Gunay E. Surgical approach to splenic hydatid cyst: single center experience. Int Surg. 2013;98(4):346–53. https://doi.org/10.9738/INTSURG-D-13-00138.1. PMID: 24229022 PMCID: PMC3829062.
29. Meimarakis G, Grigolia G, Loehe F, Jauch KW, Schauer RJ. Surgical management of splenic echinococcal disease. Eur J Med Res. 2009;14(4):165–70. https://doi.org/10.1186/2047-783x-14-4-165. PMID: 19380289 PMCID: PMC3401006.
30. Mejri A, Arfaoui K, Ayadi MF, Aloui B, Yaakoubi J. Primitive isolated hydatid cyst of the spleen: total splenectomy versus spleen saving surgical modalities. BMC Surg. 2021;21(1):46. https://doi.org/10.1186/s12893-020-01036-8. PMID: 33472623 PMCID: PMC7819254.
31. Oussama B, Makrem M, Karim A, Brahim K, Mohamed BM, Samy B. Surgical treatment and outcomes of hydatid cyst of the spleen. Open J Gastroenterol. 2014;4(2(2014)) https://doi.org/10.4236/ojgas.2014.42009.
32. Brunetti E, Kern P, Vuitton DA. Expert consensus for the diagnosis and treatment of cystic and alveolar echinococcosis in humans. Acta Trop. 2010;114(1):1–16. https://doi.org/10.1016/j.actatropica.2009.11.001. PMID: 19931502.
33. Derici H, Tansug T, Reyhan E, Bozdag AD, Nazli O. Acute intraperitoneal rupture of hydatid cysts. World J Surg. 2006;30(10):1879–1883; discussion 1884–5. https://doi.org/10.1007/s00268-005-0699-0. PMID: 16847712.
34. Lencioni R, Franchini C. Hydatid disease, abdominal. In: Baert AL, editor. Encyclopedia of diagnostic imaging. Berlin/Heidelberg: Springer; 2008. p. 894–8.
35. Öztürk G, Uzun MA, Özkan ÖF, Kayaalp C, Tatlı F, Eren S, Aksungur N, Çoker A, Bostancı EB, Öter V, Kaya E, Taşar P. Turkish HPB Surgery Association consensus report on hepatic cystic echinococcosis (HCE). Turk J Surg. 2022;38(2):101–20. https://doi.org/10.47717/turkjsurg.2022.5757. PMID: 36483170 PMCID: PMC9714645.
36. Turgut AT, Akhan O, Bhatt S, Dogra VS. Sonographic spectrum of hydatid disease. Ultrasound Q. 2008;24(1):17–29. https://doi.org/10.1097/RUQ.0b013e318168f0d1. PMID: 18362529.
37. Zalaquett E, Menias C, Garrido F, Vargas M, Olivares JF, Campos D, Pinochet N, Luna A, Dahiya N, Huete Á. Imaging of hydatid disease with a focus on extrahepatic involvement. Radiographics. 2017;37(3):901–23. https://doi.org/10.1148/rg.2017160172. PMID: 28493801.
38. Alvela-Suárez L, Velasco-Tirado V, Belhassen-Garcia M, Novo-Veleiro I, Pardo-Lledías J, Romero-Alegría A, Pérez del Villar L, Valverde-Merino MP, Cordero-Sánchez M. Safety of the combined use of praziquantel and albendazole in the treatment of human hydatid disease. Am J Trop Med Hyg. 2014;90(5):819–22. https://doi.org/10.4269/ajtmh.13-0059. PMID: 24615131 PMCID: PMC4015571.
39. Pandey S, Malhotra HS, Garg RK, Malhotra KP, Kumar N, Rizvi I, Jain A, Kohli N, Verma R, Sharma P, Uniyal R, Pandey S. Quantitative assessment of lesion load and efficacy of 3 cycles of albendazole in disseminated cysticercosis: a prospective evaluation. BMC Infect Dis. 2020;20(1):220. https://doi.org/10.1186/s12879-020-4891-5. PMID: 32171260 PMCID: PMC7071677.
40. Shams-Ul B, Arif S, Malik A, Khaja A, Dass T, Naikoo Z. Role of albendazole in the management of hydatid cyst liver. Saudi J Gastroenterol. 2011;17(5):343–7. https://doi.org/10.4103/1319-3767.84493. PMID: 21912062 PMCID: PMC3178923.
41. Hoteit A, Alimoradi M, Aljoubaei M, Fakhruddin N, Hammoud N. Primary splenic hydatid cyst treated with laparoscopic spleen preserving surgery: a case report. Int J Surg Case Rep. 2020;74:95–8. https://doi.org/10.1016/j.ijscr.2020.08.010. PMID: 32829017 PMCID: PMC7452423.
42. Hilmani S. Brain hydatid cyst. In: Rodriguez-Morales AJ, editor. Current topics in echinococcosis. InTech; 2015.
43. Neumayr A, Tamarozzi F, Goblirsch S, Blum J, Brunetti E. Spinal cystic echinococcosis—a systematic analysis and review of the literature: part 1. Epidemiology and anatomy. PLoS Negl Trop Dis. 2013;7(9):e2450. https://doi.org/10.1371/journal.pntd.0002450. PMID: 24086783 PMCID: PMC3783062.

44. Siyadatpanah A, Brunetti E, Emami Zeydi A, Moghadam YD, Agudelo Higuita NI. Cerebral cystic echinococcosis. Case Rep Infect Dis. 2020;2020:1754231. https://doi.org/10.1155/2020/1754231. PMID: 32181029 PMCID: PMC7066420.
45. El-Shamam O, Amer T, El-Atta MA. Magnetic resonance imaging of simple and infected hydatid cysts of the brain. Magn Reson Imaging. 2001;19(7):965–74. https://doi.org/10.1016/s0730-725x(01)00413-1. PMID: 11595368.
46. Kızılca O, Altaş M, Senol U, Oztek MA. Hydatid disease located in the cerebellomedullary cistern. Case Rep Med. 2014;2014:271365. https://doi.org/10.1155/2014/271365. PMID: 24782895 PMCID: PMC3982244.
47. Fakhouri F, Ghajar A, Mahli N, Shoumal N. Giant hydatid cyst in the posterior fossa of a child. Asian J Neurosurg. 2015;10(4):322–4. https://doi.org/10.4103/1793-5482.162719. PMID: 26425166 PMCID: PMC4558813.
48. Nasir MI, Gupta VK, Latawa A, Bhardwaj S. Intracranial hydatid cyst: removal by Dowling's technique of hydrodissection. J Indian Assoc Pediatr Surg. 2019;24(3):227–8. https://doi.org/10.4103/jiaps.JIAPS_206_18. PMID: 31258280 PMCID: PMC6568156.
49. Hamdani M, Bdioui A, Belkacem O, Jelidi N, Baccouche A, Majdoub W, Belhaj Salah SH. Isolated renal hydatid cyst misdiagnosed and treated as a cystic renal tumor: a diagnostic pitfall. Afr J Urol. 2021;27:75. https://doi.org/10.1186/s12301-021-00176-4.
50. Leventoğlu E, Deniz EB, Kara Ulu N, Polat M, Fidan K. A rare isolated kidney cyst: hydatid cyst. Cureus. 2023;15(5):e38566. https://doi.org/10.7759/cureus.38566. PMID: 37284360 PMCID: PMC10239546.
51. Misra A, Mandal S, Das M, Mishra P, Mitra S, Nayak P. Isolated renal hydatid disease: varied presentations, treatments, dilemmas, and the way ahead: case report series. Afr J Urol. 2021;27(1):83. https://doi.org/10.1186/s12301-021-00183-5.
52. Rexiati M, Mutalifu A, Azhati B, Wang W, Yang H, Sheyhedin I, Wang Y. Diagnosis and surgical treatment of renal hydatid disease: a retrospective analysis of 30 cases. PLoS One. 2014;9(5):e96602. https://doi.org/10.1371/journal.pone.0096602. PMID: 24796329 PMCID: PMC4010549.
53. Samet A, Mseddi MA, Hamza M, Walha M, Bahri I, Hadjslimane M. A case of an isolated hydatid cyst in the right kidney. Urol Case Rep. 2020;28:101047. https://doi.org/10.1016/j.eucr.2019.101047. PMID: 31709149 PMCID: PMC6833461.
54. Venyo AK-G. Hydatid cyst of the kidney and ureter: a review and update of the literature. J Clin Res Rep. 2021;8(5) https://doi.org/10.31579/2690-1919/188.
55. Mehta P, Prakash M, Khandelwal N. Radiological manifestations of hydatid disease and its complications. Trop Parasitol. 2016;6(2):103–12. https://doi.org/10.4103/2229-5070.190812. PMID: 27722098 PMCID: PMC5048696.
56. Srinivas MR, Deepashri B, Lakshmeesha MT. Imaging spectrum of hydatid disease: usual and unusual locations. Pol J Radiol. 2016;81:190–205. https://doi.org/10.12659/pjr.895649. PMID: 27231490 PMCID: PMC4868106.
57. Kumar S, Singla S, Ahmad A, Kurian N, Chidipotu VR. Isolated renal hydatid disease: laparoscopic approach to an uncommon entity (case report). Pan Afr Med J. 2021;38:175. https://doi.org/10.11604/pamj.2021.38.175.27993. PMID: 33995782 PMCID: PMC8077677.
58. Dehkordi AB, Sanei B, Yousefi M, Sharafi SM, Safarnezhad F, Jafari R, Darani HY. Albendazole and treatment of hydatid cyst: review of the literature. Infect Disord Drug Targets. 2019;19(2):101–4. https://doi.org/10.2174/1871526518666180629134511. PMID: 29956639.
59. John J, Marais B. Primary renal hydatid in children. EMJ. 2022; https://doi.org/10.33590/emj/22-00032.
60. Siles-Lucas M, Casulli A, Cirilli R, Carmena D. Progress in the pharmacological treatment of human cystic and alveolar echinococcosis: compounds and therapeutic targets. PLoS Negl Trop Dis. 2018;12(4):e0006422. https://doi.org/10.1371/journal.pntd.0006422. PMID: 29677189 PMCID: PMC5931691.
61. Akbulut S. Incidentally detected hydatid cyst of the adrenal gland: a case report. World J Clin Cases. 2016;4(9):269–72. https://doi.org/10.12998/wjcc.v4.i9.269. PMID: 27672642 PMCID: PMC5018624.
62. Akbulut S, Yilmaz M. Hydatid cyst of the adrenal gland: is radical surgery necessary for recurrent hydatid disease? Case Rep Surg. 2018;2018:6452402. https://doi.org/10.1155/2018/6452402. PMID: 29682390 PMCID: PMC5842735.
63. Bouchaala H, Mejdoub I, Mseddi MA, Kammoun O, Rebai N, Slimen MH. Giant primary hydatid cyst of the adrenal gland: a rare case report. Urol Case Rep. 2021;36:101580. https://doi.org/10.1016/j.eucr.2021.101580. PMID: 33643845 PMCID: PMC7889794.
64. Staouni IB, El Bouardi N, Ferhi M, Haloua M, Alami B-e, Lamrani YA, Boubbou M, Maâroufi M. Primary hydatid cyst of adrenal gland: case report. Radiol Case Rep. 2022;17(9):3188–90. https://doi.org/10.1016/j.radcr.2022.05.084.
65. Yuksel M, Demirpolat G, Sever A, Bakaris S, Bulbuloglu E, Elmas N. Hydatid disease involving some rare locations in the body: a pictorial essay. Korean J Radiol. 2007;8(6):531–40. https://doi.org/10.3348/kjr.2007.8.6.531. PMID: 18071284 PMCID: PMC2627456.
66. Aggouri Y, Ibn majdoub Hassani K. Primary adrenal hydatid cyst. Pan Afr Med J. 2013;16:153. https://doi.org/10.11604/pamj.2013.16.153.3550.
67. Hamamci EO, Besim H, Korkmaz A. Unusual locations of hydatid disease and surgical approach. ANZ

J Surg. 2004;74(5):356–60. https://doi.org/10.1111/j.1445-1433.2004.02981.x. PMID: 15144257.
68. Kothiya PK, Gupta V, Sarawagi R, Jayashankar E, Sharma J, Wani H, Balaji K, Roshny J. Isolated primary hydatid cyst of the pancreas: management challenges of a cystic masquerade. Ann Hepatobiliary Pancreat Surg. 2022;26(4):401–6. https://doi.org/10.14701/ahbps.22-031. PMID: 35995585 PMCID: PMC9721249.
69. Makni A, Jouini M, Kacem M, Safta ZB. Acute pancreatitis due to pancreatic hydatid cyst: a case report and review of the literature. World J Emerg Surg. 2012;7(1):7. https://doi.org/10.1186/1749-7922-7-7. PMID: 22445170 PMCID: PMC3325852.
70. Mitrovic M, Tadic B, Kovac J, Grubor N, Milosavljevic V, Jankovic A, Khatkov I, Radenkovic D, Matic S. Pancreatic hydatid cyst misdiagnosed as mucinous cystadenoma: CT and MRI findings. Medicina (Kaunas). 2020;56(3) https://doi.org/10.3390/medicina56030124. PMID: 32183010 PMCID: PMC7143194.
71. Ahmed Z, Chhabra S, Massey A, Vij V, Yadav R, Bugalia R, Kankaria J, Jenaw RK. Primary hydatid cyst of pancreas: case report and review of literature. Int J Surg Case Rep. 2016;27:74–7. https://doi.org/10.1016/j.ijscr.2016.07.054. PMID: 27552034 PMCID: PMC4995534.
72. Wu Y, Gong J, Xiong W, Yu X, Lu X. Primary pancreatic hydatid cyst: a case report and literature review. BMC Gastroenterol. 2021;21(1):164. https://doi.org/10.1186/s12876-021-01753-1. PMID: 33849455 PMCID: PMC8045313.
73. Akbulut S, Yavuz R, Sogutcu N, Kaya B, Hatipoglu S, Senol A, Demircan F. Hydatid cyst of the pancreas: report of an undiagnosed case of pancreatic hydatid cyst and brief literature review. World J Gastrointest Surg. 2014;6(10):190–200. https://doi.org/10.4240/wjgs.v6.i10.190. PMID: 25346801 PMCID: PMC4208043.
74. Canbak T, Acar A, Kıvanç AE, Başak F, Kulalı F, Baş G. Sinistral portal hypertension due to pancreatic hydatid cyst. Turkiye Parazitol Derg. 2017;41(4) https://doi.org/10.5152/tpd.2017.4899. PMID: 29318995.
75. Ozsay O, Gungor F, Karaisli S, Kokulu I, Dilek ON. Hydatid cyst of the pancreas causing both acute pancreatitis and splenic vein thrombosis. Ann R Coll Surg Engl. 2018;100(7):e178–80. https://doi.org/10.1308/rcsann.2018.0111. PMID: 29909684 PMCID: PMC6214065.
76. Imam A, Khoury T, Weis D, Khalayleh H, Adeleh M, Khalaileh A. Laparoscopic cystectomy for pancreatic echinococcosis: a case report and literature review. Ann Hepatobiliary Pancreat Surg. 2019;23(1):87–90. https://doi.org/10.14701/ahbps.2019.23.1.87. PMID: 30863816 PMCID: PMC6405372.
77. Butt A, Khan J. The maverick disease: cystic echinococcosis in unusual locations: a ten year experience from an endemic region. Cureus. 2019;11(10):e5939. https://doi.org/10.7759/cureus.5939. PMID: 31799081 PMCID: PMC6860736.
78. Kumar A, Kumar A, Gaurav K, Chandra G, Tiwary AK, Bhagat S, Sarawgi M. A rare case of isolated hydatid cyst of breast. Int J Surg Case Rep. 2015;7c:115–8. https://doi.org/10.1016/j.ijscr.2014.10.093. PMID: 25582084 PMCID: PMC4336384.
79. Shahriarirad R, Erfani A, Eskandarisani M, Rastegarian M, Sarkari B. Uncommon locations of cystic echinococcosis: a report of 46 cases from Southern Iran. Surg Res Pract. 2020;2020:2061045. https://doi.org/10.1155/2020/2061045. PMID: 33015320 PMCID: PMC7520003.
80. Shekh ZA. Hydatid disease—an unusual cause of a breast cyst: case report. Int J Surg Case Rep. 2020;68:178–9. https://doi.org/10.1016/j.ijscr.2020.02.066. PMID: 32182579 PMCID: PMC7090097.
81. Kassahun Tadele A, Israel Korga T, Melis Nisiro A, Abebe AS. Rare case report on hydatid cyst of breast. Pathol Lab Med Int. 2022:33–6. https://doi.org/10.2147/PLMI.S380081.
82. Mujawar P, Suryawanshi KH, Nikumbh DB. Cytodiagnosis of isolated primary hydatid cyst of breast masquerading as a breast neoplasm: a rare case report. J Cytol. 2015;32(4):270–2. https://doi.org/10.4103/0970-9371.171248. PMID: 26811577 PMCID: PMC4707791.
83. Sheikh S, Akhter R, Bhat S, Wani R. Primary hydatid disease of breast: a case report. J Parasit Dis. 2017;41(3):908–11. https://doi.org/10.1007/s12639-017-0903-9. PMID: 28848301 PMCID: PMC5555947.
84. Arora BB, Maheshwari M, Devgan N, Arora DR. Hydatid cyst of the ovary: a case report. JMM Case Rep. 2015;2(4):e000042. https://doi.org/10.1099/jmmcr.0.000042.
85. Marwah V, Kedia S, Mittal P, Dutta S, Shriya R. Laparoscopic Salpingo-oophorectomy for left-sided primary ovarian hydatid cyst. J Minim Invasive Gynecol. 2021;28(4):743–5. https://doi.org/10.1016/j.jmig.2020.08.487. PMID: 32891824.
86. Mohammed AA, Arif SH. Hydatid cyst of the ovary—a very rare type of cystic ovarian lesion: a case report. Case Rep Womens Health. 2021;31:e00330. https://doi.org/10.1016/j.crwh.2021.e00330. PMID: 34123734 PMCID: PMC8173026.
87. Nazari Z, Torabizadeh J. Primary hydatid cyst of the fallopian tube: a case report. Caspian J Intern Med. 2014;5(3):179–81. PMID: 25202448 PMCID: PMC4143742.
88. Uchikova E, Pehlivanov B, Uchikov A, Shipkov C, Poriazova E. A primary ovarian hydatid cyst. Aust N Z J Obstet Gynaecol. 2009;49(4):441–2. https://doi.org/10.1111/j.1479-828X.2009.01015.x. PMID: 19694705.
89. Zhao Q, Luo J, Zhang Q, Leng T, Yang L. Laparoscopic surgery for primary ovarian and retroperitoneal hydatid disease: a case report.

Medicine (Baltimore). 2018;97(3):e9667. https://doi.org/10.1097/md.0000000000009667. PMID: 29505009 PMCID: PMC5779778.
90. Celik A, Turanli M, Kutun S, Delibasi T, Mengi N, Comert E, Aslan S, Altundag K, Cetin A. Unusual location of hydatid cyst: soft tissue mass in the neck. Eur Arch Otorrinolaringol. 2006;263(12):1147–50. https://doi.org/10.1007/s00405-006-0122-y. PMID: 16858578.
91. Çiçekli Ö, Akgül T. An unusual cause of mass localized on vastus lateralis muscle in childhood: hydatid cyst. Int J Surg Case Rep. 2015;6c:179–81. https://doi.org/10.1016/j.ijscr.2014.09.038. PMID: 25544487 PMCID: PMC4334633.
92. Dudkiewicz I, Salai M, Apter S. Hydatid cyst presenting as a soft-tissue thigh mass in a child. Arch Orthop Trauma Surg. 1999;119(7–8):474–5. https://doi.org/10.1007/s004020050025. PMID: 10613244.
93. Guraya SY, Alzobydi AH, Guraya SS. Primary extrahepatic hydatid cyst of the soft tissue: a case report. J Med Case Rep. 2012;6:404. https://doi.org/10.1186/1752-1947-6-404. PMID: 23181425 PMCID: PMC3526461.
94. Iynen I, Sogut O, Guldur ME, Kose R, Kaya H, Bozkus F. Primary hydatid cyst: an unusual cause of a mass in the supraclavicular region of the neck. J Clin Med Res. 2011;3(1):52–4. https://doi.org/10.4021/jocmr495w. PMID: 22043272 PMCID: PMC3194026.
95. Tekin R, Avci A, Tekin RC, Gem M, Cevik R. Hydatid cysts in muscles: clinical manifestations, diagnosis, and management of this atypical presentation. Rev Soc Bras Med Trop. 2015;48(5):594–8. https://doi.org/10.1590/0037-8682-0197-2015. PMID: 26516970.
96. Di Cataldo A, Petrillo G, Trombatore C, Palmucci S, Agodi A, Barchitta M, Quattrocchi A, Crimi N, Fichera S, Puleo S. Echinococcosis—rare locations and uncommon clinical manifestations. Current topics in echinococcosis. IntechOpen; 2015.
97. Sharma D, Wadhawan G, Vyas K. Unusual locations, presentations of hydatid cyst and their laparoscopi c management. Int J Surg Sci. 2020;4(3):214–8. https://doi.org/10.33545/surgery.2020.v4.i3d.494.
98. Sachar S, Goyal S, Goyal S, Sangwan S. Uncommon locations and presentations of hydatid cyst. Ann Med Health Sci Res. 2014;4(3):447–52. https://doi.org/10.4103/2141-9248.133476. PMID: 24971224 PMCID: PMC4071749.
99. Gavriilidis P, Ananiadis A, Theodoulidis V, Barbanis S. Primary retroperitoneal echinococcal cyst. BMJ Case Rep. 2012;2012 https://doi.org/10.1136/bcr-2012-006465. PMID: 23162021 PMCID: PMC4544887.
100. Hamed Al Taei T, Ali Al Mail S, Hajjaj Al Thinayyan A, Alsetrawi A. Renal hydatid cyst: a case report. Radiol Case Rep. 2022;17(6):2063–6. https://doi.org/10.1016/j.radcr.2022.03.036. PMID: 35464801 PMCID: PMC9018802.
101. Jerraya H, Gaja A, Khalfallah M, Dziri C. Diaphragmatic hydatid disease: a diagnostic challenge for the radiologist. Tunis Med. 2017;95(4):307–9. PMID: 29492938.
102. Reza HAM, Rreza G, Nastaran B, Mousa M. Renal hydatid cyst; a rare infectious disease. Oxf Med Case Rep. 2019;2019(3):omz011. https://doi.org/10.1093/omcr/omz011. PMID: 30949349 PMCID: PMC6440275.
103. Yusuf Sherwani A, Sarmast A, Malik A, Shah M, Q-Parray F, A-Wani M, Dangroo S. Primary retroperitoneal hydatid cyst: a rare case report. Acta Med Iran. 2015;53(7):448–51. PMID: 26520634.

Artificial Intelligence and Decision Support Applications in Liver Hydatid Disease: Detection, Classification, and Complication Prediction

13

Mucahit Karaduman, Muhammed Yildirim,
and Sami Akbulut

13.1 Introduction

13.1.1 Overview of Liver Hydatid Disease

Hydatid disease, also known as echinococcosis (hydatidosis), is a zoonotic infection caused by the larval stages of cestodes (tapeworms) belonging to the genus *Echinococcus* [1]. It is widely prevalent across the globe. Although the genus *Echinococcus* is primarily responsible for echinococcosis, the most common species affecting humans are *E. granulosus*, which causes cystic echinococcosis, and *E. multilocularis*, which leads to alveolar echinococcosis [1]. It disproportionately affects populations in endemic regions, particularly in areas like the Middle East, South America, Central Asia, and parts of Eastern Europe [2]. The liver is the most frequently affected organ, with approximately 70–85% of cases involving hepatic involvement [3, 4]. Transmission typically occurs through the ingestion of parasite eggs via contaminated water, food, or direct contact with infected animals, especially dogs and sheep [5].

The clinical presentation of the disease can vary significantly, ranging from asymptomatic cases detected incidentally to severe, life-threatening complications. In the early stages, cysts are often asymptomatic [6], but as they grow, they may cause symptoms such as right upper quadrant pain, hepatomegaly, nausea, or signs of cholestasis [7, 8]. In more advanced stages, the cyst may rupture into the biliary tract, peritoneal cavity, or thoracic cavity, and in some instances leakage of cyst fluid can trigger anaphylaxis. Additional serious complications, including superinfection, fistula formation, and compression of neighboring structures, further contribute to morbidity [9, 10]. This chapter comprehensively explores the application of artificial intelligence (AI) and clinical decision support systems in the management of liver hydatid disease, with a focus on improving diagnostic accuracy, treatment planning, and complication prediction. Given that the disease predominantly affects the liver, the chapter specifically empha-

M. Karaduman
Department of Software Engineering, Malatya Turgut Ozal University, Malatya, Türkiye
e-mail: mucahit.karaduman@ozal.edu.tr

M. Yildirim
Department of Computer Engineering, Malatya Turgut Ozal University, Malatya, Türkiye
e-mail: muhammed.yildirim@ozal.edu.tr

S. Akbulut (✉)
Department of Surgery and Liver Transplantation Institute, Inonu University Faculty of Medicine, Malatya, Türkiye

Department Biostatistics and Medical Informatics, Inonu University Faculty of Medicine, Malatya, Türkiye

S. Akbulut (ed.), *Hydatid Disease*, https://doi.org/10.1007/978-3-031-97277-5_13

sizes liver hydatid disease. This focus is essential as hepatic involvement not only presents unique diagnostic and therapeutic challenges but also demands a nuanced approach to clinical decision-making and patient management.

13.1.2 Challenges in Diagnosis and Management

Although the pathophysiology and life cycle of liver hydatid disease are well defined, its diagnosis is often delayed due to nonspecific symptoms or asymptomatic presentation. Imaging techniques play a pivotal role in diagnosis and staging, serving as the gold standard. Ultrasonography (US) is the first-line imaging modality due to its widespread availability and low cost. The World Health Organization (WHO) has developed the CE1–CE5 (CE: cystic echinococcosis) classification based on ultrasonographic features, which also guides treatment choices [11]. However, the operator dependency of US and interobserver variability reduce its reliability, particularly in borderline or atypical cases [12]. Computed tomography (CT) and magnetic resonance imaging (MRI) provide higher resolution, especially in complex, calcified, or ruptured cysts. Magnetic resonance cholangiopancreatography (MRCP) is particularly useful in evaluating biliary involvement [12, 13]. However, access to these advanced imaging techniques is often limited in endemic rural areas. Additionally, serological tests such as enzyme-linked immunosorbent assay (ELISA) and indirect hemagglutination test (IHA), although useful in diagnosis, have variable sensitivity depending on the cyst stage and the host's immune response [14].

Treatment options cover a wide range: medical therapy, percutaneous drainage techniques (puncture aspiration injection re-aspiration [PAIR, standard catheterization, modified catheterization [MoCAT], percutaneous evacuation techniques [PEVAC]), as well as radical or conservative surgery [15–17]. All these approaches require accurate cyst staging and individualized risk assessment. Nonetheless, recurrence, incomplete resection, and iatrogenic dissemination remain significant concerns. Predicting complications poses another critical challenge. Factors that increase the risk of complications include cyst diameter greater than 7.5–10 cm, multivesicular structure, central location, biliary involvement, and delayed presentation [18–20]. Early identification of these risk groups allows for personalized treatment and improves outcomes.

13.2 Radiological Diagnosis of Hepatic Hydatid Disease

13.2.1 Role of US, CT, and MRI

Radiological imaging methods play a central role in the diagnosis, staging, and treatment planning of hepatic hydatid disease. Among the available imaging techniques, US is the most commonly preferred and widely used modality, especially in endemic regions. This preference is based on its cost-effectiveness, portability, and lack of ionizing radiation. US allows the evaluation of the size of cystic structures, internal architecture, presence of the germinative membrane, echogenicity, septations, and calcifications. Additionally, the basic data for the CE classification system recommended by the WHO-Informal Working Group on Echinococcosis (WHO-IWGE) are obtained through ultrasonography [21].

Recent studies have shown that DL-based models have successfully automated CE staging from US images. In particular, models developed using CNN architectures have demonstrated performance comparable to that of expert radiologists, and in some cases have provided more consistent results [22].

CT is used as a complementary method to US, particularly offering detailed evaluation in multivesicular and calcified cysts, such as CE2 and CE3b. CT provides high spatial resolution for defining cyst wall calcifications, internal septa, and the density characteristics of cyst fluid. It is essential for surgical planning as it can detail the relationship between the cyst, hepatic vascular structures, and the biliary system.

AI-based classification systems developed on CT images have been shown to reduce observer dependency, offering more consistent and reproducible results. DL architectures form the basis of models that perform CE staging.

MRI is preferred in cases where biliary involvement or potential complications (such as rupture or superinfection) need to be evaluated. Magnetic resonance cholangiopancreatography (MRCP) sequences are particularly valuable for demonstrating the connection between the cyst and bile ducts. Moreover, MRI is noted for its safety in children, pregnant women, and patients with contrast allergies, as it does not involve ionizing radiation.

In some studies focusing on the use of AI in hydatid disease, an AI-based model developed to detect subtypes of hepatic hydatid disease showed that segmentation results based on plain CT data were in significant agreement with MRI imaging findings. This suggests that MRI can play a complementary role in AI-supported diagnostic processes.

13.2.2 WHO-IWGE and Gharbi Classifications

To standardize the imaging-based evaluation of hepatic hydatid cysts, two main classification systems have been developed.

13.2.2.1 WHO-IWGE Classification [23–25]

This ultrasonographic classification system, developed by the WHO, categorizes hydatid cysts into six types:

- **CL (Cystic Lesion)**: A nonspecific anechoic lesion without a well-defined cyst wall.
- **CE1**: A single-chambered, anechoic cyst that contains hydatid sand, representing the early stage of active infection.
- **CE2**: A multichambered, multiseptated cyst containing daughter cysts, exhibiting a "honeycomb" appearance.
- **CE3a**: A transitional cyst characterized by the detachment of the endocyst membrane, giving a "water lily" appearance.
- **CE3b**: A solid cyst containing degenerative daughter cysts.
- **CE4–CE5**: Respectively, these are inactive cysts that have either undergone degenerative changes or become completely calcified.

This classification directly influences treatment decisions: CE1 and CE3a lesions are suitable for percutaneous treatment, while CE2 and CE3b typically require surgical intervention. CE4 and CE5 stages are generally managed with conservative follow-up.

13.2.2.2 Gharbi Classification [26]

The Gharbi classification, introduced in 1981, categorizes hepatic hydatid cysts based on their ultrasonographic morphology into five types:

- **Type I**: A pure, anechoic, thin-walled, simple cyst.
- **Type II**: A cyst containing detached membranes and showing sedimentation.
- **Type III**: A multilocular cyst containing septa or daughter cysts.
- **Type IV**: A lesion with heterogeneous echogenicity, presenting as a solid or complex mass.
- **Type V**: A cyst with a thick, calcified wall, usually inactive.

This classification was widely used before the WHO system and is still referenced in some centers today. One advantage of the Gharbi classification is its relatively lower complexity, as it consists of fewer classes, making it advantageous for training artificial intelligence algorithms. However, compared to the WHO-IWGE system, it has more limited guidance in clinical decision-making processes.

13.2.3 Diagnostic Pitfalls in Atypical or Complicated Hydatid Cysts

According to recent literature, despite advancements in imaging techniques, the diagnosis of complicated or atypical hepatic hydatid cysts still poses significant diagnostic challenges. Complications such as rupture, superinfection, internal hemorrhage, or extensive calcification can alter the imaging characteristics of cysts, leading to potential confusion with pyogenic abscesses, necrotic tumors, or simple hepatic cysts.

Transition cysts in the CE3a stage often exhibit both active and inactive appearances, frequently causing classification errors. In CE4–CE5 stages, the solid and calcified contents of cysts may be mistaken for solid liver tumors, especially when using low-resolution imaging techniques. Additionally, mucinous cystadenocarcinoma, biliary cysts, and metastatic cystic lesions can result in false-positive diagnoses during CE staging.

Some new ML-based approaches have been tested as decision support systems for classifying these cases. However, existing studies have not identified a validated AI model specifically designed for atypical cases. While segmentation accuracy is generally satisfactory in typical CE1–CE2 cases, a noticeable performance drop is observed in complicated cases.

Operator dependency, image quality, and the possibility of overlooking diagnostic findings (such as double contours, membrane detachment, internal septations, or calcification patterns) due to lack of experience can delay diagnosis. In low-resource areas, both equipment shortages and insufficient radiologist training exacerbate these issues.

Therefore, a prominent trend in the literature is the integration of AI-supported classification systems into routine clinical practice. These technologies automatically analyze images, reducing human error, providing decision support in identifying atypical variants, and making the diagnostic process faster and more reliable. However, for widespread implementation, more multicenter, prospective, and real-world data-based studies are required.

13.3 Rationale for Using AI-Based Technologies

Liver hydatid disease is a parasitic disease with a complex clinical course that requires a multidisciplinary approach. In diagnosis, staging, and treatment planning, decision-making variability can occur due to the heterogeneity of radiological images, the anatomical location of lesions, and patient-specific differences. To reduce this variability and ensure more consistent clinical decisions, the integration of artificial intelligence-based technologies into healthcare applications is becoming increasingly important. In particular, deep learning (DL) models have the potential to contribute to image interpretation and complication risk prediction in line with the WHO-IWGE classification.

DL architectures, developed with algorithms such as convolutional neural networks (CNN), are considered useful tools in clinical decision support processes for image-based analyses, including surgical planning, determining treatment strategies, and predicting complication risks [22, 27].

Moreover, machine learning (ML) algorithms such as Random Forest [28] and Support Vector Machine can analyze radiological variables such as age, cyst diameter, localization, and calcification, as well as serological and clinical parameters. These methods enable the creation of classification models for predictive assessments such as complication prediction, treatment response, and recurrence risk.

These AI-based technologies make significant contributions in terms of standardizing clinical processes, shortening decision-making time, and reducing human errors. However, before their widespread clinical use, their effectiveness and reliability need to be validated through multicenter and large-sample studies.

13.4 Fundamentals of AI in Medicine

13.4.1 Definitions

In medicine, AI refers to the integration of computer systems that mimic cognitive functions specific to human intelligence into clinical applications. These systems can perform tasks such as image recognition, pattern discovery, decision-making, and predictive analysis. AI is particularly valuable in processing large medical datasets, as it can identify complex relationships that are often imperceptible to the human eye, thereby supporting clinical decision-making.

ML consists of algorithms capable of learning from data. These algorithms can recognize data patterns and improve their performance over time without being explicitly programmed. ML can be applied to both structured data (such as laboratory results, age, cyst diameter) and unstructured data (like CT images and ultrasound data). In clinical practice, ML algorithms are used in areas such as predicting cyst stages, forecasting complications, and calculating recurrence risk.

DL, a subfield of ML, offers multilayered data processing capabilities based on neural network architectures. The CNN architecture, in particular, stands out in radiological image analysis. Studies have shown that CNN-based models can be effectively used to classify CE stages, perform cyst segmentation, and detect internal structural details (such as membrane detachment, calcification, and the presence of daughter cysts). Additionally, models utilizing transfer learning—leveraging pre-trained models—have demonstrated effective results even with limited data.

13.4.2 Overview of Supervised vs. Unsupervised Learning

ML algorithms can be broadly classified into two main groups: supervised learning and unsupervised learning.

- **Supervised Learning:** In this approach, data is presented to the model along with predefined labels. For example, algorithms are trained using classes such as CE1–CE5 stages or rupture status. Algorithms like Support Vector Machine, Random Forest, and Decision Tree are developed based on this structure and are applied in tasks such as CE staging and complication prediction.
- **Unsupervised Learning:** This approach allows for pattern discovery without any labeling of the data. Using algorithms like K-means and principal component analysis (PCA), clusters are formed based on cyst morphology independently of CE stages, suggesting potential new classifications. These methods have shown promise, especially in identifying atypical cyst formations and predicting cases with a high risk of complications.

In clinical applications, supervised learning methods are predominantly used when gold standard labels are available (e.g., cyst types confirmed by a radiologist). The majority of diagnostic AI applications rely on this structured learning approach.

13.4.3 Imaging-Based AI Applications in Parasitic Liver Diseases

AI systems offer significant advantages in automating imaging-based diagnostic processes and developing clinical decision support systems. CNN architectures, particularly used for medical findings, have been applied in research to classify CE stages in both US and CT images. These systems have been shown to reduce interobserver variability, especially in the interpretation of transitional stages [29].

Studies based on CT imaging have demonstrated the potential of AI models to predict CE stages by analyzing parameters such as cyst wall thickness, internal structure, and the presence of calcification. These models are trained using transfer learning techniques. Various CNN architectures like ResNet [30], MobileNet [27], GoogLeNet [31], EfficientNet [32], and VGGNet [9, 33] have been reported in the literature as suitable for classification tasks.

Radiomic analyses, by integrating features such as tissue density, shape, and edge sharpness extracted from images into AI models, contribute to a more precise assessment of patient prognosis. These methods support clinical decisions related to surgical eligibility, treatment response, and follow-up planning.

Moreover, in low-resource settings, AI-supported mobile US systems are reported to have the potential to perform CE classification without requiring expert intervention. These systems not only enhance diagnostic accuracy but also have the potential to improve standardization and resource optimization in healthcare services.

13.5 AI-Based Detection and Classification of Hydatid Cysts

13.5.1 Cyst Identification via Image Segmentation (e.g., CNNs, U-Net)

One of the most prominent applications of AI in medical imaging is image classification and segmentation. Accurate identification of hydatid cysts within the complex anatomy of the liver is critically important for both diagnosis and surgical planning. In this context, DL-based models stand out as key tools for addressing segmentation problems. In the current literature, U-Net-based models are commonly used in segmentation applications, while other similar architectures such as SegNet, V-Net, and ResUNet are also employed [34, 35].

In some studies focused on the automatic segmentation of hepatic hydatid cysts from CT and MRI images, modified versions of these models have been utilized. However, these applications often remain at the experimental stage.

AI systems working with US images have been investigated in a limited number of studies to determine CE stages. Due to the typically small sample sizes in these studies, comparative performance data with radiologists are restricted.

The use of multimodal data (e.g., combining T2-weighted MRI with contrast-enhanced CT) has been theoretically proposed in some studies; however, there is no direct application reported that utilizes an integrated AI system trained with such an approach. Similarly, creating masks for cyst volume and localization after segmentation has been minimally described.

In some models, post-processing algorithms (such as conditional random fields—CRF and morphological operations) have been implemented to make the outputs more anatomically realistic. These advancements facilitate the delineation of surgical margins during preoperative planning, reduce the risk of complications, and increase standardization in case analysis.

Overall, segmentation-based AI applications in liver hydatid disease are promising. However, clinically validated models with external validity and multicenter verification are still limited. Progress in this area will depend on developing systems trained with extensive datasets that encompass the entire CE stages.

13.5.2 Automatic Stage Classification (Gharbi/WHO Stages)

Accurate detection and automatic staging of hydatid cysts play a crucial role in shaping the therapeutic approach. Classification systems based on AI analysis of imaging data are particularly utilized for distinguishing CE stages according to the WHO-IWGE and Gharbi classifications. DL-based models identify CE1–CE5 cysts by extracting features from images, which are then transferred to the final decision-making mechanism through various classifier algorithms. These classifiers include Softmax, Random Forest [28], Support Vector Machine [36], k-Nearest Neighbors (KNN) [37], Naive Bayes [38], and Decision Tree [39].

Recent models reported in the literature have demonstrated high classification accuracy in distinguishing active, homogeneous cysts such as CE1 and CE2 from degenerative cysts like CE4 and CE5. However, CE3a and CE3b stages present more challenges both clinically and algorithmically due to their transitional morphological characteristics. These stages tend to have higher classification errors and lower overall accuracy.

Key factors influencing model performance include class imbalance, annotation inconsistencies, and low-quality imaging data. To address these challenges, some studies have experimented with methods such as attention mechanisms, ensemble modeling, data augmentation, and cost-sensitive learning. Particularly in the differentiation between CE3a and CE3b, attention-based CNN architectures have shown improved outcomes.

The integration of AI-supported automatic staging systems with a clinical decision support system in clinical practice accelerates the diagnostic process and reduces interobserver variability. In some systems, implementing a confidence threshold (uncertainty estimation) ensures that cases classified with low confidence are automatically referred to expert radiologists, helping to prevent false negatives. This approach enhances clinical safety by optimizing human–machine collaboration.

13.5.3 Comparison with Radiologist Performance

Comparing the performance of AI systems with that of radiologists is crucial when evaluating clinical validity and reliability, particularly in the context of liver hydatid cysts. Recent studies, including those leveraging CNN, have shown that AI-based models can achieve accuracy rates comparable to those of moderately experienced radiologists, especially in the classification of liver hydatid cyst subtypes such as CE1 and CE2. In the study conducted by Shang et al. [33], AI models demonstrated significant proficiency in distinguishing between different cystic echinococcosis stages using ultrasound imaging data. The study highlighted that CNN, when trained on large datasets of liver hydatid cyst images, could accurately differentiate early-stage cysts (CE1 and CE2) with precision comparable to moderately experienced radiologists. This demonstrates the potential of AI to supplement radiological expertise, especially in settings with limited access to highly experienced professionals.

However, in morphologically complex stages like CE3b and CE4, the performance of AI systems is generally limited. While AI can rapidly analyze large volumes of imaging data, its interpretation accuracy decreases in atypical or complex cases. This underlines the need for human oversight, particularly when dealing with challenging cyst morphologies.

When compared to highly experienced radiologists, AI systems may exhibit lower accuracy, particularly in interpreting atypical cases. Nonetheless, augmented intelligence models that integrate AI algorithms enhance human–machine collaboration, thereby improving diagnostic accuracy and consistency in reporting. For these models, explainability and interpretability are essential at the decision point, not only for clinical applications but also for increasing user confidence.

Explainable AI (XAI) incorporates techniques that aim to make the reasoning behind model decisions more understandable. In this context, tools like Gradient-weighted Class Activation Mapping (Grad-CAM), Local Interpretable Model-Agnostic Explanations (LIME), and SHapley Additive exPlanations (SHAP) visualize the image features on which the model's decisions are based. These visualization techniques enhance the reliability of the system and facilitate user acceptance. In clinical settings, such explanatory mechanisms play a critical role, particularly in ensuring the transparency of AI-based decisions [27].

In classification systems based on CE staging, some models have reported accuracy rates ranging from 82.45% to 94% [27, 29, 35–37]. In contrast, highly experienced radiologists have demonstrated accuracy rates of up to 89.5% [30]. This discrepancy highlights that clinical experience remains a decisive factor, especially in evaluating complex cases.

Since AI models can perform segmentation and classification tasks within seconds, they provide a substantial time-saving advantage in centers with a high patient volume. When used for second opinions or prioritization purposes, these systems can significantly streamline the radiological workflow. In conclusion, hybrid approaches that integrate AI with human expertise not only increase diagnostic speed but also reduce interobserver variability, leading to more reliable reporting.

13.6 Complication Prediction Using AI Models

13.6.1 Predicting Rupture, Superinfection, and Biliary Communication

Complication prediction in liver hydatid disease is a critical aspect that directly impacts disease management and the determination of treatment strategies. The potential of AI-based systems to predict complication development based on imaging data and patient characteristics opens a new avenue in clinical decision support applications.

Recent studies have developed models to predict conditions such as rupture, superinfection, and biliary communication by integrating radiological findings with clinical and serological parameters. These systems have particularly identified cysts in CE3b and CE4 stages as having a higher likelihood of carrying complication risks.

While AI-supported models exhibit high diagnostic accuracy in early-stage cysts such as CE1 and CE2, they tend to show lower accuracy rates in transitional stages like CE3a–CE3b and atypical cases. Nevertheless, these systems significantly enhance process efficiency, especially in high-volume centers, by rapidly analyzing large imaging datasets and performing tasks such as automated segmentation, abnormality detection, and volumetric measurements.

When compared with radiologists, AI systems demonstrate accuracy levels comparable to those of inexperienced or moderately experienced radiologists. However, experienced radiologists still outperform AI, particularly in complex cases involving calcified lesions, nested cysts, and suspected biliary fistulas.

From a clinical success perspective, the most effective approach is the "augmented intelligence" model, where AI is integrated with human expertise. This structure not only speeds up diagnosis but also reduces interobserver variability, contributing to more standardized decision-making processes.

XAI tools, particularly visualization techniques like Grad-CAM, Local Interpretable Model-Agnostic Explanations (LIME), and SHapley Additive exPlanations (SHAP), play a key role in this integration. These tools help users understand which image regions the model focuses on during decision-making, thereby increasing the reliability of decisions and facilitating clinical acceptance.

In conclusion, AI systems have shown promising results in complication prediction; however, there remain differences between AI and experienced radiologists, particularly in some subgroups. To minimize this gap, hybrid approaches that rely on confidence threshold-based model guidance and human–machine collaboration are increasingly emphasized.

13.6.2 Integrating Clinical, Laboratory, and Imaging Data

Predicting complication risk in liver hydatid disease requires a multidimensional approach that not only relies on radiological findings but also integrates clinical symptoms, laboratory values, and imaging data. AI-based models developed within this framework offer individualized risk prediction by ensuring data integration from multiple sources.

Multilayered ML algorithms, with their ability to process diverse data types simultaneously, are capable of patient-specific predictions of complications (e.g., rupture, superinfection, biliary fistula). In these systems, feature engineering and data fusion techniques are employed to integrate the following parameters:

- Demographic Information: Age, gender, history of residing in endemic areas
- Clinical History: Previous hydatid surgery, immunosuppressive status, comorbidities
- Laboratory Markers: WBC, ALP, GGT, total/direct bilirubin, eosinophil level
- Radiological Data: CE stage, number of cysts, size, location, wall calcification, membrane detachment
- Radiomic Features: Entropy, homogeneity, variance, morphological characteristics

Models incorporating such parameters perform dynamic risk scoring based on complication likelihood, guiding follow-up decisions (e.g., conservative management, percutaneous drainage, surgical intervention).

In some studies, prediction models have been developed using regression and classification algorithms. Although their performance has been found promising in specific subgroups, multicenter and prospective studies are necessary before widespread clinical implementation.

Additionally, some studies have designed complication-specific sub-models, such as algorithms dedicated solely to predicting rupture risk or superinfection likelihood. Integrating the outputs of these sub-models into clinical workflows can serve as decision support systems in grading surgical urgency, planning follow-up frequency, and predicting the need for intensive care.

In conclusion, AI approaches that combine multimodal and multivariable data sources represent advanced systems that enhance patient safety, standardize clinical decisions, and support personalized medical management. The integrative potential of AI in managing hydatid disease signifies a new paradigm, where diagnostic processes become faster, treatment planning more effective, and clinical efficiency significantly improved.

13.6.3 ML and DL Algorithms: Verified Models in Reviewed Literature

Although various AI algorithms have been proposed for the diagnosis and complication prediction of hydatid disease, only a limited number of models have been directly validated for practical use. Below is a summary of ML algorithms used specifically for predicting complications of hydatid disease, based on the reviewed studies:

- **DL Methods**: DL methods are employed for classification, segmentation, and feature extraction, particularly in determining disease stages and cyst structure in hydatid disease. Among the DL architectures, CNNs are the most commonly used. They have been applied to tasks such as CE stage classification, cyst structure segmentation, and morphological feature identification from liver images. The ResNet architecture is widely used for distinguishing between CE stages and extracting features from images [40]. The U-Net architecture, featuring an encoder-decoder structure, is particularly favored for segmentation tasks [34]. It has proven successful in defining CE stage boundaries and differentiating cyst volumes. The GoogLeNet-based Inception architecture is a CNN model capable of multiscale feature extraction. Its ability for parametric optimization makes it suitable for CE classification and subtype differentiation [31].
- **ML Methods**: ML methods are techniques that deliver rapid results compared to traditional methods, facilitating classification and diagnostic processes. (i) Support Vector Machine: Utilized in feature-based classification tasks. It is effective in enhancing the separation between CE stages and managing high-dimensional data. (ii) KNN: Often used as a supportive classifier, especially in distinguishing inactive stages. (iii) Decision Tree: Offers a rule-based classification approach, making it particularly valuable for interpretable diagnostic systems. Its high explainability is advantageous for clinical applications. (iv) Deep Learning with Ensemble Approach Model (EDAM): Combines DL and ensemble modeling principles to enhance image prediction and patient-specific classification. It contributes significantly to CE stage differentiation and diagnostic classification processes [41, 42].
- **XAI Methods**: XAI methods include visualization and interpretation techniques that make model outputs more understandable. Grad-CAM: Provides visual attention maps, highlighting regions of focus within the images. LIME: Analyzes the data points on which individual predictions are based, offering local decision explanations [27]. These XAI methods enhance the interpretability of

model decisions, making them more transparent and clinically acceptable.
- These models find application in CE staging, image classification, segmentation, complication prediction, and the development of diagnostic support systems. The choice of model depends on several factors, including imaging modality, data quality, labeling standards, and classification purpose. Therefore, it is essential to consider the specific advantages and limitations of each algorithm tailored to the clinical scenario.

In conclusion, while DL and ML models offer substantial potential, their integration into clinical practice requires careful evaluation. Selecting the appropriate algorithm for a given diagnostic context remains crucial for ensuring accuracy and reliability.

13.7 Clinical Decision Support Systems in Hydatid Disease

13.7.1 Risk Stratification Tools

AI-powered clinical decision support systems are tools designed to enhance personalized risk stratification and improve clinical decision-making in liver hydatid disease. These systems integrate clinical, laboratory, and imaging data to automatically generate risk scores related to scenarios such as cyst-associated complications, recurrence risk, or treatment failure.

For example, a clinical decision support system can combine factors such as CE stage, cyst morphology, localization, bilirubin levels, eosinophilia, and clinical symptoms to provide patient-specific predictions for rupture or biliary fistula risk. These tools can be integrated into electronic health records and radiology workflows, offering real-time alert mechanisms as well as guideline-based recommendations to support the treatment process.

In endemic regions, these systems enable more effective patient management by guiding decisions between medical therapy, percutaneous intervention, or surgical procedures. As a result, complication rates can be reduced, and healthcare resources can be utilized more efficiently.

13.7.2 Surgical Decision Aids: AI in Preoperative Planning

Surgical treatment is indispensable, especially in complicated and advanced-stage hydatid cysts. However, making decisions between options such as conservative cystotomy, pericystectomy, or hepatic resection is complex and depends on numerous parameters. AI-supported surgical decision support systems can offer several benefits in this context:

- **3D Liver Models from CT or MRI Data**: These models visualize the relationships between the cyst and vascular or biliary structures, facilitating surgical planning.
- **Prediction of Intraoperative Complications:** By analyzing morphological criteria such as cyst wall thickness, calcification, septation, and multilocular structure, potential intraoperative complications can be anticipated.
- **Supervised Learning Algorithms:** These algorithms learn from past data to recommend the optimal surgical approach (e.g., open vs. laparoscopic, radical vs. conservative).

These systems not only enhance surgical precision but also shorten operation time, reduce blood loss, and lower the risk of postoperative complications. Increasingly, they are being integrated into preoperative assessments at academic centers and multidisciplinary surgical boards.

13.7.3 AI-Assisted Post-treatment Surveillance

Post-treatment surveillance is crucial in managing hydatid disease to detect recurrence, residual cysts, and late complications (e.g., bile leakage, abscess) at an early stage. Although traditional methods can achieve this, they are often costly and prone to interpretation variability.

AI-based clinical decision support systems can process the following data to develop personalized follow-up strategies:

- **Initial Cyst Characteristics**: Including size, location, and morphological features
- **Surgical Technique and Margin Safety**: Detailing the type of intervention performed and its precision
- **Postoperative Imaging Findings**: Including early detection of residual cavities or complications
- **Serial Serological Markers**: Tracking changes over time to monitor recurrence or progression

By analyzing these data points, AI algorithms can identify high-risk patients who require early follow-up while reducing unnecessary imaging frequency in low-risk individuals, thereby improving healthcare efficiency. Additionally, algorithms that automatically compare serial US or MRI images can detect residual cavities or emerging complications at an early stage. These systems offer a personalized, dynamic, and data-driven approach to postoperative follow-up, enhancing patient safety and cost-effectiveness.

13.8 Limitations and Challenges

13.8.1 Dataset Limitations and Annotation Bias

Despite advancements in AI applications, one of the most significant limitations in liver hydatid disease is the lack of high-quality, diversified, and standardized datasets. Most of the developed models are based on single-center studies, limited sample sizes, and specific patient populations. This reliance on restricted data sources reduces the generalizability of the models and limits their effectiveness in real clinical scenarios, especially due to variations in imaging protocols and clinical presentations.

Annotation bias also represents a substantial methodological limitation. Typically, CE staging is based on the diagnoses of a limited number of expert radiologists. Due to interobserver variability, this can introduce noise during the model training process. This issue is particularly pronounced in transitional stages, such as CE3a and CE4, where subjective interpretation is more prevalent, leading to decreased model accuracy. To overcome these challenges, it is essential to employ multicenter, multi-observer, and consensus-based annotation protocols. This approach would enhance the quality of training data and make models more robust and suitable for clinical applications.

13.8.2 Interpretability of Black-Box Models

High-performance AI algorithms, particularly DL architectures, are often referred to as "black-box" models due to their inability to transparently explain decision-making processes. This lack of interpretability is one of the primary barriers that undermine clinical confidence.

CNN-based models, for instance, can perform classifications without indicating which areas of the images they focus on, making it difficult for clinical experts to assess the accuracy of the decision process. Therefore, interpretability has become as crucial as accuracy in the context of medical AI applications.

XAI tools, such as Grad-CAM, enhance the transparency of model decisions by visualizing the specific features or regions that influence the output. Although these tools help make decision-making processes more understandable, their integration into routine clinical workflows remains limited.

Future models should not only focus on prediction accuracy but also on the ability to explain their reasoning. Enhancing interpretability will accelerate the integration of AI into clinical systems and improve the effectiveness of multidisciplinary decision-making processes.

13.9 Future Directions and Research Opportunities

13.9.1 Prospective Multicenter Datasets

There is a continuous need for prospective, multicenter, and high-quality datasets to develop AI models that are generalizable, reliable, and reproducible. Currently, most existing models are based on retrospective, single-center studies with limited patient data, which restricts their clinical applicability and robustness.

To advance in this field, it is essential to standardize CE stage labeling among experts, harmonize imaging protocols, and define clinical outcomes using a unified terminology. International collaborations conducted in endemic regions are particularly valuable as they bring together high-resolution imaging data, expert annotations, and long-term follow-up information, creating a robust environment for model development.

Such datasets not only facilitate algorithmic comparisons but also provide a reference framework to objectively evaluate model performance. Furthermore, the proliferation of open-access data platforms promotes transparency and scientific productivity in the field.

13.9.2 Real-Time AI Systems for Intraoperative Use

In the management of hydatid disease, which is endemic in many regions, the use of real-time AI-supported systems represents a significant focus area for the future. AI systems designed for intraoperative use can analyze laparoscopic video and ultrasound images to:

- Predict cyst rupture risk
- Accurately delineate resection margins, which is particularly important in alveolar echinococcosis
- Detect residual cyst material or biliary connections

These systems are often integrated with augmented reality (AR) technology, which merges digital information with the real-world environment. By projecting preoperative imaging data into the surgeon's field of view, these systems enhance the understanding of relationships with vascular and biliary structures. Through this technology, objects can be visualized in three dimensions, improving both surgical planning and training.

Such technologies have become adaptable to hydatid disease management by reconfiguring algorithms previously tested in oncological surgeries. These real-time decision support systems enhance the surgeon's intraoperative assessment capacity, reduce the risk of complications, and improve operational efficiency.

13.10 Highlights

- AI enhances imaging-based accuracy and reproducibility in the automatic detection, staging, and complication prediction of liver hydatid disease.
- DL architectures have achieved high accuracy in cyst segmentation and WHO classification on CT images; in some studies, these algorithms have been tested against radiologist performance.
- ML algorithms have been successfully applied to predict rupture risk at the CE3b stage, biliary communication, and postoperative complications.
- AI-supported clinical decision support systems have contributed to improving clinical decision-making through 3D modeling in surgical planning, intraoperative risk prediction, and postoperative follow-up systems.
- Imaging-based models are utilized not only for staging but also for segmentation, treatment planning, and patient follow-up.
- Key limitations include limited labeled datasets, interobserver variability, lack of interpretability (black-box problem), and heterogeneous imaging protocols, all of which

restrict model generalizability and clinical integration.

13.11 Conclusion

AI has evolved into a multidimensional clinical tool for the diagnosis, staging, complication prediction, and treatment planning of liver hydatid disease. DL algorithms can accurately differentiate CE stages from imaging data, facilitate surgical planning through segmentation, and contribute to patient management by predicting potential complications.

Current models automatically perform CE1–CE5 staging in accordance with the WHO classification, particularly accelerating decision-making in borderline stages such as CE3 and CE4. Moreover, these systems assist clinicians in identifying cysts that have contact with biliary structures during preoperative planning.

ML systems based on radiomic features aim to predict complications such as rupture, superinfection, recurrence, or bile leakage in advance, thereby enhancing risk management and patient safety.

The contribution of AI to healthcare is not limited to high-accuracy predictions; it also encompasses guiding clinical decisions through user-friendly interfaces and interpretability support. However, the practical integration of this potential into clinical practice requires algorithms trained on multicenter, standardized datasets and a transparent explanation of decision rationale.

For real-world success, it is crucial to adhere to ethical principles, ensure data security, adopt open data policies, and develop systems that can be integrated into local healthcare infrastructures. Therefore, multidisciplinary collaborations, prospective data generation, and continuous algorithm updates are essential to making AI systems a sustainable and effective tool in the management of liver hydatid disease.

Acknowledgments The English language and grammar of this manuscript have been reviewed and enhanced with the assistance of ChatGPT and Quillbot.

References

1. Wen H, Vuitton L, Tuxun T, Li J, Vuitton DA, Zhang W, McManus DP. Echinococcosis: advances in the 21st century. Clin Microbiol Rev. 2019;32(2) https://doi.org/10.1128/cmr.00075-18. PMID: 30760475 PMCID: PMC6431127.
2. Hogea MO, Ciomaga BF, Muntean MM, Muntean AA, Popa MI, Popa GL. Cystic echinococcosis in the early 2020s: a review. Trop Med Infect Dis. 2024;9(2) https://doi.org/10.3390/tropicalmed9020036. PMID: 38393125 PMCID: PMC10891927.
3. Drocchi G, Santucciu C, Mastrandrea S, Sanguedolce F, Madonia M. Diagnosis and treatment of unusual multiorgan Echinococcus hydatid cysts. Curr Urol. 2024;18(1):75–7. https://doi.org/10.1097/cu9.0000000000000130. PMID: 38505159 PMCID: PMC10946642.
4. Azizoğlu M, Aydogdu B, Kamci TO, Arslan S, Basuguy E, Bilici S, Okur MH. Analysis of complications of a neglected disease: 13 years of experience with liver hydatid cysts in a high-volume hospital. Medicina (Kaunas). 2024;60(10) https://doi.org/10.3390/medicina60101696. PMID: 39459483 PMCID: PMC11509507.
5. Bhutani N, Kajal P. Hepatic echinococcosis: a review. Ann Med Surg (Lond). 2018;36:99–105. https://doi.org/10.1016/j.amsu.2018.10.032. PMID: 30450204 PMCID: PMC6226561.
6. Ferrer Inaebnit E, Molina Romero FX, Segura Sampedro JJ, González Argenté X, Morón Canis JM. A review of the diagnosis and management of liver hydatid cyst. Rev Esp Enferm Dig. 2022;114(1):35–41. https://doi.org/10.17235/reed.2021.7896/2021. PMID: 34034501.
7. Pradhan A, Mallick B, Dash A, Nanda D. Clinical presentation and outcome of children with hydatid disease: a retrospective cross-sectional study from a tertiary care hospital in eastern India. J Parasit Dis. 2022;46(1):230–5. https://doi.org/10.1007/s12639-021-01440-8. PMID: 35299908 PMCID: PMC8901903.
8. Nunnari G, Pinzone MR, Gruttadauria S, Celesia BM, Madeddu G, Malaguarnera G, Pavone P, Cappellani A, Cacopardo B. Hepatic echinococcosis: clinical and therapeutic aspects. World J Gastroenterol. 2012;18(13):1448–58. https://doi.org/10.3748/wjg.v18.i13.1448. PMID: 22509076 PMCID: PMC3319940.
9. Abdulla SH, Soran HM. Comparing and evaluating machine learning techniques for liver-hydatid-cyst classification; 2024. p. 307–21.
10. Abhishek KS, Shrimali T, Tak V, Nag VL, Banerjee S, Khera S. Primary pelvic echinococcosis though uncommon but not rare. Trop Parasitol. 2021;11(1):56–9. https://doi.org/10.4103/tp.TP_64_19. PMID: 34195064 PMCID: PMC8213113.

11. Solomon N, Fields PJ, Tamarozzi F, Brunetti E, Macpherson CNL. Expert reliability for the World Health Organization standardized ultrasound classification of cystic echinococcosis. Am J Trop Med Hyg. 2017;96(3):686–91. https://doi.org/10.4269/ajtmh.16-0659. PMID: 28070008 PMCID: PMC5361546.
12. Alshoabi SA, Alkalady AH, Almas KM, Magram AO, Algaberi AK, Alareqi AA, Hamid AM, Alhazmi FH, Qurashi AA, Abdulaal OM, Aloufi KM, Alsharif WM, Alsultan KD, Omer AM, Gareeballah A. Hydatid disease: a radiological pictorial review of a great neoplasms mimicker. Diagnostics (Basel). 2023;13(6) https://doi.org/10.3390/diagnostics13061127. PMID: 36980435 PMCID: PMC10047450.
13. Shah OJ, Robbani I, Zargar SA, Yattoo GN, Shah P, Ali S, Javaid G, Shah A, Khan BA. Hydatid cyst of the pancreas. An experience with six cases. JOP. 2010;11(6):575–81. PMID: 21068489
14. Erganis S, Sarzhanov F, Al FD, Cağlar K. Comparison of methods in the serologic diagnosis of cystic echinococcosis. Acta Parasitol. 2024;69(2):1122–31. https://doi.org/10.1007/s11686-024-00840-z. PMID: 38551763 PMCID: PMC11182860.
15. Öztürk G, Uzun MA, Özkan ÖF, Kayaalp C, Tatlı F, Eren S, Aksungur N, Çoker A, Bostancı EB, Öter V, Kaya E, Taşar P. Turkish HPB Surgery Association consensus report on hepatic cystic echinococcosis (HCE). Turk J Surg. 2022;38(2):101–20. https://doi.org/10.47717/turkjsurg.2022.5757. PMID: 36483170 PMCID: PMC9714645.
16. Akhan O. Percutaneous treatment of liver hydatid cysts: to PAIR or not to PAIR. Curr Opin Infect Dis. 2023;36(5):308–17. https://doi.org/10.1097/qco.0000000000000956. PMID: 37548385.
17. Schipper HG, Laméris JS, van Delden OM, Rauws EA, Kager PA. Percutaneous evacuation (PEVAC) of multivesicular echinococcal cysts with or without cystobiliary fistulas which contain non-drainable material: first results of a modified PAIR method. Gut. 2002;50(5):718–23. https://doi.org/10.1136/gut.50.5.718. PMID: 11950823 PMCID: PMC1773202.
18. Erkmen F, Yilmaz M, Yonder H, Tatli F, Ozgonul A, Karaca E, Batibay E, Uzunkoy A. Complications and management of patients with liver hydatid cyst: a single center experience. Hepatol Forum. 2025;6(2):52–6. https://doi.org/10.14744/hf.2024.2024.0052. PMID: 40248675 PMCID: PMC11999902.
19. Ramia JM, Figueras J, De la Plaza R, García-Parreño J. Cysto-biliary communication in liver hydatidosis. Langenbeck's Arch Surg. 2012;397(6):881–7. https://doi.org/10.1007/s00423-012-0926-8. PMID: 22374106.
20. Collado-Aliaga J, Romero-Alegría Á, Alonso-Sardón M, Muro A, López-Bernus A, Velasco-Tirado V, Muñoz Bellido JL, Pardo-Lledias J, Belhassen-García M. Complications associated with initial clinical presentation of cystic echinococcosis: a 20-year cohort analysis. Am J Trop Med Hyg. 2019;101(3):628–35. https://doi.org/10.4269/ajtmh.19-0019. PMID: 31359859 PMCID: PMC6726966.
21. Abbasi B, Akhavan R, Ghamari Khameneh A, Darban Hosseini Amirkhiz G, Rezaei-Dalouei H, Tayebi S, Hashemi J, Aminizadeh B, Darban Hosseini Amirkhiz S. Computed tomography and magnetic resonance imaging of hydatid disease: a pictorial review of uncommon imaging presentations. Heliyon. 2021;7(5):e07086. https://doi.org/10.1016/j.heliyon.2021.e07086. PMID: 34095581 PMCID: PMC8166760.
22. Akbari MN, Azizi A. Enhancing hydatid cyst classification with deep learning and convolutional neural networks using CT scans. Afghan J Infect Dis. 2024;2(1):9–16. https://doi.org/10.60141/AJID/V.2.I.1.2.
23. Siles-Lucas M, Uchiumi L, Tamarozzi F. 'No cyst, no echinococcosis': a scoping review update on the diagnosis of cystic echinococcosis after the issue of the WHO-IWGE Expert Consensus and current perspectives. Curr Opin Infect Dis. 2023;36(5):333–40. https://doi.org/10.1097/qco.0000000000000941. PMID: 37466103 PMCID: PMC10487401.
24. Brunetti E, Tamarozzi F, Macpherson C, Filice C, Piontek MS, Kabaalioglu A, Dong Y, Atkinson N, Richter J, Schreiber-Dietrich D, Dietrich CF. Ultrasound and cystic echinococcosis. Ultrasound Int Open. 2018;4(3):E70–e78. https://doi.org/10.1055/a-0650-3807. PMID: 30364890 PMCID: PMC6199172.
25. Srinivas MR, Deepashri B, Lakshmeesha MT. Imaging spectrum of hydatid disease: usual and unusual locations. Pol J Radiol. 2016;81:190–205. https://doi.org/10.12659/pjr.895649. PMID: 27231490 PMCID: PMC4868106.
26. Gharbi HA, Hassine W, Brauner MW, Dupuch K. Ultrasound examination of the hydatic liver. Radiology. 1981;139(2):459–63. https://doi.org/10.1148/radiology.139.2.7220891. PMID: 7220891.
27. Yildirim M. Image visualization and classification using hydatid cyst images with an explainable hybrid model. Appl Sci. 2023;13(17):9926. https://doi.org/10.3390/app13179926.
28. Peng J, Duan Z, Guo Y, Li X, Luo X, Han X, Luo J. Identification of candidate biomarkers of liver hydatid disease via microarray profiling, bioinformatics analysis, and machine learning. J Int Med Res. 2021;49(3):300060521993980. https://doi.org/10.1177/0300060521993980. PMID: 33787392 PMCID: PMC8020228.
29. Abdukeyim N, Liu Z, Yan C. CT Images Classification of Hepatic Hydatid Disease by Different Methods. In: Proceedings of the 2019 6th International Conference on Systems and Informatics (ICSAI). 2019. 2–4 Nov 2019. p. 398–403.
30. Yang Y, Cairang Y, Jiang T, Zhou J, Zhang L, Qi B, Ma S, Tang L, Xu D, Bu L, Bu R, Jing X, Wang H, Zhou Z, Zhao C, Luo B, Liu L, Guo J, Nima Y, Hua

G, Wa Z, Zhang Y, Zhou G, Jiang W, Wang C, De Y, Yu X, Cheng Z, Han Z, Liu F, Dou J, Feng H, Wu C, Wang R, Hu J, Yang Q, Luo Y, Wu J, Fan H, Liang P, Yu J. Ultrasound identification of hepatic echinococcosis using a deep convolutional neural network model in China: a retrospective, large-scale, multicentre, diagnostic accuracy study. Lancet Digit Health. 2023;5(8):e503–14. https://doi.org/10.1016/s2589-7500(23)00091-2. PMID: 37507196.

31. Tegshee T, Dorjsuren T, Lee S, Batjargal D. A study on staging cystic echinococcosis using machine learning methods. Bioengineering (Basel). 2025;12(2) https://doi.org/10.3390/bioengineering12020181. PMID: 40001700 PMCID: PMC11852189.
32. Yildirim M, Eroğlu O, Eroğlu Y, Çinar A, Cengil E. COVID-19 detection on chest X-ray images with the proposed model using artificial intelligence and classifiers. N Gener Comput. 2022;40(4):1077–91. https://doi.org/10.1007/s00354-022-00172-4.
33. Shang F, Song T, Wang Z, Wu M, Yan C, Wang X. Clinical application of artificial intelligence in the ultrasound classification of hepatic cystic echinococcosis. Am J Trop Med Hyg 2024; 111(1): 93–101. DOI: https://doi.org/10.4269/ajtmh.23-0519. PMID: 38806047 PMCID: PMC11229658.
34. Chen J, Li K, Wang Z, Wang Z, Huang J. HES-UNet: a U-net for hepatic echinococcosis lesion segmentation. arXiv preprint arXiv:241206530 (2024).
35. Xin S, Shi H, Jide A, Zhu M, Ma C, Liao H. Automatic lesion segmentation and classification of hepatic echinococcosis using a multiscale-feature convolutional neural network. Med Biol Eng Comput. 2020;58(3):659–68. https://doi.org/10.1007/s11517-020-02126-8. PMID: 31950330.
36. Wu M, Yan C, Sen G. Computer-aided diagnosis of hepatic cystic echinococcosis based on deep transfer learning features from ultrasound images. Sci Rep. 2025;15(1):607. https://doi.org/10.1038/s41598-024-85004-9. PMID: 39753933 PMCID: PMC11698856.
37. Gul Y, Muezzinoglu T, Kilicarslan G, Dogan S, Tuncer T. Application of the deep transfer learning framework for hydatid cyst classification using CT images. Soft Comput. 2023;27(11):7179–89.
38. Cengil E, Çınar A, Yıldırım M. A hybrid approach for efficient multi-classification of white blood cells based on transfer learning techniques and traditional machine learning methods. Concurr Comput Pract Exp. 2022;34(6):e6756. https://doi.org/10.1002/cpe.6756.
39. Lissandrin R, Cicerone O, Vola A, D'Alessandro G, Frassini S, Manciulli T, Famularo S, De Silvestri A, Viganò J, Quaretti P, Ansaloni L, Brunetti E, Maestri M. Machine learning to predict the decision to perform surgery in hepatic echinococcosis. HPB (Oxford). 2025;27(4):480–8. https://doi.org/10.1016/j.hpb.2024.12.014. PMID: 39755481.
40. Nijiati M, Tuerdi M, Damola M, Yimit Y, Yang J, Abulaiti A, Mutailifu A, Aihait D, Wang Y, Zou X. A deep learning radiomics model based on CT images for predicting the biological activity of hepatic cystic echinococcosis. Front Physiol. 2024;15:1426468. https://doi.org/10.3389/fphys.2024.1426468. PMID: 39175611 PMCID: PMC11338923.
41. Manciulli T, Brunetti E. Stepping stones and challenges in the use of artificial intelligence in the diagnosis of echinococcosis. Lancet Digit Health. 2023;5(11):e750–1. https://doi.org/10.1016/s2589-7500(23)00183-8. PMID: 37770336.
42. Wang Z, Bian H, Li J, Xu J, Fan H, Wu X, Cao Y, Guo B, Xu X, Wang H, Zhang L, Zhou H, Fan J, Ren Y, Geng Y, Feng X, Li L, Wei L, Zhang X. Detection and subtyping of hepatic echinococcosis from plain CT images with deep learning: a retrospective, multicentre study. Lancet Digit Health. 2023;5(11):e754–62. https://doi.org/10.1016/s2589-7500(23)00136-x. PMID: 37770335.

Post-treatment Follow-Up, Surveillance, and Prevention Strategies in Hydatid Disease

14

Sami Akbulut

14.1 Introduction

Echinococcosis, also known as hydatid disease or hydatidosis, is caused by *Echinococcus* species and is a significant zoonoticparasitic infection prevalent in endemic regions such as the Mediterranean, Middle East, South America, and parts of Asia [1]. The most common forms of the disease are cystic echinococcosis, caused by *Echinococcus granulosus*, and alveolar echinococcosis, caused by *Echinococcus multilocularis* [2, 3]. It primarily affects the liver and lungs, but can also be found in other organs, including the spleen, kidneys, thyroid, breast, adrenal gland, retroperitoneum, pancreas, soft tissue, bone, and the central nervous system [4, 5]. The disease follows a slow-growing pattern, with cysts enlarging over months to years, often remaining asymptomatic until they reach a critical size or cause complications.

Due to its chronic nature and potential for recurrence, effective follow-up and surveillance are crucial to prevent disease progression, detect complications, and assess the long-term success of treatment [6]. Long-term follow-up is especially important in high-risk patients, such as those with multiple cysts, incomplete treatment response, or previous recurrence [7, 8]. This chapter outlines the follow-up protocols after surgical treatment (radical, conservative), radiological intervention (puncture aspiration injection reaspiration [PAIR], puncture aspiration injection reaspiration drainage [PAIRD], modified catheterization technique [MoCaT], percutaneous evacuation [PEVAC]), medical therapy (albendazole, mebendazole, prazyquentel), and wait and watch with an emphasis on imaging, serological monitoring, and multidisciplinary management [9–11].

14.1.1 Post-treatment Follow-Up and Surveillance

14.1.1.1 Goals of Follow-Up

The primary objectives of post-treatment follow-up in hydatid disease are:

- Early detection of recurrence: Preventing progression through timely intervention
- Monitoring residual lesions: Assessing the evolution of treated cysts
- Evaluating complications: Including biliary fistula, secondary bacterial infections, or anaphylactic reactions
- Assessing treatment response: Determining if further therapeutic intervention is required

S. Akbulut (✉)
Department of Surgery and Liver Transplantation Institute, Inonu University Faculty of Medicine, Malatya, Türkiye

Department Biostatistics and Medical Informatics, Inonu University Faculty of Medicine, Malatya, Türkiye

S. Akbulut (ed.), *Hydatid Disease*, https://doi.org/10.1007/978-3-031-97277-5_14

- Long-term patient care: Preventing reinfection and ensuring appropriate patient education

14.1.1.2 Follow-Up Protocols by Treatment Type

Surgical Treatment

Surgical management of hydatid disease is divided into conservative and radical approaches, each requiring distinct follow-up strategies [12]. Conservative surgical treatment aims to empty and sterilize the cyst cavity while managing the residual space [12]. Partial cystectomy, where the portion of the cyst wall outside the liver parenchyma is removed, is a common conservative approach [12, 13]. Various techniques, such as unroofing, marsupialization, tube drainage, capitonnage, and omentoplasty, can be performed intraoperatively to manage the remaining cavity and prevent postoperative complications [13, 14]. Radical surgical treatment involves complete removal of the cyst without opening it, minimizing the risk of spillage [13, 15]. This includes pericystectomy (total cystectomy), where only the pericyst is excised, or hepatectomy, where the pericyst and surrounding hepatic parenchyma are resected [13, 15]. Radical surgery significantly reduces recurrence risk but necessitates long-term hepatic function monitoring and imaging to assess liver regeneration and detect late complications [12, 15]. Given these different surgical approaches, follow-up must be tailored accordingly. Conservative surgical patients require close radiological monitoring to evaluate cavity resolution and detect secondary infections or biliary complications [15]. In addition, long-term uninterrupted or cyclic interrupted anthelmintic benzimidazoles is recommended for this group to minimize the risk of recurrence and ensure complete parasite eradication. Radical surgical patients need long-term hepatic function surveillance and imaging to ensure disease-free status. If no cystic lesions remain following radical surgery, prolonged anthelmintic benzimidazoles therapy is not necessary [16]. Although there are opposing views in the literature, we believe that this approach is appropriate [12].

- **Imaging Surveillance:** Postoperative imaging is crucial for assessing treatment response and detecting early recurrence. The recommended follow-up protocol begins with the first imaging assessment at 1 month post-treatment using ultrasonography (US). Subsequent checks should be conducted at the 3rd, 6th, and 9th months within the first year. After the first year, follow-up should continue every 6 months for at least 3 to 5 years, especially in endemic areas. This structured imaging protocol ensures that both early and late recurrences are detected promptly. If there is suspicion of recurrence or if the US findings are inconclusive, contrast-enhanced computed tomography (CT) should be considered to obtain more detailed information. In cases characterized by a high risk of recurrence, complicated cysts, or incomplete cyst evacuation, magnetic resonance imaging (MRI) may offer valuable insights into residual disease, thereby enhancing post-treatment surveillance [13, 17, 18].
- **Serological Surveillance**: Serological tests, including enzyme-linked immunosorbent assay (ELISA) or indirect hemagglutination (IHA), are essential tools in the follow-up of hydatid disease. Baseline serology should always be obtained before initiating any treatment protocol, as it provides a reference for post-treatment comparisons. Following treatment, serial serological assessments allow for tracking therapeutic response. Serological tests, however, tend to lose sensitivity over time, making them less reliable for detecting late recurrences. If antibody titers decline steadily, this suggests effective treatment. In contrast, persistently high or rising titers may indicate residual disease or recurrence [13, 17]. Notably, if serology turns negative post-treatment and later becomes positive again, this is a strong indicator of disease recurrence, necessitating further evalua-

tion with imaging techniques [13]. Given these limitations, serological monitoring should always be interpreted in conjunction with imaging studies [19]. Serological tests should be conducted every 3–6 months in the first year, followed by annual assessments for long-term surveillance [17].

- **Clinical Evaluation**: Patients should be regularly assessed for symptoms of recurrence, including abdominal pain, fever, allergic reactions, or general malaise. Signs of biliary obstruction, such as jaundice and pruritus, may suggest biliary fistula or cyst rupture into the biliary tree.
- **Biliary Tree Assessment:** Given the high likelihood of cystobiliary fistula formation, especially in liver hydatid disease, post-surgical patients with persistent liver enzyme elevation, jaundice, or cholangitis should undergo magnetic resonance cholangiopancreatography (MRCP) for detailed biliary evaluation. Endoscopic retrograde cholangiopancreatography (ERCP) may be required for therapeutic intervention in cases of obstructive biliary involvement.
- **Recurrence Prevention Strategies**: The postoperative recurrence rate varies significantly depending on the treatment method employed. While conservative surgery are associated with a local recurrence rate of 20–25%, radical surgery significantly reduces this risk, with reported recurrence rates ranging from 0% to 6.4%. In fact, the recurrence rate in the conservative surgery group is 5.9 times higher compared to the radical surgery group [13, 20]. When recurrence is detected, treatment strategies should mirror primary treatment protocols. For small, unilocular cysts, percutaneous treatment together with anthelmintic therapy should be prioritized, while complex cysts may require repeat surgery after anthelmintic therapy [13]. Preventive measures include meticulous surgical technique, intraoperative scolicidal agent use, and uninterrupted or cyclic interrupted adjuvant anthelmintic benzimidazole therapy for at least 3–6 months postoperatively. Patients with multiple cysts or prior recurrences may benefit from longer courses of adjunctive anthelmintic therapy.

Percutaneous Treatment (PAIR, PAIRD, MoCaT, PEVAC)

Primarily PAIR, as well as PAIRD, MoCaT, and PEVAC, are minimally invasive techniques that require structured follow-up to evaluate the therapeutic response and detect complications [9–11]. While PAIR is effective in CE1 and CE3a cysts, catheter-based drainage techniques is more suitable for larger CE2 and CE3b cysts, those with biliary communication, and postoperative collections [13, 21–23]. PAIR is typically recommended for cysts smaller than 10 cm, while catheter-based drainage techniques is more suitable for cysts larger than 10 cm or in complex cases where simple aspiration alone is inadequate [13, 17, 24]. Percutaneous therapy is effective for cysts with liquid or viable contents, but CE3b cysts with solid structures or exophytic growth have a higher risk of recurrence, often requiring surgical intervention. If recurrence occurs after PAIR, repeating the procedure alongside anthelmintic therapy is an option, but persistent failure necessitates surgical management. Follow-up after PAIR should incorporate clinical, radiological, and serological assessments to confirm cyst resolution and detect recurrence. High-risk patients require prolonged monitoring to ensure effective disease control and prevent complications.

- **Imaging Surveillance**: Patients should undergo US evaluation at 1, 3, 6, 9, and 12 months post-procedure to assess cyst involution and determine whether the cystic cavity is shrinking as expected. Progressive reduction in cyst size and increased echogenicity within the lesion indicate a favorable response, whereas persistent or enlarging cysts may suggest treatment failure or recurrence.
- **Serological Surveillance**: Serological monitoring with ELISA or IHA should be

conducted periodically, as decreasing antibody titers suggest successful therapy, while persistently high or rising titers may indicate residual disease or recurrence. However, due to potential false positives and prolonged antibody persistence, serology should always be interpreted alongside imaging findings.

- **Re-evaluation and Repeat Percutaneous Treatment**: In cases where cyst regression is inadequate or if imaging shows persistent viable cysts, repeat percutaneous treatment sessions may be necessary. The decision for additional treatment should be based on a combination of imaging, serological results, and clinical symptoms. Long-term follow-up extending beyond 1 year is recommended for patients with large or multiple cysts, particularly those at high risk of recurrence.

Medical Therapy

Medical treatment serves as either an adjunct to surgical or percutaneous treatment procedures or as a primary therapy in patients who are not surgical candidates [13, 15]. It is primarily based on benzimidazole derivatives such as albendazole and mebendazole, which act by inhibiting microtubule formation in *Echinococcus* larvae [25]. Medical therapy is indicated in patients with inoperable cysts due to high surgical risk, multiple organ involvement, peritoneal hydatidosis, or when cysts are small (<5 cm) and asymptomatic [17, 26]. In selected cases, a "watch and wait" strategy may be employed, particularly for inactive or degenerating cysts detected on imaging [17]. These patients require close monitoring through serial US and serological testing to ensure the cyst remains inactive and does not progress to a symptomatic or viable stage [17]. The decision between active treatment and surveillance depends on cyst characteristics, patient condition, and risk of complications.

- **Imaging surveillance**: US should be performed every 3 months during treatment to assess cyst size, wall thickness, and internal echogenicity. A stable or shrinking cyst with progressive calcification suggests effective therapy, while persistent cyst activity may indicate treatment failure. After therapy completion, imaging intervals should be extended to every 6 months for the first 2 years, followed by annual scans if no recurrence is detected.
- **Liver function tests**: Patients receiving prolonged benzimidazole therapy should undergo routine hematological and biochemical assessments. Monthly complete blood counts are necessary to detect neutropenia, a known side effect of prolonged high dose albendazole. Hepatotoxicity should be closely monitored, and any significant enzyme elevation may require dose adjustment or discontinuation. Tests should be conducted monthly during treatment and for at least 3 months post-therapy to ensure hepatic recovery. Patients with underlying liver disease or prolonged treatment duration require closer monitoring. Patients should also be assessed for gastrointestinal symptoms and other adverse effects that could impact adherence to therapy.
- **Serological monitoring**: ELISA or IHA should be performed at baseline, during therapy, and after treatment completion. A decline in antibody titers suggests a favorable response, whereas persistently elevated or rising titers may indicate ongoing infection or recurrence. Since antibodies can persist long after successful treatment, serological findings should always be interpreted in conjunction with imaging results.

14.1.1.3 Imaging Strategies in Follow-Up

- **US**: US is the primary imaging tool for post-treatment follow-up due to its availability, affordability, and noninvasive nature. It enables dynamic assessment of cyst morphology, including changes in size, wall thickness, and internal structure. It is also useful for identifying postoperative biliary complications such as fistulae. However, its limited ability to assess deeply seated cysts and extrahepatic involvement necessitates additional imaging techniques in specific cases.

- **CT**: CT provides superior anatomical detail, making it the imaging method of choice when complications such as rupture, superinfection, or peritoneal dissemination are suspected. Contrast-enhanced CT is particularly useful in differentiating active cysts from calcified or degenerated lesions by assessing the presence of cyst wall enhancement. It is especially beneficial in the evaluation of pulmonary, bone, and multi-organ involvement, where US has limited penetration. Due to its radiation exposure, CT is generally reserved for cases where US findings are inconclusive or when a more detailed assessment is required for treatment planning.
- **MRI**: MRI is the preferred imaging modality for detecting cystobiliary communication and assessing complex intrahepatic or peritoneal hydatid cysts. MRCP plays a crucial role in evaluating bile duct involvement, which is a common postoperative complication. MRI is also advantageous in differentiating cystic lesions from other hepatic masses and provides superior contrast resolution in soft tissues. However, its use is generally limited due to higher costs and longer imaging times, making it a secondary option when US and CT are inconclusive.
- **Positron Emission Tomography (PET-CT):** PET-CT is increasingly recognized as a valuable tool in the follow-up of patients with alveolar echinococcosis, particularly in distinguishing viable cysts from inactive ones by detecting metabolic activity through fluorodeoxyglucose (FDG) uptake. This advanced imaging technique proves especially useful in cases where conventional follow-up methods, such as US or CT, are inconclusive or when serological markers remain elevated despite negative imaging findings. Furthermore, PET-CT is beneficial in monitoring the potential extrahepatic spread of the disease, which is particularly relevant for immunocompromised patients [27–30]. Although PET-CT is not routinely recommended for follow-up due to its high cost, limited accessibility, poor popularity, rare availability in primary centers, longer scanning time, and expensive examination, it is widely accepted as one of the most prolific imaging techniques and remains a valuable diagnostic asset for both the present and future, particularly when conventional follow-up methods prove inadequate [31].

14.1.1.4 Surveillance After Liver Transplantation for Alveolar Echinococcosis

Liver transplantation is a life-saving procedure for patients with end-stage alveolar echinococcosis, particularly those with extensive hepatic involvement and extrahepatic spread. Due to the aggressive and infiltrative nature of *E. multilocularis*, post-transplant follow-up requires a comprehensive, long-term surveillance strategy to prevent disease recurrence and manage complications.

- **Imaging Surveillance:** Routine imaging with US, CT, or MRI is essential to monitor for graft function, detect potential disease recurrence, and assess complications such as biliary strictures or vascular complications. PET-CT may be considered in cases with suspected recurrence or ambiguous findings on conventional imaging.
- **Serological Monitoring:** Post-transplant patients should undergo regular serological testing, including ELISA and IHA, to track disease activity. Persistent or rising antibody titers may indicate recurrence and warrant further imaging assessment. However, in immunosuppressed patients, standard serological tests may yield false-negative results. In such cases, Western blot testing is considered a more reliable alternative for disease monitoring.
- **Immunosuppressive Therapy Considerations:** Given that alveolar echinococcosis behaves like a parasitic malignancy, immunosuppressive regimens must be carefully adjusted. Calcineurin

inhibitors and corticosteroids remain the cornerstone of immunosuppression, while albendazole is often prescribed to suppress residual parasite activity and prevent relapse. Maintaining a balance between anthelmintic benzimidazoles and immunosuppression is essential. While some centers do not routinely use albendazole therapy post-transplant, most recommend treatment for up to 2 years to minimize recurrence risk.

- **Medical Treatment:** In patients undergoing liver transplantation due to alveolar echinococcosis, the use of immunosuppressive therapy may inadvertently encourage the reactivation of residual larvae or lead to the emergence and enlargement of metastases. Although the current literature is limited, some reports suggest that, even under continuous immunosuppression, benzimidazole treatment may successfully control the progression of alveolar echinococcosis metastases. As a result, the combination of liver transplantation with prolonged benzimidazole therapy may contribute to improved survival outcomes, regardless of whether metastases are present in other organs. Nonetheless, there remains a lack of consensus on the optimal protocol, including the appropriate dosage and duration of benzimidazole therapy post-transplantation. This uncertainty underscores the need for further clinical research to develop evidence-based guidelines for managing alveolar echinococcosis in liver transplant recipients [17, 32, 33].
- **Long-Term Follow-up:** Patients require lifelong monitoring to assess graft function, detect late disease recurrence, and evaluate potential complications related to immunosuppressive therapy. Multidisciplinary collaboration between hepatologists, transplant surgeons, infectious disease specialists, and radiologists is crucial for optimal patient outcomes.

14.1.1.5 Future Perspectives

The integration of advanced technologies, particularly artificial intelligence, into hydatid disease follow-up is a promising development. AI-assisted imaging techniques can improve the accuracy of cyst classification, particularly in borderline cases where distinguishing viable from inactive cysts remains challenging. This is especially relevant in post-transplant patients, where alveolar echinococcosis can metastasize to other organs or mimic hepatic tumors. In such cases, artificial intelligence-driven imaging analysis may help differentiate true recurrence from tumor-like lesions, particularly in Stage 4–5 calcified hydatid cysts that resemble malignancies. Deep learning algorithms applied to USG, CT, and MRI may enhance the detection of recurrence and improve risk stratification. Additionally, AI-driven automated analysis of serological patterns could refine disease monitoring, especially in immunosuppressed patients where conventional tests may yield false-negative results. Future studies should focus on validating artificial intelligence-based approaches in clinical practice to optimize hydatid disease surveillance, particularly in cases where standard imaging and serology are inconclusive.

In conclusion, effective follow-up after hydatid disease treatment is crucial for preventing recurrence, managing complications, and optimizing long-term outcomes. A structured, multidisciplinary approach combining imaging, serology, and clinical evaluation enhances patient care and early complication detection. In high-risk patients, extended follow-up is particularly important to ensure sustained disease control and improve overall prognosis.

14.1.2 Prevention and Control Strategies of Hydatid Disease

Prevention and control of echinococcosis require comprehensive strategies tailored to both cystic echinococcosis and alveolar echinococcosis, as their transmission dynamics and risk factors differ significantly [1, 34–42].

14.1.2.1 Cystic Echinococcosis Prevention

CE is primarily controlled by preventing the transmission of the parasite from infected livestock to dogs and subsequently to humans. Key prevention measures include:

- Regular deworming of dogs using praziquantel, ideally every 4 to 6 weeks, to break the transmission cycle.
- Managing dog populations through registration and controlled breeding to reduce stray dog numbers.
- The EG95 vaccine has been extensively demonstrated effective in reducing *E. granulosus* transmission to sheep, providing 95–99% protection.
- Limiting areas where dogs are allowed, especially near livestock.
- Preventing dogs from feeding on the carcasses of infected sheep.
- Controlling stray dog populations.
- Restricting home slaughter of sheep and other livestock to minimize exposure.
- Avoiding the consumption of food or water that may have been contaminated by fecal matter from dogs.
- Ensuring proper hand hygiene, especially after handling dogs and before preparing food.
- Teaching children the importance of hand washing to prevent infection.
- Improving hygiene practices during livestock slaughter, including safe disposal of infected offal.

14.1.2.2 Alveolar Echinococcosis Prevention

Alveolar echinococcosis prevention involves reducing contact between domestic dogs and wildlife hosts, such as foxes and rodents. Key strategies include:

- Avoiding contact with wild animals such as foxes, coyotes, and stray dogs, and their fecal matter.
- Reducing fox populations and minimizing human-fox contact, especially in endemic regions.
- Applying anthelmintic bait to wildlife, particularly in areas with high *E. multilocularis* prevalence.
- Implementing targeted culling and monitoring of wild carnivore populations.
- Preventing domestic dogs from hunting or feeding on wild rodents.
- Avoiding the encouragement of wildlife near human settlements.

14.1.2.3 Strengthening Prevention and Control Measures

Strengthening prevention and control requires a multifaceted approach, including:

- Integrating vaccination of lambs with dog deworming schedules to optimize herd immunity.
- Applying the One Health approach, integrating human, animal, and environmental health strategies.
- Establishing public health campaigns to educate at-risk populations on prevention practices.

14.1.2.4 Surveillance and Monitoring

Monitoring echinococcosis in animal populations remains a challenge due to the asymptomatic nature of the infection in livestock and dogs. Key strategies include:

- Regular inspection of carcasses at slaughter to detect cyst presence.
- Conducting periodic serological surveys in high-risk areas to monitor transmission dynamics.
- Developing mathematical models to predict infection trends and evaluate control strategies.
- Surveillance programs targeting wildlife reservoirs, particularly in endemic areas.
- Periodic mass screening of at-risk populations to identify early cases.
- Utilizing data from sporadic surveys and integrating them into public health strategies for better disease tracking.

By combining preventive measures, regular monitoring, and community education, the burden of echinococcosis can be significantly reduced, particularly in endemic regions where both cystic echinococcosis and alveolar echinococcosis pose ongoing public health challenges.

Acknowledgments The English language and grammar in this chapter have been improved with the help of ChatGPT and QuillBot.

References

1. Hogea MO, Ciomaga BF, Muntean MM, Muntean AA, Popa MI, Popa GL. Cystic echinococcosis in the early 2020s: a review. Trop Med Infect Dis. 2024;9(2) https://doi.org/10.3390/tropicalmed9020036. PMID: 38393125 PMCID: PMC10891927.
2. Eckert J, Deplazes P. Biological, epidemiological, and clinical aspects of echinococcosis, a zoonosis of increasing concern. Clin Microbiol Rev. 2004;17(1):107–35. https://doi.org/10.1128/cmr.17.1.107-135.2004. PMID: 14726458 PMCID: PMC321468.
3. Badwaik N, Gharde P, Shinde RK, Tayade H, Navandhar PS, Patil M. Hydatid cyst or echinococcosis: a comprehensive review of transmission, clinical manifestations, diagnosis, and multidisciplinary treatment. Cureus. 2024;16(7):e63713. https://doi.org/10.7759/cureus.63713. PMID: 39099980 PMCID: PMC11294710.
4. Gun E, Etit D, Buyuktalanci DO, Cakalagaoglu F. Unusual locations of hydatid disease: a 10-year experience from a tertiary reference center in Western Turkey. Ann Diagn Pathol. 2017;29:37–40. https://doi.org/10.1016/j.anndiagpath.2017.01.011. PMID: 28807340.
5. Yuksel M, Demirpolat G, Sever A, Bakaris S, Bulbuloglu E, Elmas N. Hydatid disease involving some rare locations in the body: a pictorial essay. Korean J Radiol. 2007;8(6):531–40. https://doi.org/10.3348/kjr.2007.8.6.531. PMID: 18071284 PMCID: PMC2627456.
6. Forooghi M, Hosseini H, Yousufzai S, Ebrahimi D, Shahrokhi R, Ebrahimi A, Abdollahzade R, Hooshmandi S, Gahromi MG, Sharifi S, Sharifi Z, Tadayon A. Ten-year experience with pediatric splenic hydatidosis: clinical profiles, surgical outcomes, and prognostic indicators. Food Waterborne Parasitol. 2025;39:e00266. https://doi.org/10.1016/j.fawpar.2025.e00266.
7. Saberi S, Naghizadeh H, Bagheri N, Saffar H, Afshar ZM, Khabiri SS. Extensive humerus and scapula hydatidosis: a unique case report and comprehensive review of management strategies. Int J Surg Case Rep. 2024;125:110503. https://doi.org/10.1016/j.ijscr.2024.110503. PMID: 39602939 PMCID: PMC11629264.
8. Lee FM, Gaszynski R, Merrett N. Recurrent hydatid disease, a case report and literature review. Int Surg J. 2021;9(1):194–9. https://doi.org/10.18203/2349-2902.isj20215156.
9. Kahriman G, Ozcan N, Dogan S, Karaborklu O. Percutaneous treatment of liver hydatid cysts in 190 patients: a retrospective study. Acta Radiol. 2017;58(6):676–84. https://doi.org/10.1177/0284185116664226. PMID: 27565632.
10. Çalta AF, Duran A. [Our experience with laparoscopic pericystectomy in liver hydatid disease]. Med J SDU. 2022;29(2) 223–8. https://doi.org/10.17343/sdutfd.1052285.
11. Botezatu C, Mastalier B, Patrascu T. Hepatic hydatid cyst—diagnose and treatment algorithm. J Med Life. 2018;11(3):203–9. https://doi.org/10.25122/jml-2018-0045. PMID: 30364592 PMCID: PMC6197524.
12. Baimakhanov Z, Kaniyev S, Serikuly E, Doskhanov M, Askeyev B, Baiguissova D, Skakbayev A, Sadykov C, Barlybay R, Seisembayev M, Baimakhanov B. Radical versus conservative surgical management for liver hydatid cysts: a single-center prospective cohort study. JGH Open. 2021;5(10):1179–82. https://doi.org/10.1002/jgh3.12649. PMID: 34622005 PMCID: PMC8485413.
13. Öztürk G, Uzun MA, Özkan ÖF, Kayaalp C, Tatlı F, Eren S, Aksungur N, Çoker A, Bostancı EB, Öter V, Kaya E, Taşar P. Turkish HPB Surgery Association consensus report on hepatic cystic echinococcosis (HCE). Turk J Surg. 2022;38(2):101–20. https://doi.org/10.47717/turkjsurg.2022.5757. PMID: 36483170 PMCID: PMC9714645.
14. Tepelenis K, Stefanou S, Stefanou C, Pappas-Gogos G, Zikos N, Spyrou S, Koulas S. Comparing two different surgical approaches (conservative and radical) in combination with Albendazole for the treatment of hydatid cyst. A 30-year retrospective study from a large series of patients. Intern Med Rev. 2019;5(4):1–15.
15. Sozuer E, Akyuz M, Akbulut S. Open surgery for hepatic hydatid disease. Int Surg. 2014;99(6):764–9. https://doi.org/10.9738/intsurg-d-14-00069.1. PMID: 25437585 PMCID: PMC4254238.
16. Akbulut S, Yilmaz M. Hydatid cyst of the adrenal gland: is radical surgery necessary for recurrent hydatid disease? Case Rep Surg. 2018;2018:6452402. https://doi.org/10.1155/2018/6452402. PMID: 29682390 PMCID: PMC5842735.
17. Brunetti E, Kern P, Vuitton DA. Expert consensus for the diagnosis and treatment of cystic and alveolar echinococcosis in humans. Acta Trop. 2010;114(1):1–16. https://doi.org/10.1016/j.actatropica.2009.11.001. PMID: 19931502.
18. Rosa DD, Figueredo E, Rosas M, Goñi F. Follow up study of symptomatic human cystic echinococcosis treatment with albendazole and praziquantel, in Uruguay. BMC Infect Dis. 2024;24(1):737.

https://doi.org/10.1186/s12879-024-09539-y. PMID: 39060987 PMCID: PMC11282602.
19. Manzano-Román R, Sánchez-Ovejero C, Hernández-González A, Casulli A, Siles-Lucas M. Serological diagnosis and follow-up of human cystic echinococcosis: a new hope for the future? Biomed Res Int. 2015;2015:428205. https://doi.org/10.1155/2015/428205. PMID: 26504805 PMCID: PMC4609352.
20. Dziri C, Dougaz W, Samaali I, Khalfallah M, Jerraya M, Mzabi R, Bouasker I, Nouira R, Fingerhut A. Radical surgery decreases overall morbidity and recurrence compared with conservative surgery for liver cystic echinococcosis: systematic review with meta-analysis. Ann Laparosc Endosc Surg. 2019;4:92. https://doi.org/10.21037/ales.2019.08.01.
21. Golemanov B, Grigorov N, Mitova R, Genov J, Vuchev D, Tamarozzi F, Brunetti E. Efficacy and safety of PAIR for cystic echinococcosis: experience on a large series of patients from Bulgaria. Am J Trop Med Hyg. 2011;84(1):48–51. https://doi.org/10.4269/ajtmh.2011.10-0312. PMID: 21212200 PMCID: PMC3005516.
22. Ormeci N. PAIR vs Örmeci technique for the treatment of hydatid cyst. Turk J Gastroenterol. 2014;25(4):358–64. https://doi.org/10.5152/tjg.2014.13018. PMID: 25254515.
23. Akhan O, Özbay Y, Ünal E, Karaagaoglu E, Çiftçi TT, Akıncı D. Long-term results of modified catheterization technique in the treatment of CE type 2 and 3b liver hydatid cysts. Cardiovasc Intervent Radiol. 2025;48(4):503–11. https://doi.org/10.1007/s00270-025-03976-1. PMID: 39953155 PMCID: PMC11958407.
24. Kuehn R, Uchiumi LJ, Tamarozzi F. Treatment of uncomplicated hepatic cystic echinococcosis (hydatid disease). Cochrane Database Syst Rev. 2024;7(7):Cd015573. https://doi.org/10.1002/14651858.Cd015573. PMID: 38994714 PMCID: PMC11240857.
25. Chai JY, Jung BK, Hong SJ. Albendazole and mebendazole as anti-parasitic and anti-cancer agents: an update. Korean J Parasitol. 2021;59(3):189–225. https://doi.org/10.3347/kjp.2021.59.3.189. PMID: 34218593 PMCID: PMC8255490.
26. Aydın Y, Ali Bilal U, Dostbil A. Current management of pulmonary hydatid cyst. Eurasian J Med. 2025;57(1):1–7. https://doi.org/10.5152/eurasianjmed.2025.24761. PMID: 40377486 PMCID: PMC12036369.
27. Graeter T, Eberhardt N, Shi R, Schmidberger J, Beer AJ, Beer M, Henne-Bruns D, Hillenbrand A, Barth TFE, Grimm J, Kratzer W, Gruener B. Hepatic alveolar echinococcosis: correlation between computed tomography morphology and inflammatory activity in positron emission tomography. Sci Rep. 2020;10(1):11808. https://doi.org/10.1038/s41598-020-68624-9. PMID: 32678174 PMCID: PMC7366930.
28. Husmann L, Deibel A, Skawran S, Ledergerber B, Muehlematter UJ, Hasse B, Huellner MW, Reiner CS, Muellhaupt B. Follow-up PET/CT of alveolar echinococcosis: comparison of metabolic activity and immunodiagnostic testing. PLoS One. 2022;17(6):e0270695. https://doi.org/10.1371/journal.pone.0270695. PMID: 35767557 PMCID: PMC9242476.
29. Husmann L, Muehlematter UJ, Grimm F, Ledergerber B, Messerli M, Kudura K, Gruenig H, Muellhaupt B, Hasse B, Huellner MW. PET/CT helps to determine treatment duration in patients with resected as well as inoperable alveolar echinococcosis. Parasitol Int. 2021;83:102356. https://doi.org/10.1016/j.parint.2021.102356. PMID: 33872794.
30. Caoduro C, Porot C, Vuitton DA, Bresson-Hadni S, Grenouillet F, Richou C, Boulahdour H, Blagosklonov O. The role of delayed 18F-FDG PET imaging in the follow-up of patients with alveolar echinococcosis. J Nucl Med. 2013;54(3):358–63. https://doi.org/10.2967/jnumed.112.109942. PMID: 23303963.
31. Ailixire SA, Sai Charan Goud K, Hari Krishna K, Wenya L. A new classification of hepatic alveolar echinococcosis (HAE) calcification and biological activity using CT and PET-CT. Radiol Infect Dis. 2019;6(2):61–7. https://doi.org/10.1016/j.jrid.2019.06.002.
32. Bresson-Hadni S, Koch S, Miguet JP, Gillet M, Mantion GA, Heyd B, Vuitton DA. Indications and results of liver transplantation for Echinococcus alveolar infection: an overview. Langenbeck's Arch Surg. 2003;388(4):231–8. https://doi.org/10.1007/s00423-003-0394-2. PMID: 12905036.
33. Zavoikin VD, Zelya OP, Tumolskaya NI. The importance of uninterrupted albendazole treatment in patients with unresectable alveolar echinococcosis undergoing liver transplantation. Transpl Infect Dis. 2020;22(4):e13291. https://doi.org/10.1111/tid.13291. PMID: 32277859.
34. Borhani M, Fathi S, Lahmar S, Ahmed H, Abdulhameed MF, Fasihi HM. Cystic echinococcosis in the Eastern Mediterranean region: neglected and prevailing! PLoS Negl Trop Dis. 2020;14(5):e0008114. https://doi.org/10.1371/journal.pntd.0008114. PMID: 32379760 PMCID: PMC7205190.
35. Craig PS, Hegglin D, Lightowlers MW, Torgerson PR, Wang Q. Echinococcosis: control and prevention. Adv Parasitol. 2017;96:55–158. https://doi.org/10.1016/bs.apar.2016.09.002. PMID: 28212791.
36. McManus DP, Zhang W. Helminthic diseases: echinococcosis. Ref Mod Biomed Sci. 2024; https://doi.org/10.1016/B978-0-12-801238-3.03103-2.
37. McManus DP, Zhang W, Li J, Bartley PB. Echinococcosis. Lancet. 2003;362(9392):1295–304. https://doi.org/10.1016/s0140-6736(03)14573-4. PMID: 14575976.
38. Menghi C, Gatta C, Arias L. Human cystic and alveolar echinococcosis. Curr Treat Options Infect Dis. 2017;9(2):210–22. https://doi.org/10.1007/s40506-017-0121-0.

39. Moro P, Schantz PM. Echinococcosis: a review. Int J Infect Dis. 2009;13(2):125–33. https://doi.org/10.1016/j.ijid.2008.03.037. PMID: 18938096.
40. Wen H, Vuitton L, Tuxun T, Li J, Vuitton DA, Zhang W, McManus DP. Echinococcosis: advances in the 21st century. Clin Microbiol Rev. 2019;32(2) https://doi.org/10.1128/cmr.00075-18. PMID: 30760475 PMCID: PMC6431127.
41. Yang Y, Cairang Y, Jiang T, Zhou J, Zhang L, Qi B, Ma S, Tang L, Xu D, Bu L, Bu R, Jing X, Wang H, Zhou Z, Zhao C, Luo B, Liu L, Guo J, Nima Y, Hua G, Wa Z, Zhang Y, Zhou G, Jiang W, Wang C, De Y, Yu X, Cheng Z, Han Z, Liu F, Dou J, Feng H, Wu C, Wang R, Hu J, Yang Q, Luo Y, Wu J, Fan H, Liang P, Yu J. Ultrasound identification of hepatic echinococcosis using a deep convolutional neural network model in China: a retrospective, large-scale, multicentre, diagnostic accuracy study. Lancet Digit Health. 2023;5(8):e503–14. https://doi.org/10.1016/s2589-7500(23)00091-2. PMID: 37507196.
42. Borhani M, Fathi S, Harandi MF, Casulli A, Ding J, Liu M, Zhang W, Wen H. Echinococcus granulosus sensu lato control measures: a specific focus on vaccines for both definitive and intermediate hosts. Parasit Vectors. 2024;17(1):533. https://doi.org/10.1186/s13071-024-06581-2. PMID: 39716337 PMCID: PMC11665232.

FSC
www.fsc.org
MIX
Papier aus verantwortungsvollen Quellen
Paper from responsible sources
FSC® C105338

If you have any concerns about our products,
you can contact us on
ProductSafety@springernature.com

In case Publisher is established outside the EU,
the EU authorized representative is:
Springer Nature Customer Service Center GmbH
Europaplatz 3, 69115 Heidelberg, Germany

Printed by Libri Plureos GmbH
in Hamburg, Germany